Estuarine Biogeochemical Dynamics of the East Coast of India

Sourav Das • Tuhin Ghosh
Editors

Estuarine Biogeochemical Dynamics of the East Coast of India

Springer

Editors
Sourav Das
School of Oceanographic Studies
Jadavpur University
Kolkata, West Bengal, India

Tuhin Ghosh
School of Oceanographic Studies
Jadavpur University
Kolkata, West Bengal, India

ISBN 978-3-030-68979-7 ISBN 978-3-030-68980-3 (eBook)
https://doi.org/10.1007/978-3-030-68980-3

This Springer imprint is published by the registered company Springer Nature Switzerland AG
The registered company address is: Gewerbestrasse 11, 6330 Cham, Switzerland

In the loving memory of

Late Prof. Ananda Deb Mukhopadhyay
(3 May 1938 – 8 October 2020)

Prof. Ananda Deb Mukhopadhyay was an eminent academician and a renowned researcher in multifarious fields of geology, oceanography, and environmental studies. He was a well-loved teacher with father-like character and guided a number of students in academia and their personal lives. His love and affection towards the student community is still cherished by many who came in touch with him. Being a professor in the Department of Geological Sciences at Jadavpur University, West Bengal, India, he founded the School of Oceanographic Studies. He has also served as the Vice Chancellor of Vidyasagar University, West Bengal, India. Till his last day, he was actively involved with the National Council of Education, India.

Preface

Estuaries, located at the interface between land and coastal oceans, are dynamic, highly productive systems that, in many cases, have been historically associated with the development of many of the great centers of early human civilization. Biogeochemistry of estuaries offers a comprehensive and interdisciplinary approach to understanding biogeochemical cycling in estuaries. Estuary and river systems play a critical role in the natural self-regulation of Earth's surface conditions by serving as a significant sink for anthropogenic CO_2. Approximately 90% of global carbon burial occurs in ocean margins, and the majority of this carbon remains buried in large delta-front estuaries. Many of the existing books in estuarine science comprise a suite of edited volumes, typically focused on specific topics in estuaries all over the world. However, the present book entitled *Estuarine Biogeochemical Dynamics of the East Coast of India* provides a unique foundation for the first time on the east coast of India. This book utilizes numerous illustrations and an extensive literature base to impart the current state-of-the-art knowledge in this field on the east coast of India adjacent to the Bay of Bengal. We collated chapters on geomorphology, carbon dynamics, bacterial population, estuarine pollution, and nutrient cycling of this region. The book also comprised the role of microbial diversity, microzooplankton variability in estuaries, CDOM dynamics of the east coast of India, and anthropogenic impacts of Indian Sundarbans (the largest mangrove forest in the world) with linkages to physical and biological processes in estuarine sciences. Consequentially, these systems have and continued to be severely impacted by anthropogenic inputs. This timely book can act as the foundational basis of elemental cycling in estuaries of the east coast of India and estuarine management issues. Estuarine and marine scientists, ecologists, biogeochemists, and environmentalists around India and other parts of the world would find interest in the present title. Intermediate to advanced level students can benefit by going through this book. This book presents both review and original study findings involving estuaries on the east coast of India. The future state of all of these estuaries may be a sensitive indicator of shifts in global weather patterns.

The book opens with an introductory chapter by the editors (Dr. Sourav Das and Prof. Tuhin Ghosh). Then, Mr. B. K. Saha (Former Senior Deputy Director-General, Geological Survey of India) presents a brief account of the geology of the east coast of India. Studies about the estuarine carbon dynamics along the east coast of India have seen tremendous growth over the past decade. Dr. Kunal Chakraborty (Scientist-E, INCOIS, Govt. of India) has reviewed these works in one of the chapters. A synthesis of previous research works on the biogeochemistry of the Mahanadi estuarine ecosystem is presented in one of the chapters by Dr. Tamoghna Acharyya (focusing on increased anthropogenic interferences). Dr. Abhra Chanda (Assistant Professor, School of Oceanographic Studies, Jadavpur University) has reviewed different pollution parameters along the east coast of India in three of the chapters (focusing on persistent organic pollutants, heavy metals, eutrophication, algal bloom, fecal coliform, organic matter, and petroleum hydrocarbon). Dr. Anirban Mukhopadhyay illustrated the variability of suspended particulate matter with the help of geo-statistical analysis. Dr. Rajdeep Roy (Scientist-E, National Remote Sensing Centre, India) has described the nutrient cycling, phytoplankton community structure, and seasonal dynamics of primary production of the estuarine waters of the east coast of India. Mercury-resistant marine bacterial population has been synthesized by Dr. Surajit Das. Dr. Biraja Kumar Sahu covered the microzooplankton studies carried out in estuaries, mangroves, and lagoons of the east coast of India. The book continues with a chapter by Dr. Sudarsanarao Pandi covering all the information and gap on CDOM-related researches carried out in estuaries and rivers draining into the Bay of Bengal. The book closes with an overview of the current understanding of biogeochemical dynamics and anthropogenic impacts on the Indian Sundarbans ecosystems by A. C. G. Henderson (Faculty, School of Geography, Politics & Sociology, Newcastle University, UK), Dr. S. Das, Prof. T. Ghosh, Dr. V. N. Panizzo, Dr. H. L. Moorhouse, Dr. L. R. Roberts, Dr. R. E. Walton, Dr. Y. Zheng, Dr. A. M. Bass, and Dr. S. McGowan.

Kolkata, India

Sourav Das
Tuhin Ghosh

Acknowledgments

We thank Dr. Rajdeep Roy, Dr. Sanjibakumar Baliar Singh, Dr. Tamoghna Acharyya, Dr. Anirban Akhand, and Dr. Abhra Chanda for their help in reviewing several chapters. We also thank Mr. Aaron Schiller, editor at Springer Nature, for his kind help and encouragement. We specially thank Dr. Andrew C. G. Henderson (School of Geography, Politics & Sociology, Newcastle University, UK) for his kind support and encouragement throughout this book project.

Contents

List of Figures

List of Tables

Chapter 1
Introduction: An Overview of Biogeochemical Cycle of Estuarine System

Sourav Das and Tuhin Ghosh

Abstract Estuaries are one of the most dynamic regions of the world where a suite of biogeochemical phenomena drives the ecosystem functions. This chapter has provided a brief overview of the role of estuaries as a fragile ecosystem and the environmental processes involved in this regime. The most significant topics which have gained impetus from the viewpoint of research are introduced in the chapter. These topics include the effect of climate change, nutrient dynamics, sedimentation, and the carbon cycle. India encompasses a long coastline, and the eastern side of the Indian peninsula has an intricate network of estuaries. This chapter has set up a background of the present book giving special emphasis on the estuaries of the east coast of India. The scope of the entire book along with the chapter summary is also included in this chapter to engage the readers. Overall, this chapter can be considered as a brief introduction to the book.

Keywords Overview of estuarine system · Coast · Estuarine nutrient dynamics · Estuarine carbon cycle · Sedimentation in estuarine ecosystems · Estuaries of India

1.1 Coast and Estuaries

Coastal environments are areas of substantial productivity and high convenience. Despite their comparatively small areal cover (only 7% of the biosphere ocean surface), coastal regions play a significant role in the global carbon cycle and defending from human impacts on ocean systems. They are contributing to 12–16% of the biosphere ocean net annual productivity and approximately responsible for more than 40% of the yearly carbon sequestration (Muller Karger et al. 2005). The coastal sector characterizes more than 85% of the biosphere fish catch (Pauly and Christensen 1995), and Costanza et al. (1997) estimate its economic value, i.e., >40% of the

S. Das (✉) · T. Ghosh
School of Oceanographic Studies, Jadavpur University, Kolkata, West Bengal, India

S. Das, T. Ghosh (eds.), *Estuarine Biogeochemical Dynamics of the East Coast of India*, https://doi.org/10.1007/978-3-030-68980-3_1

value of the world ecosystem services. Alongi et al. (1998) quantified the coastal area as a zone of severe human impact on the marine environs, and more than 60% of the world population survive along the shoreline. The present book is based on the biogeochemical characteristics of major coastal and estuarine regions (adjoining to the Bay of Bengal) of the east coast of India.

However, the estuary is an integral sector of the coastal environment. An estuary is a transitional zone between the fluvial and marine environments and the outfall province of the river. The estuarine ecosystem is the most dynamic and delivers a direct source of natural resources to humankind. These ecosystems are also used for industrial, commercial, and recreational purposes. Bianchi (2007) defines an estuarine system as a resource for commercially valuable estuarine species.

Estuaries also provide shelter and food for the juvenile stages of the economically important species. More than 85% of the world's land surface is joined to the sea by rivers (Ludwig and Probst 1998). Present-day estuaries were shaped during the last 5000–5500 years in a stable interglacial period. Only 12–15% of the world's 180 prominent rivers that empty freely from the source to the ocean form different types of estuaries (Wang et al. 2007). The Ganges River – one of the largest rivers (~2510 km in length) in the world – stands in the third position (first one is Amazon and second is the Congo River) for draining freshwater to the Bay of Bengal from terrestrial through the Hooghly estuary, one of the estuarine systems presented in the book.

1.2 Estuarine Processes

Estuaries represent a biogeochemically active zone since it receives massive inputs of terrestrial organic matter and nutrient and exchanges large amounts of matter and energy with the open ocean (Bouillon et al. 2007). In tropical estuaries, where the water temperature is more or less stable, a number of plants and animals are less affected. Tides are necessary for healthy estuaries as they flush the systems and provide nutrients to keep the food webs functional. However, the tides create incessantly changing conditions of exposure to air and inundation to water. Water circulation is important because it transports animals and plants; mixes nutrients, oxygen, and sediments; and removes wastes. Estuarine circulation, river and groundwater discharge, tidal flooding, re-suspension events, and exchange flow with adjacent areas (Leonard and Luther 1995) all constitute important physical variables that exert some level of control on estuarine biogeochemical cycles.

1.3 Impact of Climate Change on Estuaries

Previous studies emphasized that disturbance in natural processes limits the estuarine health and viability (Goldberg 1995), making them all the more vulnerable to the consequences of climate change. Some of the critical potential impacts of climate change on estuaries may result from changes in physical mixing characteristics caused by changes in freshwater runoff (Scavia et al. 2002). A globally intensified hydrological cycle and regional alterations in runoff all comprehensively indicate changes in coastal water quality. Freshwater inflows into estuaries influence water residence time, nutrient delivery, vertical stratification, salinity, and control of phytoplankton growth rates. Increased freshwater inflows decrease the water residence time and enhance vertical stratification and vice versa (Moore et al. 1997). The effects of altered residence times can have significant impacts on phytoplankton populations, which have the potential to increase fourfold per day. Changes in the timing of freshwater delivery to estuaries could lead to a decoupling of the juvenile phases of many estuarine and marine fishery species from the available nursery habitat. Increased water temperature also affects a suite of microbial processes such as nitrogen fixation and denitrification in estuaries (Lomas et al. 2001). Apart from that, extreme weather events such as cyclones and flooding are likely to be a future threat to estuarine systems (Nicholls et al. 2007).

1.4 Estuarine Nutrient Dynamics

Ecosystem responses depend on several critical physicochemical characteristics and processes. The transport, transformation, retention, and export of nutrients in estuarine ecosystems are strongly influenced by estuary size (surface area), depth, volume, flushing rate, water residence time, tidal exchange, vertical mixing, and stratification. Riverine input influences estuarine hydrography by creating salinity gradients and stratification and assures large transport of silt, organic material, and inorganic nutrients to the estuaries. The open marine areas impose large-scale physical and chemical forcing on the estuarine ecosystem due to tide and wind-generated water exchange (Berner and Berner 1996; Flindt et al. 1999).

During the last six decades, global riverine C, N, and P inputs (inorganic nutrient flux) into the ocean have trebled due to regional climate, geology, and human activities, whereas the dissolved silicate stemming from natural sources was significantly reduced (Turner et al. 2005; Lohrenz et al. 2002). Changed ratios of the essential nutrients Si, N, and P entail changes of the plankton community and the biogeochemical cycles (Smith et al. 2003). These enhanced inputs of C, N, and P are due to increasing population density in the areas of major river drainage basins and close oceanic coastlines; socio-economic development and changes in land-use practices; enhanced discharges of industrial, agricultural, and municipal waste into continental margin waters via the river; groundwater discharges; and atmospheric transport (Meybeck and Vörösmarty 2005). Engineering projects (damming of the rivers) have an opposite effect by altering the hydrological regime of most of the world's major rivers. The current state of knowledge indicates that the impact of dams on estuarine ecosystems is profound, complex, varied, multiple, and mostly negative (Adams et al. 2002). By storing or diverting water, dams alter the natural distribution and timing of stream flows. This alteration in land use, in turn, changes sediment and nutrient regimes and alters water temperature and chemistry, with consequent ecological and economic impacts.

1.5 Estuarine Carbon Cycle

Recent observations have shown that river-estuary systems release a significant amount of CO_2 into the atmosphere in addition to the commonly recognized fluvial export of inorganic/organic matter (Borges et al. 2006; Hofmann et al. 2009). Over 97% of the runoff has been classified as the $Ca(HCO_3)_2$ type, making HCO_3^-, Ca^{2+}, SO_4^{2-}, and SiO_2 the dominant dissolved constituents in global surface river waters (Mehrbach et al. 1973; Bianchi 2007). Rivers are generally net heterotrophic, resulting in greater consumption of HCO_3^- by river phytoplankton (Bianchi et al. 2004). High decomposition rates in estuarine systems may result in the export of DIC that rivals that of riverine export to coastal waters (Wang and Cai 2004). Estuarine and freshwater systems are close to equilibrium with atmospheric CO_2 and are influenced by temperature and salinity (Wetzel 2001).

The abiotic source of CO_2 in rivers and estuaries is the photochemical mineralization of dissolved organic carbon (DOC). This process happens either by direct photo-oxidation of DOC to CO_2 by solar UV radiation (Granéli et al. 1996) or by cleavage of DOC molecules into low molecular weight compounds available for bacterial metabolism (Bertilsson and Stefan 1998). Allochthonous DOC is preferentially photomineralized, while autochthonous DOC is preferentially mineralized by heterotrophic microbes (Obernosterer and Benner 2004). Finally, CO_2 in estuarine water also is the result of respiration by heterotrophic organisms. CO_2 loss from estuaries is supported largely by microbial decomposition of OC produced in coastal wetlands (Cai 2011).

Annually, world rivers transport large quantities of C to coastal seas (0.9 Gt C, out of which 40% is organic and 60% inorganic) (Etcheber et al. 2007). Particulate organic carbon (POC) in estuarine systems is derived from a multitude of sources. The autochthonous POC includes phytoplankton, submerged vegetation, benthic diatoms and cyanobacteria, and periphyton living on stems of emergent plants, whereas the allochthonous sources of POC consist of marginal marsh and swamp vegetation, marine- or river-borne phytoplankton and detritus, and beach, shoreline, and wind-blown material (Schlesinger 1997).

Several studies showed that nearshore ecosystems such as estuaries, mangrove waters, salt marsh waters, and coral reefs are assumed to be a net source of CO_2 (Cai et al. 2004; Borges et al. 2006; Abril and Borges 2005). In general, estuaries are CO_2-supersaturated as a result of the respiration of the riverine OC input. The overall source of CO_2 from nearshore ecosystems has been evaluated to ~0.50 Pg C y−1, mainly related to the emission of CO_2 to the atmosphere from estuaries (~0.36 Pg C y−1) (Chen and Borges 2009). Bouillon et al. (2008) reported oversaturation of CO_2 in different mangrove forests surrounding waters, suggesting that this surface water is a significant source of CO_2 to the atmosphere. The direction and magnitude of air-water CO_2 exchanges strongly depend on the type of ecosystem at the coast (Borges et al. 2006), the ocean currents dominating at a respective coast (Liu et al. 2000), and the geographical latitude (Liu et al. 2000; Borges et al. 2006). Isotopic signatures from sediments suggest that most of the terrestrial POCs are degraded in estuaries (Hedges et al. 1997) under the strong net heterotrophic nature of the system (Gattuso et al. 1998; Hopkinson and Smith 2005).

Estuarine water and wetlands are the dominant natural source of CH_4 all over the globe and emit between 100 and 230 Tg CH_4 y^{-1} globally (Denman et al. 2007) and are expected to remain largely unchanged in the future. Bange et al. (1994) reported that up to 75% of total oceanic CH_4 emissions are from estuarine and coastal areas that contribute around 2% of global atmospheric CH_4 emission.

1.6 Sedimentation in Coastal/Estuarine Ecosystems

Sedimentation (as well as erosion) is an elementary phenomenon of nature dealing with loose sediments within the transporting cycle from source to sink locations. Sedimentation of the coastal environment is often associated with human interference in the physical system, such as land-use changes (Walling 2006), construction of artificial structures, or the dredging of sediment from the bed to increase the flow depth or width. Sedimentation affects the navigation, shoreline erosion and stability, migration of shoals, fate of nutrients, contaminants such as heavy metals and pesticides, turbidity, and the primary productivity of estuaries (Wolanski 1995).

Estuaries are recognized as a trap for fine, cohesive coastal sediments (FitzGerald and Knight 2005; Syvitski et al. 2005). Coastal sedimentary processes are inherently dynamic, but significant changes have been related to anthropogenic activities in upland areas (Syvitski et al. 2005). Estuaries are associated with rivers or other forms of runoff from land. They are the immediate recipients of sediment carried by those rivers, as manifest by the formation of river deltas. In the long term, sediment build-up in estuaries is limited by a dynamic balance between the effects of tides, waves, and rivers on sediment inputs and outputs in different parts of an estuary. Estuarine export of fine and coarse sediment is poorly understood because of the tidal pumping mechanism involved in estuarine sediment import (Wolanski et al. 2006). The Hooghly is a macro-tidal estuary that receives high sediment loads from the river Ganges, majorly filled with fluvial sediments. Particle settling in this estuary can be enhanced by the change from fresh to saltwater, the rise and fall of the water level with the tides, and the presence of turbidity maxima during slack tides.

1.7 Estuaries of India

The length of the total coastline in India is approximately 7515 km, and estuaries cover about 27,000 square km area. The Indian subcontinent has 160 minor, 45 medium, and 14 major rivers with an overall catchment region of 3.12×10^6 square km. The combined length of all rivers is 4.5×10^4 km. Kumar et al. (2006) stated

Table 1.1 Major estuaries in the east coast of India

States in the east coast of India	Estuaries	Average discharge (m^3/s/day)
West Bengal	Ganges delta	35,217
Odisha	Mahanadi	2100
Andhra Pradesh	Godavari	3500
	Krishna	2100
	Pennar	200
Tamil Nadu	Cauvery	600
	Ponnaiyar	21
	Vaigai	28

Source: http://iomenvis.nic.in

Table 1.2 The status of important ecologically sensitive areas along the east coast of India

Sensitive area	State	Area (sq km)	Coast length	Status of inlet/ channels	Importance
Indian Sundarban	West Bengal (40%), Bangladesh	4260	85 km inland	Cluster of 102 islands	World's largest mangrove, erosion
Chilika (Asia's largest lagoon)	Odisha	906–1155	64	North shifting/ opening and closing inlets	Aqua catch, dolphins, birds
Kolleru (Asia's largest freshwater body)	Andhra Pradesh	308–954	Inland lake	60 km long river Upputeru joining sea and lake	Aquaculture, agriculture, birds
Pulicat Lake (Asia's second largest lagoon)	Andhra Pradesh (84%)/Tamil Nadu (16%)	350–450	60	Three inlets – Tupili Palem, Pulicat, Rayadoruvu	Mangroves, aqua culture
Golf of Mannar	Tamil Nadu	10,500	63.22	21 islands, coral reefs, estuaries, mangrove	Coral hub

Source: Mishra (2016)

that about 26% of the total population survive within 100 km from the coast in India. There are five major rivers, namely, Ganges, Mahanadi, Godavari, Krishna, and Cauvery, on the east coast (Table 1.1) and Narmada and Tapti on the west coast. The Ganges estuary is considered the largest estuary on the east coast of India as well as in India. Geological processes along with various physical processes like wave, wind, current, tide, and also sediment influx play a crucial role in the formation of the coastal domain and for the development of the estuarine system as well to characterize the water quality of the coastal water of India. The coast is experiencing varying tidal ranges of 4.3 m at Sagar Island (the mouth of Hooghly River), 1.3 m at Kakinada, 1.0 m at Chennai, and 0.9 m at Pondicherry (Nayak and Hanamgond 2010). Besides, there are several ecologically sensitive zones along the east coast of India (Table 1.2). The Ganges estuary, also known as the Hooghly-

Matla estuarine systems, deserves a special mention in this regard. Millions of people are directly and indirectly dependent on this estuarine system, which is also one of the most polluted estuaries of this country as well. This estuarine system also exhibits unique carbon dynamics. The eastern part of the estuary essentially acts as a source of CO_2, whereas the other counterpart acts as transient sinks, especially in the post-monsoon season (Akhand et al. 2016). Due to its vast north to south as well as east to west stretch, the spatial variability in the carbon dynamics depends on an array of biogeochemical and metrological factors.

1.8 Scope of the Book

The biogeochemical properties from upland to coastal margin ecosystems of the northeastern flank of the Indian Ocean (east coast of India) are least understood. There are various research carried out along the east coast of India, while one or two estuaries encompassed most of those studies. Several small to large estuaries, like the Hooghly, Mahanadi, Godavari, Krishna, and Cauvery (Table 1.1), intersperse the east coast of India facing the Bay of Bengal (BoB). This coastal region is one of the most dynamic regimes in the world. A significant amount of river discharge from several perennial estuaries vis-à-vis the effect of monsoon and frequent depression and tropical cyclones makes the estuarine and coastal waters of the east coast of India unique from various perspectives. These physical forcing and extreme atmospheric events exert a substantial impact on the biogeochemistry of the water column in the east coast estuaries and nearshore waters. The purpose of the present work (book) is to understand the water quality parameters and their biogeochemical interaction within the estuarine ecosystem of the east coast of India.

Therefore, it is necessary to identify the principal anthropogenic activities impacting the estuaries of the east coast of India, including the physicochemical variability, nutrient fluxes, organic carbon loading, and heavy metal pollution. It is worth examining these in detail to develop effective management strategies to mitigate their impacts. Moreover, this book addresses a few critical global change problems on the regional estuarine ecosystem, emphasizing their interactions with water quality.

Because of the complex nature of the processes (geologic or anthropogenic) occurring in the coastal zone of the east coast of India, a multi-disciplinary effort is necessary to find a holistic solution for their implementable management options, and the present book is a step forward in that direction (Table 1.3).

Table 1.3 Chapter outline of the present book

Chapter no.	Outline of the chapters
Chapter 2	A brief account of the geology of the east coast of India is presented to describe the geological processes which play an important role in the development of the different type of estuarine systems
Chapter 3	Discusses the present state of the art of several biogeochemical parameters in the estuaries along the east coast of India under normal conditions and the impact of different physical forcing in modulating these parameters. Moreover, it defines a stark difference between the northern estuaries of the east coast and the southern estuaries concerning biogeochemical characteristics
Chapter 4	Discusses the understanding of estuarine carbon dynamics along the east coast of India
Chapter 5	Synthesizes previous research works on the biogeochemistry of the Mahanadi estuarine ecosystem concerning increased anthropogenic interferences
Chapter 6	Presents a detailed study on the mercury-resistant marine bacterial (MRMB) population along the Odisha coast and the effect of the physicochemical parameters on the population dynamics
Chapter 7	The chapter focuses on the persistent organic pollutant (POP) accumulation in the sediments, water column, and selected biotas in the estuaries along the east coast of India
Chapter 8	Discusses the present state of the art of heavy metal contamination in the water column, sediments, and marine organisms adjoining the estuaries of the east coast of India as well as assesses the human health risk
Chapter 9	Confers on the suspended particulate matter concentration variability along the northwestern coastal waters of Bay of Bengal to explain the regionalized coastal phenomenon affecting the environmental state by geo-statistical analysis
Chapter 10	Discusses all the principal findings observed concerning pollutions (eutrophication, algal bloom, fecal coliform, organic matter, and petroleum hydrocarbon) in the estuaries and nearshore waters along the east coast of India
Chapter 11	Describes the nutrient cycling, phytoplankton community structure, and seasonal dynamics of primary production of the estuarine waters of the east coast of India
Chapter 12	The chapter describes all the information on microzooplankton studies carried out in estuaries, mangroves, and lagoons to give a holistic view of the microzooplankton of the east coast of India
Chapter 13	This chapter addresses the impact of physical forcing mechanisms on spatiotemporal variation of biological productivity aided by nutrient dynamics in the coastal waters of the Bay of Bengal (east coast of India)
Chapter 14	The chapter describes all the information and gap on CDOM related researches carried out in estuaries of the east coast of India
Chapter 15	The present chapter presents an overview of our current understanding of biogeochemical dynamics and anthropogenic impacts on the Indian Sundarbans ecosystems

References

Akhand A, Chanda A, Manna S, Das S, Hazra S, Roy R, Choudhury SB, Rao KH, Dadhwal VK, Chakraborty K, Mostofa KMG, Tokoro T, Kuwae T, Wanninkhof R (2016) A comparison of CO2 dynamics and air-water fluxes in a river-dominated estuary and a mangrove-dominated marine estuary. Geophys Res Lett 43(22):11726–11735

Abril G, Borges AV (2005) Carbon dioxide and methane emissions from estuaries. In: Greenhouse gas emissions—fluxes and processes. Springer, Berlin, Heidelberg, pp 187–207

Adams JB, Bate GC, Harrison TD, Huizinga P, Taljaard S, Van Niekerk L, Plumstead EE, Whitfield AK, Wooldridge TH (2002) A method to assess the freshwater inflow requirements of estuaries and application to the Mtata Estuary, South Africa. Estuaries 25(6):1382

Alongi DM, Sasekumar A, Tirendi F, Dixon P (1998) The influence of stand age on benthic decomposition and recycling of organic matter in managed mangrove forests of Malaysia. J Exp Mar Biol Ecol 225(2):197–218

Bange HW, Bartell UH, Rapsomanikis S, Andreae MO (1994) Methane in the Baltic and North Seas and a reassessment of the marine emissions of methane. Glob Biogeochem Cycles 8(4):465–480

Berner EK, Berner RA (1996) Global environment water, air, and geochemical cycles. Prentice-Hall, Englewood Cliffs

Bertilsson S, Stefan LJ (1998) Photochemically produced carboxylic acids as substrates for freshwater bacterioplankton. Limnol Oceanogr 43(5):885–895

Bianchi TS (2007) Biogeochemistry of estuaries. Oxford University Press on Demand, New York

Bianchi TS, Filley T, Dria K, Hatcher PG (2004) Temporal variability in sources of dissolved organic carbon in the lower Mississippi River. Geochim Cosmochim Acta 68(5):959–967

Borges AV, Schiettecatte LS, Abril G, Delille B, Gazeau F (2006) Carbon dioxide in European coastal waters. Estuar Coast Shelf Sci 70(3):375–387

Bouillon S, Dehairs LS, Borges AV (2007) Biogeochemistry of the Tana estuary and delta (Northern Kenya). Limnol Oceanogr 52:46–57

Bouillon S, Borges AV, Castañeda-Moya E, Diele K, Dittmar T, Duke NC, Kristensen E, Lee SY, Marchand C, Middelburg JJ, Rivera-Monroy VH (2008) Mangrove production and carbon sinks: a revision of global budget estimates. Glob Biogeochem Cycles 22(2), GB2013, https://doi.org/10.1029/2007GB003052

Cai WJ (2011) Estuarine and coastal ocean carbon paradox: CO_2 sinks or sites of terrestrial carbon incineration? Annu Rev Mar Sci 3:123–145

Cai WJ, Dai M, Wang Y, Zhai W, Huang T, Chen S, Zhang F, Chen Z, Wang Z (2004) The biogeochemistry of inorganic carbon and nutrients in the Pearl River estuary and the adjacent Northern South China Sea. Cont Shelf Res 24(12):1301–1319

Chen CTA, Borges AV (2009) Reconciling opposing views on carbon cycling in the coastal ocean: continental shelves as sinks and near-shore ecosystems as sources of atmospheric CO_2. Deep-Sea Res II Top Stud Oceanogr 56(8–10):578–590

Costanza R, d'Arge R, De Groot R, Farber S, Grasso M, Hannon B, Limburg K, Naeem S, O'neill RV, Paruelo J, Raskin RG (1997) The value of the world's ecosystem services and natural capital. Nature 387(6630):253–260

Denman SE, Tomkins NW, McSweeney CS (2007) Quantitation and diversity analysis of ruminal methanogenic populations in response to the antimethanogenic compound bromochloromethane. FEMS Microbiol Ecol 62(3):313–322

Etcheber H, Taillez A, Abril G, Garnier J, Servais P, Moatar F, Commarieu MV (2007) Particulate organic carbon in the estuarine turbidity maxima of the Gironde, Loire and Seine estuaries: origin and lability. Hydrobiologia 588(1):245–259

FitzGerald DM, Knight J (eds) (2005) High resolution morphodynamics and sedimentary evolution of estuaries, vol 8. Springer Science & Business Media, Dordrecht

Flindt MR, Pardal MA, Lillebø AI, Martins I, Marques JC (1999) Nutrient cycling and plant dynamics in estuaries: a brief review. Acta Oecol 20(4):237–248

Gattuso JP, Frankignoulle M, Wollast R (1998) Carbon and carbonate metabolism in coastal aquatic ecosystems. Annu Rev Ecol Syst 29(1):405–434
Goldberg ED (1995) Emerging problems in the coastal zone for the twenty-first century. Mar Pollut Bull 31(4–12):152–158
Granéli W, Lindell M, Tranvik L (1996) Photo-oxidative production of dissolved inorganic carbon in lakes of different humic content. Limnol Oceanogr 41(4):698–706
Hedges JI, Keil RG, Benner R (1997) What happens to terrestrial organic matter in the ocean? Org Geochem 27(5–6):195–212
Hofmann DJ, Butler JH, Tans PP (2009) A new look at atmospheric carbon dioxide. Atmos Environ 43(12):2084–2086
Hopkinson CS, Smith EM (2005) Estuarine respiration: an overview of benthic, pelagic, and whole system respiration. In: Respiration in aquatic ecosystems, pp 122–146. http://iomenvis.nic.in
Kumar KR, Sahai AK, Kumar KK, Patwardhan SK, Mishra PK, Revadekar JV, Kamala K, Pant GB (2006) High-resolution climate change scenarios for India for the 21st century. Curr Sci 90(3):334–345
Leonard LA, Luther ME (1995) Flow hydrodynamics in tidal marsh canopies. Limnol Oceanogr 40(8):1474–1484
Liu KK, Atkinson L, Chen CTA, Gao S, Hall J, Macdonald RW, McManus LT, Quinones R (2000) Exploring continental margin carbon fluxes on a global scale. EOS Trans Am Geophys Union 81(52):641–644
Lohrenz SE, Redalje DG, Verity PG, Flagg C, Matulewski KV (2002) Primary production on the continental shelf off Cape Hatteras, North Carolina. Deep-Sea Res II 49:4479–4509
Lomas MW, Glibert PM, Clougherty DA, Huber DR, Jones J, Alexander J, Haramoto E (2001) Elevated organic nutrient ratios associated with brown tide algal blooms of Aureococcus anophagefferens (Pelagophyceae). J Plankton Res 23(12):1339–1344
Ludwig W, Probst JL (1998) River sediment discharge to the oceans; present-day controls and global budgets. Am J Sci 298(4):265–295
Mehrbach C, Culberson CH, Hawley JE, Pytkowicx RM (1973) Measurement of the apparent dissociation constants of carbonic acid in seawater at atmospheric pressure 1. Limnol Oceanogr 18(6):897–907
Meybeck M, Vörösmarty C (2005) Fluvial filtering of land-to-ocean fluxes: from natural Holocene variations to Anthropocene. Compt Rendus Geosci 337(1–2):107–123
Mishra SP (2016) Estuaries and lateral channel development along east coast of India. Int J Adv Res 4(12):2360–2371
Moore DE, Lockner DA, Ma S, Summers R, Byerlee JD (1997) Strengths of serpentinite gouges at elevated temperatures. J Geophys Res 102:14787–14801
Muller Karger FE, Varela R, Thunell R, Luerssen R, Hu C, Walsh JJ (2005) The importance of continental margins in the global carbon cycle. Geophys Res Lett 32(1), L01602, https://doi.org/10.1029/2004GL021346
Nayak GN, Hanamgond PT (2010) Chapter on India in Encyclopedia of the World's coastal landforms. Springer, Dordrecht, pp 1065–1070
Nicholls RJ, Wong PP, Burkett VR, Codignotto J, Hay J, McLean R, Ragoonaden S, Woodroffe CD (2007) Coastal systems and low-lying areas. In: Parry ML, Canziani OF, Palutikof JP, van der Linden PJ, Hanson CE (eds) Climate change 2007: impacts, adaptation and vulnerability. Contribution of Working Group II to the fourth assessment report of the Intergovernmental Panel on Climate Change. Cambridge University Press, Cambridge, UK, pp 315–356
Obernosterer I, Benner R (2004) Competition between biological and photochemical processes in the mineralization of dissolved organic carbon. Limnol Oceanogr 49(1):117–124
Pauly D, Christensen V (1995) Primary production required to sustain global fisheries. Nature 374(6519):255–257
Scavia D, Field JC, Boesch DF, Buddemeier RW, Burkett V, Cayan DR, Fogarty M, Harwell MA, Howarth RW, Mason C, Reed DJ (2002) Climate change impacts on US coastal and marine ecosystems. Estuaries 25(2):149–164

Schlesinger WH (1997) Biochemistry: an analysis of global change, 2nd edn. Academic Press, New York, p 588
Smith RA, Alexander RB, Schwarz GE (2003) Natural background concentrations of nutrients in streams and rivers of the conterminous United States. Environ Sci Technol 37:3039–3047
Syvitski JP, Vörösmarty CJ, Kettner AJ, Green P (2005) Impact of humans on the flux of terrestrial sediment to the global coastal ocean. Science 308(5720):376–380
Turner RE, Rabalais NN, Swenson EM, Kasprzak M, Romaire T (2005) Summer hypoxia in the northern Gulf of Mexico and its prediction from 1978 to 1995. Mar Environ Res 59:65–77
Walling DE (2006) Human impact on land–ocean sediment transfer by the world's rivers. Geomorphology 79(3–4):192–216
Wang ZA, Cai WJ (2004) Carbon dioxide degassing and inorganic carbon export from a marsh-dominated estuary (the Duplin River): a marsh CO_2 pump. Limnol Oceanogr 49(2):341–354
Wang D, Chen Z, Wang J, Xu S, Yang H, Chen H, Yang L, Hu L (2007) Summer-time denitrification and nitrous oxide exchange in the intertidal zone of the Yangtze Estuary. Estuar Coast Shelf Sci 73(1–2):43–53
Wetzel RG (2001) Limnology, 3rd edn. Academic Press, San Diego
Wolanski E (1995) Transport of sediment in mangrove swamps. In: Asia-Pacific symposium on mangrove ecosystems. Springer, Dordrecht, pp 31–42
Wolanski E, Williams D, Hanert E (2006) The sediment trapping efficiency of the macro-tidal Daly Estuary, tropical Australia. Estuar Coast Shelf Sci 69(1–2):291–298

Chapter 2
Geological Setup of the East Coast of India

Bijan Kumar Saha

Abstract Precambrian crystalline highlands, quaternary sediments, marine deposits, and alluvium occupy the east coast of India and its hinterland. The Precambrian comprising charnockite, khondalite, granite gneiss, and schist are exposed in the central and southern part of the east coast. The major part of the east coast pertains to the low-lying deltaic and estuarine environment at an average elevation of 1–3 m above sea level. The sand dunes on the east coast, at places, are well developed, having considerable width, notably a height of 18 m along Ganjam, Orissa, and 17 m along Baruva, Andhra Pradesh. Neotectonic faults have resulted in upliftment/subsidence of seabed and adjoining coast at several places, e.g. the one across Pulicat Lagoon and offshore extension of east-west faults and vertical upliftment of the seabed in Palk Bay, Tamil Nadu. The east-west trending lineaments associated with thermal springs on the Palk Bay coast are deep-seated. Near Baruva, Mid-Holocene transgression coincided with neotectonic upliftment and led to the evolution of the barrier-lagoon coast. Geological and physical processes play a crucial role in the development of different types of estuaries. The Hooghly estuary is one of the significant estuarine systems of India, having a long tidal intrusion and characterized by season-wise and location-wise variation in salinity and dissolved oxygen levels.

Keywords East coast of India · Bay of Bengal · Eastern continental shelf · Neotectonics · Active faults · Precambrian · Eastern Ghats · Promontories and cliff · Delta · Estuaries · Lagoon · Rocky coast · Aeolian sand dunes · Red sediments · Sandy beaches · Placer minerals · Erosion · Coast configuration · Salinity

2.1 Introduction

The coastal zone extends from the coastal plains to continental shelves, approximately matching the region that has been alternately flooded and exposed during the sea-level fluctuations of the late Quaternary period. Thereby, the coastal domain covers parts of the area above and below the present sea level.

B. K. Saha (✉)
Formerly, Geological Survey of India, Kolkata, West Bengal, India

School of Oceanographic Studies, Jadavpur University, Kolkata, West Bengal, India

S. Das, T. Ghosh (eds.), *Estuarine Biogeochemical Dynamics of the East Coast of India*, https://doi.org/10.1007/978-3-030-68980-3_2

The east coast of India has a long coastline from Ganga Delta in the northeast to Kanyakumari in the southwest, stretching over the Indian states of West Bengal, Orissa, Andhra Pradesh, and Tamil Nadu from north to south. The coast configuration varies from NE-SW direction in the northern and central part (north of 15° latitudes) of the east coast to N-S in the southern sector, especially from Chennai to Point Calimere, and swerves southwest to Kanyakumari along Palk Bay and Gulf of Mannar. The coastline along the Ganga Delta trends in the E-W direction. The east-west trending sand shoals/coral ridges around Rameswaram-Dhanushkodi area with a few of them submerged, known as Adam's Bridge connecting Jaffna Peninsula (Sri Lanka), separate Palk Bay from the Gulf of Mannar. Thick mangrove forests of Sunderbans and Pichavaram, as well as around Godavari and Mahanadi deltas, have emerged as protective barriers.

The Indian subcontinent witnessed climate changes since the last glacial maximum (LGM). During LGM, the climate was marked by weak southwest monsoon with reduced rainstorms and reduced fluvial activity and domination of the north-east monsoon with scanty rainfall in Bengal Basin. At the end of LGM, the north-east monsoon weakened, and the intensity of the southwest monsoon became stronger. The maximum southwest monsoon precipitation was recorded during the early Holocene to 5500 years BP, and the reduced minimum precipitation was in the late Holocene (around 3500 years BP). Cool and dry phases witnessed the widespread deposition of sediments by rivers and also the active dune-building processes along the coast, whereas warm and wet monsoonal phases experienced large-scale fluvial erosion (Kale 2004). The changes in monsoon precipitation, wind pattern, sea level, runoff, and sediment load had shaped the configuration of the coast.

Keeping in view the overall perspective of the book, this chapter presented a brief account of the geology of the east coast of India.

2.2 Geology and Tectonics of the East Coast of India

It is postulated that the east coast of India came into existence after its separation from Australia and Antarctica (Subramanya 1994). Since the last century, the peninsular shield or the stable continental region has been witnessing moderate seismicity, indicating tectonic activity or reactivation of Precambrian faults in the area (Murthy 2006).

Presently, the east coast can be categorized as (a) stable coast, (b) coastal tract with active faults, and (c) pre-Holocene seismic imprints (Banerjee et al. 2001). Coastal areas that are relatively free from the late Quaternary faulting lie from Bheemunipatnam to Pudimadaka on either side of Vishakhapatnam. Further south, a 380 km stretch of stable coast occurs along the Gulf of Mannar, where no indication of an active fault is recorded (Banerjee et al. 2001). Beach ridges, sand dunes, Quaternary aeolinite and grainstone, Neogene grainstone, and limestone mostly cover this stretch of coast, underlain by Precambrian garnetiferous charnockite.

There are areas of active faults in the eastern coastal belt. Seismological section delineated a series of faults cutting across up to Jurassic sediments, up to ~4.5 km beneath, from off Paradip to off west of Ganga Delta (surface seabed map off Paradip: GSI 2005). In Mahanadi offshore Precambrian basement about 3 km beneath near the coast and 8.5 km beneath in the seaward, which is faulted at several places, has been picked up from coastline up to ~75 km seaward (surface seabed map off Chilika-1; GSI 2002). Jagannathan et al. (1983) have established the horst-graben structure of the sedimentary formations of lower Cretaceous to upper Miocene off Puri-Konarak.

Shallow seismic surveys by the Geological Survey of India (GSI) picked up signatures of faults in the different offshore domains of the east coast. On the coast near Baruva, Vaz et al. (1998) recorded lateral shifting of coastline along the E-W trending fault. Offshore extension of this fault is also reported (Banerjee et al. 2001). Mid-Holocene transgression coincided with this neotectonic upliftment of the southern part along the lineament had led to the evolution of the barrier-lagoon coast near Baruva. Transgressed shoreline on the western side of Vasishta Godavari River mouth can be due to subsidence caused by neotectonism. The signatures of neotectonism have been reported both on land and offshore in the area. Other factors are divergence of littoral current caused by the perpendicular oriented shoreline and the presence of river mouth bar (Mahapatra et al. 2002). Vaz and Banerjee (1997) discussed the block faults in the basement across Pulicat Lagoon. This lagoon might have reached its maximum spread in the Mid-Holocene period. The submergence of coastal forests along Palk Bay in recent times may be due to relative vertical movements along E-W lineaments, as reported by Vaz et al. (2006). A landward shift of the coast and the formation of a localized bay north of River Vellar and Manamelkudi lineaments (~10° N lat.) support the movement of the southern block of the lineaments. Offshore extension of E-W faults and vertical upliftment of the seabed in the southern block formed the shore perpendicular 9 km spit off Manamelkudi. The east-west trending lineaments in association with thermal springs on the Palk Bay coast are deep-seated and neotectonically activated. A linear alignment of hot springs far away from the volcanic arc is indicative of recent movements along fault zones. Water samples collected from thermal springs located close to the Palk Bay coast show higher values of Na (568 ppm) and Cl (1015 ppm), indicating possible mixing of seawater with thermal water at shallow levels. The geomorphic shape of the Dhanushkodi coast has changed. The southern part of erstwhile Dhanushkodi Town subsided during the year 1948–1949 (Vaz et al. 2007). Analysis of magnetic data of Cauvery Basin indicated that the basin is controlled with two major E-W faults extending from land into offshore at north off Pondicherry and at north off Vedaranyam (Murthy et al. 2006). Diamond Harbour lies in the delta of river Ganga (about 50 km upstream of the coast), where subsidence rates vary up to 4 mm/year (Goodbred and Kuel 2000).

The east coast of India and its hinterland can be categorized as (a) highlands predominantly constituting Precambrian crystalline, (b) sedimentaries and raised matured dunes, and (c) coastal plains composed of unconsolidated aeolian and marine deposits.

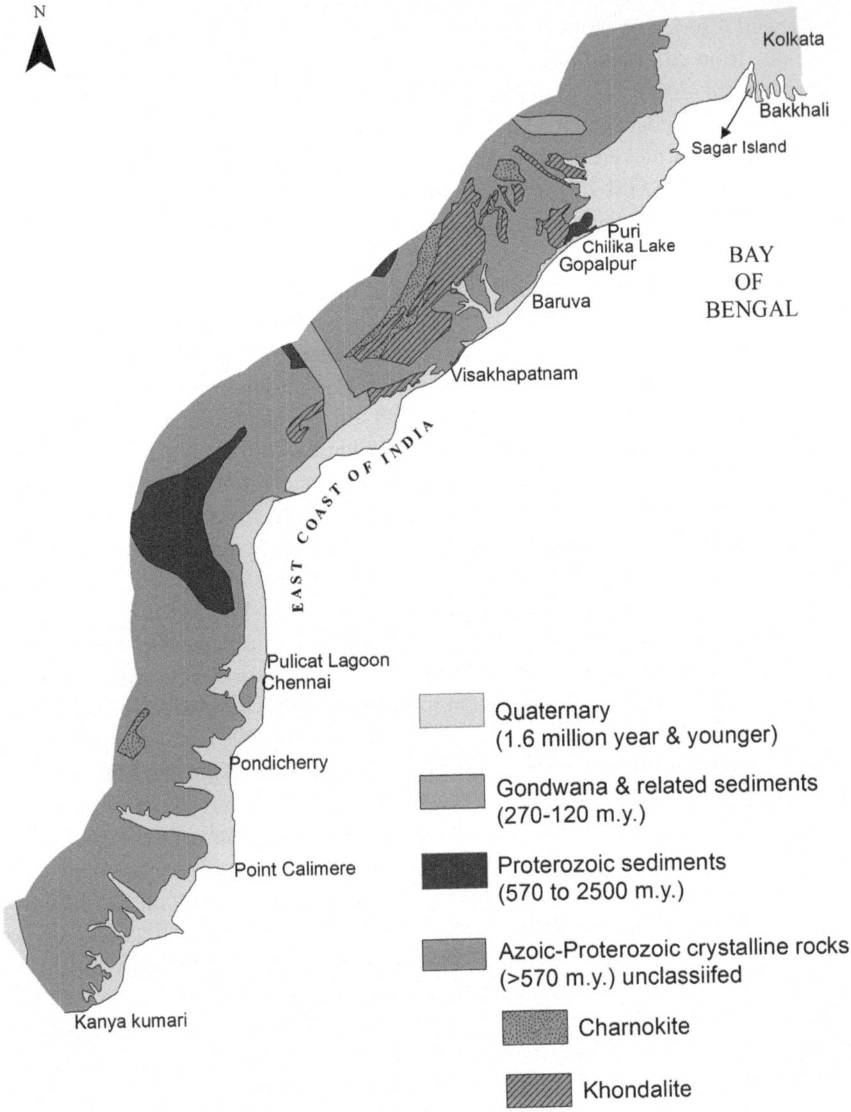

Fig. 2.1 Outline geological map of east coast of India: compiled after map published by Geological Survey of India (1968) and after map by Acharyya (2004)

The Precambrian rocks comprising charnockite, khondalite, granite gneiss, and schist with emplacement of dykes occupy mostly the central and southern parts of the east coast of India (Fig. 2.1). The rocky coastline around Visakhapatnam stretches intermittently for about 100 km and exposes rocks of khondalite of Eastern Ghats Mobile Belt. Promontories and cliffs are present in northern Andhra Pradesh and the Tamil Nadu coast. At Erayyapalem, 30 km north of Visakhapatnam, a cliff is present

Fig. 2.2 Rocky cliff near Visakhapatnam Light House, Andhra Pradesh, India

right on the coastline. Khondalite is the predominant rock type in the Visakhapatnam area, with the presence of charnockite in the hinterland. In Visakhapatnam, there is a rocky cliff on the sea face (Fig. 2.2). In Jodugullapalem, Visakhapatnam, there is a beautiful rocky (khondalite) coast (Fig. 2.3). Here, the wave-cut platform/coastal bench/shore platform is a flat-topped rock. It is a gently sloping surface having considerable width. A small detached part (farthest in the photograph towards/in the sea) is headland eroded by seawater action (wave) crashing against the rock, which caused them later to collapse and form free-standing stacks (Fig. 2.3). A sea arch (Fig. 2.4) has been developed near Thotlakonda, about 10 km north of Visakhapatnam. Sea arch/natural arch have been formed due to wave erosion with an opening underneath. Soft rock materials got eroded because of continuous seawater (wave and current) action. Near the Gopalpur coast, there are patches of khondalite. On the western flank of Chilika Lake, the hill made up of this rock is prominent with patchy charnockite occurrences. The geological map (1968), published by GSI, describes the rest of the Eastern Ghats rocks as unclassified crystalline. A geological map compiled by Aftalion et al. (1988) shows the principal areas of khondalite and charnockite occurrences. Khondalite is exposed along the coast from near Puri to Vishakhapatnam and slightly beyond south with patches of charnockite about 100 km inland. In the hinterland of Chennai, charnockite is exposed, like St. Thomas hills in the south of Chennai. Tamil Nadu, along with the Union Territory of Pondicherry coast, is known for Precambrian crystalline rocks overlain by Gondwana,

Fig. 2.3 Rocky coast near Jodugullapalem, Andhra Pradesh, India

Fig. 2.4 Sea arch near Thotlakonda, Andhra Pradesh, India

Cretaceous, and Tertiary sediments with a cover of soils and alluvium (Subramanian and Selvan 2001). The city of Chennai is on a broad sandy coastal plain, formed by Holocene progradation of beach with several coast-parallel beach ridges.

The deltas, formed by the rivers like Cauvery, Pennar, Krishna, Godavari, Mahanadi, Subarnarekha, and Ganges, are the most significant features on the east coast. A substantial part of the east coast pertains to the low-lying deltaic and estuarine environment at an average elevation of 1–3 m above sea level. The studies brought out progradation of delta in the river mouths of Hooghly (60 m/year), Mahanadi (10 m/year), Krishna and Godavari (15 m/year), and Cauvery (5 m/year) (Nayak and Hanamgond 2010). The Cauvery delta stretches from near Pondicherry in the north to Vedaranyam in the south.

There are prominent geomorphic elements on the east coast like beaches, spits, aeolian dunes, wetlands, estuaries, tidal inlets, and lagoons. The dunes on the coast, at places, are well developed, having considerable width. The height of the dune is 18 m near Ganjam and 17 m along Baruva. In the central and southern parts of the east coast, the dunes are stabilized and paved the way for habitation. North of Baruva, the dunes are migrating towards land. On either side of Baruva, there are shore-linear patches of red sediments, sandwiched between the beach in the east and pediplains (composed of highly deformed metamorphic rocks of khondalite group with soil cover) in the west, occurring from Bhavanapadu to Ichchapuram. The width of these red sediments is about 10 km north of Baruva (Vaz et al. 2002). The red sediments attain a height of 15–30 m in the area south of Baruva. Red dunes also exist along the coastal tract between Visakhapatnam and Bheemunipatnam (Vaz et al. 1998). These low-lying loose and unconsolidated dunes have beautiful exposures near Bheemunipatnam, located 20 km northeast of Visakhapatnam (Fig. 2.5). These are migrating dunes with denudation, and the presence of gullies is attributed to the badland topography. The bottom-most sand unit is fluvial, whereas the overlying counterparts are aeolian in origin. The sediments of aeolian parts underwent transport during the lowered sea level of the last glacial maximum. Iron-bearing minerals in the sand, because of geochemical activity, resulted in the red colouring of the sediments. Calcareous concretions are also present, which yielded a radiometric age of Mid-Holocene. The growth of the spit, enclosing Kakinada Bay, is due to enhanced sediment discharge from the Godavari River. Krishna delta is growing with the addition of spits and bars. In the Krishna delta, 4–6 m above mean sea level, beach ridges occur even 20 km inland (Nayak and Hanamgond 2010).

The Orissa coast is under strong littoral drift and longshore current from south to north (Mohanty et al. 2008). There are significant shoreline changes near the Mahanadi estuary. The Orissa coast is prograding and depositional in nature, having estuaries, mangrove forest of Bhitarkanika, Chilika lagoon, and sandy beaches (Mohanty et al. 2008). The Gopalpur coast of Ganjam is characterized by a steep slope beach (5°–8°) with a well-developed berm. Along the Ganjam coast, there are three rows of dunes: neo dune, old dune, and ancient dune (Chakraborty et al. 2019). Major aeolian processes in the coastal domain were more active during LGM, during which period aeolian deposit formation took place in the coast of Southeast Asia, east coast of India, and Sri Lanka and considered as an oldest aeolian unit

Fig. 2.5 Red sand dunes in Bheemunipatnam, Andhra Pradesh, India

(Bowler et al. 1995). The ancient dune unit of the Ganjam coast, occurring about 700 m inland, might have been formed during the last glacial maximum. Three aeolian dune units along the Ganjam coast might have been formed by the three successive dune-building phases during LGM (18,000 years BP), Mid-Holocene (4500 years BP), and medieval period (1400–1800 AD) or modern period (Dash and Adhikari 2008). The red colour of an ancient dune is due to the highly oxidized dark red colour sediments, which formed due to in situ weathering of iron-bearing heavy minerals (Chakraborty et al. 2019) (Fig. 2.6).

The West Bengal coast, lying in the northeastern part of India, is characterized by the presence of the largest tide-dominated Hooghly estuary with numerous channels, creeks, and mangroves of Sunderbans. This part of the coast can be distinguished into two geological environments, having contrasting tidal range, beach morphology, and provenance of sediments: (1) coast lying east of Hooghly River (downstream part of river Ganga) mouth from Sagar Island to Bakkhali – Henry's Island and further east – which is under meso- to macro-tidal environment with erosion and accretion in phases, where sediments are mostly derived from Ganga River system and (2) coast lying west of Hooghly River from Shankarpur to Udaipur (Subarnarekha River mouth), which is under meso-tidal environment with strong erosive phases and not much of sediment input from Hooghly River. The Junput coast, lying along the western flank of Hooghly River mouth, may be categorized in between where sediments are derived from the Hooghly River (Chakraborty et al. 2017). The author has studied the entire coast from Subarnarekha River mouth to Saptamukhi River mouth

Fig. 2.6 Ancient dune of Ganjam coast, Orissa, India

since the 1970s. Presently, in several places, due to anthropogenic activities, dunes are being damaged and flattened, and only a few are left for studies. Sagar Island, Bakkhali, and Henry's Island are part of the subaerial delta plain of the Ganga Delta system as well as part of the Hooghly estuarine environment.

The gently sloping (1.3^0–0.5^0) 84–528 m wide beach of Bakkhali is covered by greyish-white fine to very fine sands with exposed palaeomud, at places, in the lower part of the intertidal zone. In the backshore area of Bakkhali coast, there are aeolian dunes like neo dunes, followed landward by a continuous belt of mature dunes and older dunes with undulating topography. A strong storm in the year 1984 has washed out 13 m high dunes in the Bakkhali area. There is a net erosion of coast of 100 to 175 m from the year 1969 to 2001. After this period, the coast is more or less stable concerning erosion and accretion with marginal seasonal changes (Saha et al. 2011). The swampy intertidal zone and part of the backshore of Henry's Island (adjoining east of Bakkhali) of the year 1969 have been eroded and submerged. The inundation was up to 1000 m inland. The present intertidal zone was a part of an erstwhile mixed forest with clayey/silt sediments. Modern sands, transported from the near-coastal seabed, are now being deposited over the clayey/silt bed of the present intertidal zone, and this has resulted in the development of a beautiful sand beach (Saha et al. 2014) (Fig. 2.7).

Sagar Island is the largest island of the Ganga deltaic system. Hooghly River flows to its west. Perhaps, the island was completely covered by mangroves until the year 1811. From the changing configuration of the island during the years 1851–1992, it appears that the island area was reduced significantly by fluvial and marine processes (Bandyopadhyay 1997) and due to slope failure and landslide (Saha and Saha 2014). The Gangasagar coast, lying in the southernmost part of Sagar Island and facing the Bay of Bengal, is 2.8 m above MSL with a width of

Fig. 2.7 Henry's Island coast, West Bengal, India

intertidal zone varying from 65 m to 382 m and slope varying from 1.51^0 to 0.42^0. The entire coast is strongly influenced by high tidal waves, especially during spring tide. The coast has experienced several erosion-accretion phases, recorded since 1969. There is a net landward shift of coastline in the eastern sector by 600–800 m and the western sector by 250–500 m.

Udaipur-Shankarpur and Junput coasts are having three distinct rows of dunes, separated by an old clayey tidal flat. These are neo dunes, followed landward by the older dune, old tidal flat, and older dune ridge. Well-preserved older dune ridge might have been formed during late Holocene higher sea level, before 3000 years BP (Chakraborty et al. 2017). After stabilization of the present sea level at around 3000 years BP, older dunes might have been formed due to strong aeolian action. Another dune unit, described as an ancient dune complex (brownish colour sands), lying about 10 km inland from the coast, is well documented from Udaipur to the west of Junput (Chakrabarty 1990). Its age and process of formation are yet to be established. Perhaps, this ancient dune might have been formed in similar environments like the red dune on the Ganjam coast and on the Bheemunipatnam coast.

2.3 Seabed Character of Eastern Continental Shelf of India, Bay of Bengal

The coastline of the east coast of India is fronted by a well-defined continental shelf of the Bay of Bengal. The Bay of Bengal is an embayment bound by the east coast of India, Bangladesh, Myanmar, Sri Lanka, and Andaman-Nicobar Islands. River systems like Ganga, Mahanadi, Godavari, and Krishna on the Indian side, the Brahmaputra in Bangladesh, and Irrawaddy in Myanmar are the principal contributors of sediments in the Bay of Bengal. There are several active channels and submarine canyons on the eastern shelf and beyond.

The width of the eastern continental shelf of India is relatively narrow compared to the western shelf. The shelf width is about 100 km off Ganga Delta, 60 km off Chilika Lake, 15–20 km off Krishna and Godavari delta, and 25–35 km off Pondicherry-Porto Novo (Mahapatra et al. 1992; Mahapatra and Hari Prasad 2005; Saha 2009). In general, terrigenous sands occupy the continental shelf mostly within 200 m isobath, except in the deltaic area, and clay/silty sediments cover the outer shelf. From Ganga Delta to Paradip offshore off the northern part of the east coast, these sands occur within 100 m isobath, and off Tuticorin Coast in the south this is within 200 m water depth, where sands are having coral fragments (surface seabed sediment map, GSI 2003). There are relict sands in the inner/middle shelf. These palaeosands are differentiated from the modern sands by their coarser size, iron staining, and dissolution pitting due to subaerial weathering. Relict sands were partly reworked by transgressed sea or covered by modern sediments (sands and clay/silt) at many places on the eastern shelf. Some of these situations are cited. The palaeosands off Bahuda River, south of Gopalpur, are exposed on the seabed, close to the river mouth. Away from the coast, the relict sands are veneered by 110 cm thick silty sediments. The radiocarbon date of these buried relict sands is 10, 690 ± 120 years BP (Bhattacharjee et al. 2007). Finer sediments cover the seabed close to the Krishna River mouth off Nizampatnam, whereas the sandy zone is 20 km away and around 100 m isobaths (Rao and Sesha Rao 2002). Off the Digha-Udaipur coast, modern sands are present up to 1.5 km, followed seaward by clayey silt (Mitra and Samadder 2003).

The northeastern part of the continental shelf of India is characterized by the presence of several deltaic islands. The major contribution of sediments to this part is through the River Ganga. One-third of the total sediment load of this river is accumulated on the shelf. A substantial load of sediments carried by Ganga River systems and the tidal regime resulted in the formation of sand bars and shoals in the inner shelf, making significant changes in the geomorphic pattern in the near coastal seabed in recent times. This part of the shelf is about 100 km wide with an average slope of less than 1°. Prominent U- to V-shaped submarine canyon known as "Swatch of No Ground" accepts sediment load in this part of the shelf. This canyon is the presumed location of the river mouth at the last glacial sea-level low. The canyon is 8–10 km wide (Sengupta et al. 1992a). The coarse sediments on the floor of the canyon and the finer sediments on the flank indicate that the canyon is active as a conduit for sediment transport.

Erosion in the Ganga Delta front is not only localized along the land-sea boundary but also in the marine domain. The delta front zone of Sagar Island offshore is being eroded, as evident from the shoreward shifting of 10 m isobath by about 3 to 8 km from the year 1985 to 2000 (Biswas et al. 2007). The rate of erosion is not uniform everywhere. Off Sagar Island, submarine erosion of sandy shoals and subsequent landward movement of materials by tidal current leading to deposition of new sandbar close to the coast are prominent. Subsidence due to compaction or neotectonic activity and strong tidal currents are also probable reasons for the disappearance of some of the earlier shoals. The signature of underwater slope failure followed by slumping is recorded in the near-coastal seabed off the southwest coast of Sagar Island (Saha and Saha 2014). Dredging, being done regularly in the main navigation channel of the Hooghly River, may also cause instability of the slope. The eastern coast of Sagar Island has been under strong cliff erosion. In the easternmost part of Bakkhali, deposition of very fine to fine sands on palaeomud in the low tide area might have been caused by shoreward transportation of sandy sediments after submarine erosion of sandbar. Near Bakkhali, the shifting of seabed sediment towards the coast is illustrated by silting up of Saptamukhi creek mouth and the appearance of new shoals in the estuary. Comparison of isobath drawn in the year 2012 with that of 1969 indicated a shoreward shifting of deeper isobath lines and steep slope along the sandy shoal all along due to submarine erosion by bottom friction of strong tidal current in this narrow entrance (Saha et al. 2011).

2.4 Impact of Late Quaternary Sea-Level Fluctuations

The sea has stood at both higher and lower levels on the east coast during the late Quaternary in respect to the present sea level. The existing coastal geomorphology has evolved during the late Quaternary marine transgression.

The signatures of the last glacial maximum and subsequent rise of sea level as well as its fluctuations during the late Quaternary are traced on the seabed from the near shelf edge to the middle shelf of the east coast. During the late Holocene marine transgression, the rise of seawater by 2–3 m above the present sea level is witnessed in the form of raised marine terraces, sea cliffs, natural bridge, and wave-cut terraces/notches/sea caves on the Precambrian rocks in the central part of the east coast in the areas of Rishkonda, Erayyapalem, Tugidam, and near Visakhapatnam. This transgressed phase was followed by a regressed one, as evidenced by the presence of palaeobeach ridges 2–5 km and further inland. Wave-cut notches are present near Erayyapalem at different levels of 1.06 m, 1.8 m, 2.73 m, and 4.73 m higher than the present sea level, which indicated imprints of higher sea levels of the last glacial cycle. Banerjee et al. (2001) recorded six terraces near Visakhapatnam at 6 m, 3.84–4.00 m, 2.3 m (radiocarbon date, 3480 ± 90 years BP), 1.8 m, 1.3 m, and 1.00 m (date 1730 ± 140 years BP) above the present sea level. The coral reefs in the Rameswaram peninsula appear to have formed around 6000 years BP.

Bathymetry and shallow seismic surveys of GSI recorded three to four distinct palaeostrand lines on the seabed, mostly between 20 and 120 m of water depths. Terraces/platforms, oolites, coarse sands, oxidized sands, peaty materials, and buried corals are the salient characteristics of strandlines/palaeoshorelines at different stages of the lowered sea (Saha 2011). The studies carried out by several workers, namely, Banerjee and Sengupta (1992), Vaz (1996, 2000), Ramamohana and Viswanath (2008), and Sreenivas et al. (1990), have given the following radiocarbon dates:

Off Visakhapatnam, corals exist at 80 m (10,790 ± 170 years BP) and 100 m (12,530 ± 170 years BP) of water depths. Off Nizampatnam Bay, the mollusc shells at 17 m of water depth yielded dates of 8200 ± 120 years BP. The corals from 115 m of water depth off Mahabalipuram and from 125 m of water depth off northeast of Karaikal dated 14,510 ± 190 years BP and 18,390 ± 220 years BP, respectively.

Off Gopalpur, the terrace/topographic highs are at 20–25 m, 30–35 m, 40–45 m, 50–55 m, 60–65 m, and 75–80 m of present water depths. Off Chennai, ooids are recorded at 60 m and 110 m water depths. Terrace/ridges around 100 m isobaths are distinct and continuous throughout, unlike the one at 120 m, perhaps coinciding with the LGM.

2.5 Placer Mineral Resources

Placers are deposits of economic heavy minerals, deposited by mechanical concentrations of detrital mineral particles after liberating on the breakdown of rocks or vein minerals. The formation of placers takes place through dynamic processes having suitable source rocks, favourable climatic conditions, weathering, transportation by the drainage system, and sorting.

Beach placer minerals are present in patches along the east coast of India. There are good concentrations of placer minerals in the Kalingapatnam area, north Andhra Pradesh coast, and in the Ganjam area, south Orissa coast. The intertidal zone, as well as the dunes of Gopalpur in Ganjam, contains a good concentration of heavy minerals having economic potential. Among the heavy minerals, ilmenite is the dominant one. Extensive dune sand deposits with heavy mineral content like ilmenite, sillimanite, garnet, zircon, rutile, and monazite occur along Chatrapur (Ganjam) coast over a linear stretch of 18 km in an area of 26 sq. km between Gopalpur in the south and Rushikulya River in the north and contain about 230 million tons of sands with 20–25% of heavy minerals (source: a pamphlet of Indian Rare Earth Ltd n.d.). The placer mineral deposits contain 9.4% ilmenite, 6.8% garnet, 0.4% rutile, 0.33% zircon, and 0.29% monazite (Rajamanickam 1999; source: Indian Mineral yearbook 1995). The heavy minerals are being exploited by IRE at its Chatrapur plant. Beach placer minerals are also present in the Tirunelveli, Ramnad, and Tanjore districts of Tamil Nadu (Rajamanickam 1999).

During the lowered sea level of Pleistocene time, heavy minerals were deposited on the exposed middle shelf area. When the sea level rose during the end of Pleistocene, these placer minerals were covered by rising water levels and are reworked, spread over the area destroyed, or partly covered by recent sediments. Offshore exploration by GSI brought out a rich incidence of placer minerals like ilmenite, sillimanite, garnet, monazite, zircon, and rutile in the inner shelf off Orissa between Gopalpur and Chilika Lake. The heavy mineral-bearing sands extend up to 10–12 km offshore and occur mostly within 30 m of water depth (Sengupta et al. 1992b, 2002; Saha 2005, 2007). The concentration is high (~15 wt. %) off Gopalpur. The continental shelf off Kalingapatnam-Baruva area has a substantial concentration (~6 wt. %) of economic heavy minerals within 15–20 m of water depth (Ravi Kumar et al. 2002; Saha 2007).

2.6 Physical Processes Along the East Coast of India

Geological processes, along with physical processes, like, wave, wind, current, and tide and sediment influx, play a critical role in the formation of the coastal domain and for the development of the estuarine system. The east coast is experiencing varying tidal ranges of 4.3 m at Sagar Island (the mouth of Hooghly River), 1.3 m at Kakinada, 1.0 m at Chennai, and 0.9 m at Pondicherry (Nayak and Hanamgond 2010).

The Hooghly estuary is one of the most important estuarine systems of India because of its large drainage area of the Himalayan originated River Ganga and a long tidal intrusion up to 280 km inland. Season-wise and location-wise changes are observed in the salinity and dissolved oxygen values from the river mouth to away. This estuary has three distinct geographical zones characterized by varying physicochemical and hydrological parameters (Bhattathiri 2001). Tidal creeks of the Hooghly estuarine system have dissected the landmass into many smaller islands. Mahanadi estuary is a tide-dominated coastal plain shallow estuary, where temperatures of surface and bottom waters do not differ much. The estuary of Rushikulya River is shallow and well mixed. The Godavari estuary is having a maximum width of 1000 m. The Vasishta Godavari estuarine system experiences seasonal stratification variation like salt wedge type in July, partly mixed type in September, and well-mixed type during February (Ramana et al. 1989). The maximum depth of the Cauvery estuary is 12 m. The Vellar estuary comprises four zones with varying salinity distribution patterns (Bhattathiri 2001). When the rivers discharge freshwater in the sea, a low salinity zone occurs at the discharge points, closer to the coast than offshore. In contrast, from in situ observations, Mahapatra and Rao (2017) recorded low salinity pool(s) in the offshore region (south of 16° N lat.) and its migration to further offshore, compared to the discharge points of Krishna and Godavari Rivers. They suggested that this anomalous low salinity pool in the offshore away from the coast would be due to orientation of coastline from 14^0 to 17^0N lat., surface circulation along with weak East India Coastal Current, and influx of low salinity water from larger rivers to the north.

Chilika Lake is a large coastal lagoon partly enclosed by a long sandy barrier with seawater entrances at the southern and northern ends. Two of the three openings of this lagoon towards the sea (as in 1973) eventually closed because of longshore current and littoral drift (Chandramohan and Nayak 1994). After opening the new mouth along the outer channel in September 2000, the average salinity near Satpada increased to 14% in December 2000, when compared to the salinity of 3–4% during the same period one decade ago. The primary threats of Chilika lagoon are siltation, reduction in salinity, weed infestation, a decline of fishery resources, and an overall loss of biodiversity (Mohanty et al. 2008). It changes from freshwater domination during southwest monsoon to seawater domination during pre- and post-monsoon. Pulicat Lake is brackish with huge intertidal flats and faced sea-level changes during the middle and late Holocene (Vaz and Banerjee 1997). The east coast is vulnerable to geohazards like a cyclone, storm surge, and tsunami though rare. Sri Lanka provides a shelter to the southwestern coast from the high wave action. The Palk Strait coast has low wave energy (Nayak and Hanamgond 2010).

2.7 Discussion and Conclusion

The east coast of India and its hinterland is having geological formations comprising of Precambrian crystalline, Quaternary sediments, marine deposits, and modern sediments. Precambrian, represented by khondalite, charnockite, and basic rocks, occurs mostly in the central and southern part of the coast, at places right on the sea face. A substantial part of the east coast encompasses deltas with estuaries and mangrove forests. There are extensive sandy beaches, often backed by a chain of beach ridges/matured dunes, aligned parallel to the coast. There are sand bars and shoals. Longshore current and littoral drift have nourished sand spits and barrier in front of lagoons. Coral reefs are rare on the east coast. There are defunct coral reefs in Rameswaram Island and Adam's Bridge. Hooghly estuary has a tidal intrusion up to 280 km inland. Erosion along the delta front of Ganga, Mahanadi, and Godavari Rivers and adjoining offshore has changed the configuration of the coast and as well as seabed character. This observation is evident, in particular, from the morpho-bathymetric changes in the Ganga Delta offshore. Submarine erosion of seabed off Sagar Island is prominent. The shoreline changes may be due to a combination of several factors like erosion-accretion, landslide, and neotectonic movements. Continental shelf, adjoining the east coast of India is veneered by terrigenous sands up to 200 m water depth. But in the deltaic domain, clay-silt covers the seabed. Gopalpur coast and its offshore are endowed with placer deposits, containing ilmenite, sillimanite, garnet, monazite, zircon, and rutile.

Though the east coast was considered to be a part of a stable continental margin, there are active faults, now recorded along the coast and offshore. Banerjee et al. (2001) opined that up to the 1960s Indian peninsular shield was considered to be practically aseismic because instrumentally recorded global seismic events listed only a few earthquakes from the Indian peninsula, not that seismicity was mild and rare. The theme-based geoscience database on coast and seabed terrain analysis are essential to classify the coast, especially in the context of the continuing growth of coastal settlements and sustainable developmental activities in the coastal domain (Saha 2009). Areas affected by active faults, coastal landslides, slope failure, and changes in seabed morphobathymetric patterns are to be mapped. Instrumental evidence of seismicity along the coast and adjoining offshore is required to be re-looked. A tsunami in December 2004 has alerted the geoscientific community about the need for theme-based geological appraisal of the east coast of India.

Acknowledgements The author acknowledges the technical help rendered by Ms. Jaya Pradhan, Senior Research Fellow, Department of Science and Technology, Government of West Bengal, for the compilation and drawing of the geological map. The author is thankful to Mr. Jayant Kumar, Director (Retd.), Geological Survey of India, who has gone through the text critically and has offered valuable suggestions.

References

Acharyya SK (2004) Simplified geological map of India: workshop on interlinking of rivers, Kolkata. Proceedings, West Bengal Academy of Science and Technology, p. 234

Aftalion M, Bowes DRB, Dempster TJ (1988) Late Proterozoic Charnockites in Orissa, India: A U-Pb and Rb-Sr Isotopic Study. J Geol 96(6):663–676. Published by: The University of Chicago Press

Bandyopadhyay S (1997) Natural environment hazards and their management: a case study of Sagar Island, India. Singap J Trop Geogr 18(1):20–45

Banerjee A, Sengupta R (1992) Evidences of low stands on the continental shelf of east coast of India: seminar on Recent geoscientific studies in the Bay of Bengal and the Andaman Sea, Kolkata in 1990. Proceedings volume Geol. Surv. India Spl. Pub. No. 29, pp. 163–170

Banerjee PK, Vaz GG, Sengupta BJ, Bagchi A (2001) A qualitative assessment of seismic risk along peninsular coast of India, south of 19° N. J Geodyn 31:481–498

Bhattacharjee D, Ghosh SK, Nagendran G, Katari A, Panda PK (2007) Foraminifera evidence of Holocene transgression in sediments column from the territorial waters off Behuda River mouth of Orissa, East coast of India: Geological Survey of India. Marine Wing, News letter, XXI(1):23–24

Bhattathiri PMA (2001) Features at some significant estuaries of India. In: Book on The Indian Ocean –a perspective, Oxford and IBH, New Delhi, vol 1, pp 271–298

Biswas NR, Roy Chaudhury S, Mullick A, Sakhare VV (2007) Some observations of the effect of sea level rise in the Ganga- Mahanadi delta region. Marine Wing, GSI News letter, XXI(1&2):12–13

Bowler JM, Jones R, Zhisheng AN (1995) Climatic change in Asia and West Pacific; its significance for green-house scenario. J Geol Soc India 32:1–21

Chakrabarty P (1990) Process response system analysis in the macro tidal estuarine and meso tidal coastal plain of eastern India. Mem Geol Soc India 22:165–181

Chakraborty D, Saha BK, Hazra S (2017) Geomorphology and sediment character of beach-dune complex along Udaipur, Shankarpur and Junput coasts, West Bengal, India. Indian J Geosci 71(2):413–428

Chakraborty D, Saha BK, Hazra S (2019) Sediment character of beach-dune complex of Gopalpur coast, Orissa. Indian J Geosci 73(1):25–38

Chandramohan P, Nayak BU (1994) A study for the improvement of Chilka lake tidal inlet, east coast of India. J Coast Res 10:909–918

Dash PC, Adhikari KN (2008) Aeolian sediments of Ganjam coast, Orissa and their palaeoclimate significance; Workshop on IGCP 464- Continental shelves during the last glacial cycle; Knowledge and applications in 2005, Visakhapatnam. Proceedings volume Geol. Surv. Ind. Spl. Pub. No 96, pp. 154–162

Geological Survey of India (1968) Map: Geology and minerals of India

Geological Survey of India (2002) Surface sea bed map off Chilka-1, Bay of Bengal

Geological Survey of India (2003) Surface sediment map

Geological Survey of India (2005) Surface seabed map off Paradip, Bay of Bengal

Goodbred SL, Kuel SA (2000) The significance of large sediment supply, active tectonism, and eustasy on margin sequence development: Late Quarternary stratigraphy and evolution of the Ganges-Brahmaputra delta. Sediment Geol 133:227–248

Indian Mineral year book (1995) Govt. of India

Indian Rare Earths Limited (n.d.) OSCOM; Pamphlet

Jagannathan CR, Ratnam C, Baishya NC, Das Gupta U (1983) Geology of offshore Mahanadi Basin. Petrol Asia J:101–104

Kale VS (2004) Palaeoclimatic and palaeohydrological changes during late Pleistocene and Holocene: the Indian scene. Proc AP Akademi Sci 8(3):167–174

Mahapatra DK, Rao AD (2017) Redistribution of low-salinity pools off east coast of India during southwest monsoon. Estuar Coast Shelf Sci 84:21–29

Mahapatra GP, Hari Prasad M (2005) Morphology of the southern continental margin of India off the area between Puri and Point Calimere in the east coast of India. Marine Wing, GSI News letter, XVIII(2) & XIX(1):63–64

Mahapatra GP, Hari Prasad M, Rao AD, Prabhakar P (2002) Recent sea level changes and shoreline movement from the study of river mouth processes of Vasistha Godavari, East coast of India: National seminar on Four decades of marine geosciences in India- A retrospect in 2001, Mangalore. Proceedings volume Geol. Surv. Ind. Spl. Pub. No. 74, pp. 108–113

Mahapatra GP, Rao BR, Biswas NR (1992) Morphology and surface sediments of the eastern continental shelf off peninsular India: Seminar on Recent geoscientific studies in the Bay of Bengal and the Andaman Sea, Kolkata in 1990. Proceedings volume Geol. Surv. Ind. Spl. Pub. No. 29, pp. 229–243

Mitra PK, Samadder AK (2003) Correlation of wave parameters with beach slope along some coastal tracts of West Bengal and Orissa. Marine Wing, GSI News letter, XVII(1):20–21

Mohanty PK, Panda US, Pal SR, Mishra P (2008) Monitoring and management of environmental changes along the Orissa coast. J Coast Res 24(2B):13–27

Murthy KSR (2006) Seismotectonics of the Indian subcontinent and the Bengal Fan. Proceedings of the National commemorative conference on tsunami, Madurai, pp 26–29

Murthy KSR et al (2006) Factors guiding tsunami surge at the Nagapattinam-Cuddalore shelf, Tamil Nadu, east coast of India. Curr Sci 90(11):1535–1538

Nayak GN, Hanamgond PT (2010) Chapter on India in Encyclopedia of the World's coastal landforms. Springer, pp 1065–1070

Rajamanickam GV (1999) Light heavy minerals on the Indian continental shelf, including beaches. In: Cronan DS (ed) Hand book of Marine mineral deposits. CRC, London, pp 13–26

Ramamohana RT, Viswanath K (2008) Integrated approach in the studies on the continental shelf on the east coast of India with reference to the last glacial cycle: Workshop on IGCP 464-Continental shelves during the last glacial cycle; Knowledge and applications. Proceedings volume Geol. Surv. Ind. Spl. Pub. No. 96, pp. 68–78

Ramana YV, Ranga RV, Reddy BRS (1989) Diurnal variation in salinity and currents in Vasishta Godavary estuary, east coast of India. Indian J Marine Sci 18:54–59

Rao BR, Sesha Rao PV (2002) Sea bed sediments of the Nizampatnam Bay and their texture. Marine Wing, GSI News letter, XVI(2):15–16

Ravi Kumar V, Khan MIA, Vaz GG, Vijay KP, Sankar J, Devdas V, Subba Rao V (2002) Offshore placer minerals off Iduvanipalem of north Andhra coast: National seminar on Four decades of marine geosciences in India- A retrospect in 2001, Mangalore. Proceedings volume Geol. Surv. Ind. Spl. Pub. No. 74, pp. 183–186

Saha BK (2007) Exploration of Multi-mineral placer occurrence and silica sands in the offshore domain of India: International Conference on The Role of Natural Resources and Environment in Sustainable Development in South and South East Asia (NESDA), at Dacca, 2003, Proceedings volume, 2007

Saha BK (2009) Geological appraisal and coastal hazard preparedness with special reference to east coast of India, Bay of Bengal. International conference on geosciences for global development, Dhaka, Bangladesh, 2009. NESDA seminar volume

Saha BK (2005) Exploration for non-living resources by Geological Survey of India in Bay of Bengal and some findings thereof: Seminar on mineral and energy resources of eastern and northeastern India by MGMI, Kolkata. Proceedings volume, pp. 75–78

Saha BK (2011) Presented paper on Geomorphic signatures of sea level fluctuations during Late Quaternary transgression in the eastern continental shelf of India and on its coast: IGCP 588 Conference on Quaternary coastal change and records of extreme marine inundation on coastal environments, Bangkok, Abstract volume, p. 14

Saha BK, Saha P (2014) Signatures of landslides in coastal and submarine domain in deltaic environment in eastern continental shelf of India, Bay of Bengal- some observations. Proceed. World Landslide Forum 3 conference, Beijing, Vol 4

Saha P, Saha BK, Hazra S (2014) Recent changes in coastal configuration of Henry's Island. Curr Sci 107(4):679–688

Saha P, Saha BK, Hazra S, Sinha Roy S (2011) Geomorphology and sediment character of Bakkhali-Frasergunj coastal belt and adjoining inner continental shelf, Bay of Bengal. Indian J Geosci 65(3):195–210
Sengupta R, Basu PC, Bandyopadhyay RR, Bandyopadhyay A, Rakshit S, Sharma B (1992a) Sediments in the continental shelf in and around the swatch of No Ground: Seminar on Recent geoscientific studies in the Bay of Bengal and the Andaman Sea, Kolkata in 1990. Proceedings volume Geol. Surv. Ind. Spl. Pub. No. 29, pp. 201–207
Sengupta R, Bhattacharyya S, Rakshit S, Das F, Roy Chaudhuri S (2002) The variable nature of placer occurrences from Gopalpur to Chilka Lake- significant observations: National seminar on Four decades of marine geosciences in India- A retrospect in 2001, Mangalore. Proceedings volume Geol. Surv. Ind. Spl. Pub. No. 74, pp.176–178
Sengupta R, Khalil SM, Rakshit S, Deb DK, Sinha JK, Mitra SK, Majumder S, Raghav S, Bhattacharya S (1992b) Multimineral Placer deposits in the inner shelf off Orissa coast: Seminar on Recent geoscientific studies in the Bay of Bengal and the Andaman Sea, Kolkata in 1990. Proceedings volume Geol. Surv. Ind. Spl. Pub. No. 29, pp. 135–143
Sreenivas R, Krishna Rao NVN, Swamy ASR (1990) Sedimentation and sea level variations in Nizampatnam Bay, East coast of India. Indian J Marine Sci 19:261–264
Subramanian KS, Selvan TA (2001) Book on geology of Tamil Nadu and Pondicherry. Geological Society of India
Subramanya KR (1994) Post-Gondwana tectonics of the Indian peninsula. Curr Sci 67(7):527–530
Vaz GG, Subba Rao V, Ravikumar V (2006) Thermal springs in Indian coastal areas of the Palk Bay: their implications in relation to lineaments, coastal morphology and seismicity. J Geol Soc India 68:593–596
Vaz GG (1996) Relict coral reefs and evidence of pre Holocene sea level standoff Mahabalipuram, Bay of Bengal. Curr Sci 71(3):240–241
Vaz GG, Banerjee PK (1997) Middle and late Holocene sea level changes in and around Pulicat lagoon, Bay of Bengal, India. Marine Geol 138:261–271
Vaz GG (2000) Age of relict coral reef from the continental shelf off Karaikal,Bay of Bengal: Evidence of Last Glacial Maximum. Curr Sci 79(2):230
Vaz GG, Hariprasad M, Rao BR, Subba Rao V (2007) Subsidence of southern part of erstwhile Dhanuskhodi Township, Tamil Nadu-evidences from bathymetry, side scan and under water videography. Curr Sci 92(5):671–675
Vaz GG, Mohapatra GP, Hariprasad M (1998) Origin and palaeoenvironmental aspects of red sediment from Bavanapadu-Ichchapuram, Andhra Pradesh. J Geol Soc India 52:463–471
Vaz GG, Mohapatra GP, Hariprasad M (2002) Geomorphology and evolution of Barrier-Lagoon coast in a part of North Andhra coast. Mem Geol Soc India 49:31–40

Chapter 3
Aquatic Biogeochemistry of the Estuarine and Coastal Waters of the Bay of Bengal: Impact of Physical Forcing and Extreme Atmospheric Events

Suchismita Pattanaik, Abhra Chanda, and Pradipta Kumar Mohapatra

Abstract The northeastern flank of the Indian ocean, i.e., the Bay of Bengal (BoB), is considered to be one of the most dynamic oceanic regions of the world. A significant amount of river discharge from several perennial estuaries vis-à-vis the effect of monsoon and frequent depression and tropical cyclones makes the estuarine and coastal waters of the BoB unique from various perspectives. These physical forcing and extreme atmospheric events exert a substantial impact on the biogeochemistry of the water column in these east coast estuaries and nearshore waters. We discussed the general biogeochemical conditions of the major estuaries along the east coast of India in this chapter with an emphasis on (i) thermal stratification induced during the monsoon and by the tropical cyclones (TCs), (ii) phytoplankton bloom and biological productivity along with their relationship with freshwater discharge and extreme weather events, (iii) variability in dissolved oxygen levels with relation to the physical forcing, and (iv) nutrient injection and cycling driven by monsoonal discharge and extreme weather events. We observed a stark difference between the northern estuaries of the east coast and the southern estuaries concerning biogeochemical characteristics.

Keywords Physical forcing · Atmospheric disturbance · Monsoon · Tides · River discharge · Tropical cyclone · Thermal stratification · Algal bloom · Nutrient · Hypoxia

---

S. Pattanaik (✉) · P. K. Mohapatra
Department of Botany and Biotechnology, Ravenshaw University, Cuttack, Odisha, India

A. Chanda
School of Oceanographic Studies, Jadavpur University, Kolkata, West Bengal, India

S. Das, T. Ghosh (eds.), *Estuarine Biogeochemical Dynamics of the East Coast of India*, https://doi.org/10.1007/978-3-030-68980-3_3

3.1 Introduction

Several small to large estuaries, like the Hooghly, Mahanadi, Godavari, Krishna, and Cauvery, intersperse the east coast of India facing the Bay of Bengal (BoB). The monsoonal rainfall drives most of these estuaries and is, therefore, referred to as monsoonal estuaries (Sarma et al. 2014). These estuaries experience significantly variable freshwater discharges annually. At the peak of monsoon, most of these estuaries receive so much freshwater from the monsoon rain and associated land runoff that they lose their estuarine character and temporarily transform into a riverine one (Sarma et al. 2009). Compared to the other estuaries of the world, Indian estuaries are different as freshwater discharge during the monsoon leads to loss of vertical salinity gradient, whereas, in most of the other estuaries of the world, freshwater discharge promotes vertical stratification (Christopher et al. 2002; Vijith et al. 2009). Such contrasting observations show that monsoon-associated freshet exerts a physical forcing in these estuaries, and several researchers observed that this freshwater discharge strongly influences the biogeochemistry and productivity of these estuaries (Acharyya et al. 2012; Sarma et al. 2009, 2011).

Apart from river discharge and monsoonal rain, tides also play a crucial role in regulating several biogeochemical parameters in these estuaries. The estuaries along the east coast of India, depending on their size and width, exhibit a wide range of tidal amplitudes, making some of the estuaries micro-tidal in nature and some act as meso-macro-tidal estuaries (Sarma et al. 2014). Unlike river discharge, tides regulate the variability of the estuarine biogeochemistry on a short-term scale, as most of these estuaries experience a semidiurnal tidal cycle (Ganguly et al. 2011; Akhand et al. 2016). Such a tidal nature implies that a regular shifting of riverine and marine influence takes place two times during a 24-hour cycle within these estuaries.

In addition to these forces, the BoB frequently experiences extreme weather events like tropical cyclones (TCs), and the wrath of such events often leaves severe impacts on life and property in the coastal periphery. These TCs are capable of significantly altering the mixing and hence the biogeochemistry of the upper layer of ocean water (Maneesha et al. 2012), and thus such effects are often seen in the coastal waters adjacent to the estuarine mouths. Manifestations like drastic enhancement in primary productivity, cooling of water surface temperature, and nutrient enrichment in the euphotic zone often take place along the east coast of India (Chacko 2017).

3.2 Aims and Objective

Several studies have observed that physical forcing like the monsoon, freshwater discharge, and tides along with extreme atmospheric events like TCs governs the aquatic biogeochemistry of the estuaries along the east coast of India. The main aim of this chapter is to collate all the principal findings and observations in this regard and, at the same time, discuss the present state of the art of several biogeochemical parameters in the estuaries along the east coast of India under normal conditions. We laid stress on the role of the abovementioned physical forcing in modulating these parameters.

3.3 Variability of the Basic Physicochemical Parameters in the Major Estuaries and Coastal Waters of the East Coast of India

3.3.1 *Hooghly-Matla Estuarine Complex*

The Hooghly River Estuary is the first deltaic offshoot of the River Ganges. Hooghly, along with several other estuaries like Saptamukhi, Thakuran, Matla, Bidya, Gosaba, Raimangal, and Haribhanga, intersects the world's largest mangrove of Sundarban (Indian part). This complex network of estuaries in the southern end of the Ganges-Brahmaputra-Meghna (GBM) delta (in the Indian state of West Bengal) forms the Hooghly-Matla Estuarine complex (Chatterjee et al. 2013). Among these estuaries, only the Hooghly Estuary and the Raimangal-Haribhanga estuaries, which flow through the west and east of this estuarine complex, carry freshwater perennially. However, all the other estuaries of this complex have lost their connection with the upper riverine reaches (Raha et al. 2012). This differential freshwater discharge has not only led to a marked difference in annual mean salinity between the Hooghly Estuary ($\approx$20.0) and the Matla Estuary ($\approx$7.1) but also led to a difference in carbonate chemistry dynamics between the two (Akhand et al. 2016). Due to higher freshwater discharge, there is a higher dominance of sediment-laden suspended particulate matter in the Hooghly Estuary, which is testified by the higher turbidity observed in the Hooghly compared to Matla (Akhand et al. 2016). Annual mean DO was more or less similar in the two estuaries (175 ± 19 μmol kg^{-1} in Hooghly and 185 ± 25 μmol kg^{-1} in Matla). However, the annual mean pH was significantly low in Hooghly (7.850–8.040) compared to Matla (8.044–8.205) (Akhand et al. 2016).

3.3.2 *Mahanadi Estuarine System*

The Mahanadi River Estuary system situated in Odisha is the third largest river system in India. The estuarine reaches of Mahanadi bifurcate into various distributaries, which in turn meets the BoB mainly through the Paradip Port and Harbours and the Chilika lagoon (Panda et al. 2006). The Mahanadi Estuary experiences perennial freshwater flow with peak discharge as high as ~45,000 $m^3\ s^{-1}$, coupled with substantial sewage discharge from the adjacent industrial setup and port activities (Panda et al. 2006; Ganguly et al. 2011). Unlike the Hooghly, this estuary is comparatively narrower toward the sea end. Salinity varies over a wide range of 1.1–26.5. Dissolved oxygen (DO) and pH vary from 5.6 to 8.05 mg l^{-1} and 7.77 to

8.40, respectively, throughout an annual cycle (Pattanaik et al. 2020a). Similar DO and pH ranges prevailed in other indirect distributaries of Mahanadi, like the Dhamra Estuary (which shelters the mangrove forest of Bhitarkanika). However, the salinity varied over a narrower range of 11.8–27.4 for being situated much closer to the BoB and thus having more marine dominance (Pattanaik et al. 2020b). Additionally, the magnitude of physical forcing, in terms of monsoonal and non-monsoonal freshwater discharge in the Mahanadi Estuary, is several times higher than that of Dhamra Estuary, thus influencing the range of salinity. Unlike many other eastern Indian estuaries, this estuary exhibited the lowest oxygen saturation coupled with the highest CO_2 efflux during the pre-monsoon (summer) season, whereas, during the monsoon and post-monsoon seasons, a reverse scenario prevailed (Ganguly et al. 2011; Pattanaik et al. 2019).

3.3.3 Godavari Estuarine System

The Godavari River, the second-longest river of India (1465 km), drains into the BoB through Andhra Pradesh and encompasses a catchment area of >3,00,000 km^2. Despite its large catchment area, the river discharge from the Godavari is low, compared to other north Indian rivers because of moderate annual average precipitation (Acharyya et al. 2012). The river water is widely used for domestic and industrial purposes by a large urban population, and consequently, enormous quantities of domestic and industrial wastes enter the river (Deshmukh and Ambore 2006). The biogeochemical processes in the Godavari estuary during the monsoonal discharge period vary from the dry period. The seasonal runoff exceeds the total volume of the estuary during the phase of peak discharge (Sarma et al. 2011). Mean salinity and pH as low as 0.08 and 6.43, respectively, were observed during monsoon, whereas, during the dry periods, these parameters exhibited values like 33.5 and 8.608, respectively (Table 3.1). The DO was under-saturated during the peak discharge period and relatively supersaturated during the dry period (Sarma et al. 2010). Chl-a concentrations varied over a wide range, and regular bloom formations occur in the months of October and November (Acharyya et al. 2012).

3.3.4 Cauvery and Krishna Estuarine Systems

The Rivers Cauvery (805 km) and Krishna (1288 km), in south India, are comparatively less studied than the northern estuaries. The variation of physicochemical parameters in these two estuaries was more or less found in the same order and range as other Indian monsoonal estuaries influenced by both tidal cycle and seasonal variation (Senthilkumar et al. 2008; Kumari and Rao 2009). However, the chl-a concentration was very high compared to the other estuaries mentioned above (Table 3.1) (Senthilkumar et al. 2008).

Table 3.1 Range of physicochemical parameters observed throughout an annual cycle in the major estuaries along the east coast of India

Estuary	Salinity	Temp. (°C)	pH	DO (mg/l)	Turbidity (NTU)	Chl-a (mg/m³)	Authors
Mahanadi	4.3–22.3	21.6–31.4	7.525–8.467	4.5–8.1	0.5–126.7	0.27–5.07	Pattanaik et al. (2020a, b) and Ganguly et al. (2011)
Dhamra	11.8–27.4	24.1–31.0	7.061–8.930	4.0–8.9	40.0–462.9	1.69–5.29	Pattanaik et al. (2019, 2020a)
Hooghly	0.1–20.1	19.7–31.4	7.433–8.316	4.0–7.3	109–402	0.52–4.98	Akhand et al. (2016) and Das et al. (2017)
Matla	8.5–29.7	18.6–31.2	7.419–8.564	4.5–8.2	75–301	0.41–4.19	Akhand et al. (2016)
Godavari	0.08–33.5	24.0–31.2	6.437–8.608	6.4–8.7	95–163	0.69–3.31	Sarma et al. (2010, 2011, 2013) and Bhalla and Sekhon (2010)
Krishna	2.6–33.2	26.6–30.0	7.300–7.430	5.3–11.7	67–141		Sarwade and Kamble (2014) and Kumari and Rao (2009)
Cauvery	16.0–34.0	24.0–32.0	7.000–8.800	4.0–5.8		6.50–12.74	Krishna et al. (2012), Vijayan et al. (2018), and Senthilkumar et al. (2008)

3.4 Characteristic Features of Principal Physical Forcing and Extreme Atmospheric Events in the East Coast of India

3.4.1 Monsoonal Rainfall

Indian subcontinent experiences two monsoons in a year. The southwest monsoon (June–September) brings in the rain for the entire country. The effect of this monsoonal rain is mainly predominant in the central and northeastern parts of the east coast of India (Bharathi and Sarma 2019). On the contrary, the northeast monsoon (October–December), which is also known as retreating monsoon, mostly influences the southeast coast of India, as the Western Ghats mountain range obstructs the return flow of monsoon toward the Arabian Sea (Deepthy and Balakrishnan 2005). India receives more than 75% of the annual total rainfall between the months of June and September. The precipitation pattern significantly varies along the east coast of India, with substantially high rainfall in the northeast (1000–2500 mm) and comparatively less in the southeast (300–500 mm) (Soman and Kumar 1990).

However, in the past decades, severe erratic and irregular patterns in southwest monsoon were noticed, often leading to higher rainfall in non-monsoon months (Thomas and Prasannakumar 2016).

3.4.2 *Freshwater Discharge*

Based on the magnitude and pattern of monsoonal rain, freshwater discharge varies over a wide range in the estuaries of India (Vijith et al. 2009; Sarma et al. 2014). During the peak of the southwest monsoon, river discharges vary between very low magnitudes like 28 $m^3 s^{-1}$ to magnitudes as high as 3505 $m^3 s^{-1}$ (Kumar et al. 2005). On the whole, the estuaries in the northern part of the east coast exhibited significantly higher freshwater discharges than those present in the southern part of the coastline, as most of the perennial rivers, which has a steady source of freshwater from the glacial Himalayas, are situated in the northern part (Bharathi and Sarma 2019). Major estuaries like Hooghly, Mahanadi, Godavari, Krishna, and Cauvery exhibit very high peak discharge rates of ~1600 $m^3 s^{-1}$, ~2100 $m^3 s^{-1}$, ~3500 $m^3 s^{-1}$, 550 $m^3 s^{-1}$, and ~700 $m^3 s^{-1}$, respectively (Kumar et al. 2005).

3.4.3 *Tropical Cyclones*

TC is a common natural calamity that develops in the BoB and eventually makes landfall on the east coast of India and the neighboring countries. In the last five decades, more than 200 cyclonic storms hit the east coast of India (Sahoo and Bhaskaran 2016). Table 3.2 shows the list of the most disastrous cyclones. In the present date, on average, 5–6 TCs generate each year in the BoB (Sahoo and Bhaskaran 2018). Several researchers argue that due to the ongoing global warming, the sea surface temperature in BoB has increased over the past decades, and often it exceeds the threshold, which is conducive for cyclone genesis. However, the frequency and intensity of the cyclones, both in terms of their size and maximum sustained wind speed, have significantly increased over the last decades and likely to continue shortly as well (Murty et al. 2016; Sahoo and Bhaskaran 2016).

3.4.4 *Tides*

The entire east coast of India, along with all the estuaries in this coastline, exhibit semi-diurnal tides of varying amplitudes throughout the year (Jithin et al. 2017). The tidal amplitude varies following the lunar cycle, with the highest amplitudes during the new moon/full moon (known as spring tide) and the lowest during the first quarter/third quarter (known as neap tide) (Das et al. 2017). The tidal amplitude

Table 3.2 Tropical cyclones that affected the east coast of India during 1980–2020

Serial number	Name of the cyclone	Date of landfall	Max. wind speed during landfall (km/h)
Tropical cyclones that affected West Bengal			
1	1981 North Indian Ocean cyclone	20/06/1981	140
2	1988 Bangladesh cyclone	21/11/1988	205
3	1997 North Indian Ocean cyclone	14/05/1997	100
4	1998 North Indian Ocean cyclone	17/05/1998	145
5	2000 North Indian Ocean cyclone	27/03/2000	75
6	2002 West Bengal cyclone	10/11/2002	100
7	Sidr	11/11/2007	215
8	Rashmi	25/10/2008	85
9	Aila	26/05/2009	120
10	Komen	26/07/2015	75
11	Roanu	19/05/2016	85
12	Mora	28/05/2017	150
13	Fani	26/04/2019	215
14	Bulbul	05/11/2019	145
15	Amphan	20/05/2020	260
Tropical cyclones that affected Odisha			
1	1999 Odisha cyclone	29/10/1999	276
2	Phailin	04/10/2013	215
3	Hudhud	07/10/2014	185
4	Titli	08/10/2018	110
5	Fani	29/10/1999	250
Tropical cyclones that affected Andhra Pradesh			
1	1990 Andhra Pradesh cyclone	04/05/1990	230
2	1998 North Indian Ocean cyclone	17/05/1998	195
3	2003 North Indian Ocean cyclone	10/05/2003	140
4	Yemyin	21/06/2007	95
5	Khai-Muk	27/04/2008	165
6	Laila	17/05/2010	120
7	Nilam	28/10/2012	100
8	Helen	19/11/2013	130
9	Lehar	19/11/2013	140
10	Hudhud	07/10/2014	215
11	Kyant	17/05/2016	130
12	Titli	08/10/2018	150
Tropical cyclones that affected Tamil Nadu			
1	1991 North Indian Ocean cyclone	17/01/1991	85
2	1992 North Indian Ocean cyclone	16/05/1992	85
3	1993 North Indian Ocean cyclone	17/06/1993	175
4	1996 North Indian Ocean cyclone	07/05/1996	120

(continued)

Table 3.2 (continued)

Serial number	Name of the cyclone	Date of landfall	Max. wind speed during landfall (km/h)
5	2000 North Indian Ocean cyclone	27/03/2000	190
6	Fanoos	07/01/2005	85
7	Nisha	25/11/2008	85
8	Jal	01/11/2010	100
9	Thane	25/12/2011	140
10	Nilam	28/102012	85
11	Madi	10/05/2013	120
12	Roanu	19/05/2016	85
13	Kyant	17/05/2016	85
14	Nada	29/11/2016	75
15	Vardah	06/12/2016	130
16	Ockhi	29/11/2017	155
17	Gaja	10/11/2018	128

Data source: Rao (1994), Mohapatra et al. (2012), Bahinipati (2014), Muthusamy et al. (2018), and Nandi et al. (2020)

varies in the estuaries of the east coast. However, estuaries on the north of the coastline exhibit higher amplitudes (>2 m), and those in the south show lower amplitudes (<2 m) (Bharathi and Sarma 2019). The Hooghly Estuary deserves a special mention, in this regard, as tidal amplitude in this estuary can be as high as 7 m, and the water current ranges between 108 and 117 cm s^{-1} during high tide and low tide, respectively, rendering it to be a macro-tidal estuary (De et al. 2011). Moreover, the Hooghly Estuary is one of the rare estuaries of earth, where tidal bores (occasional intense nonlinear wave propagation) take place due to its typical geometry and tidal amplitude (Bonneton et al. 2016).

3.5 Influence of Physical Forcing on the Variability of Biogeochemistry

3.5.1 Thermal Stratification in the Water Column Induced by Monsoon and TCs

Bharathi and Sarma (2019) reported that the estuaries of the east coast of India remain comparatively warmer (~30.8 °C) than that of the west coast estuaries (~27.1 °C) during the monsoon months. Reddy and Rao (1994) observed a typical monsoon induced stratification in the Gautami-Godavari Estuary. From the onset to the mid of monsoon, the water column remains well stratified, in terms of salinity, due to freshwater discharge. However, in the latter end of monsoon months, the salinity profile becomes uniform due to massive freshwater discharge making the entire water column full of freshwater. On the contrary, the difference in surface and bottom temperatures within the estuary remains substantially high (~4 to ~4.5 °C) all around the monsoon season, compared to monsoon months, with low temperature prevailing in the bottom of the estuary, which shows that monsoon favors strong thermal stratification in this estuary. However, such monsoon-induced thermal stratification is not prominent in the Hooghly Estuary. The dominant effect of tidal mixing and the typical geomorphology of the estuary could be the principal reasons behind the absence of such stratification (Manna et al. 2013).

The TCs, depending on the seasons, exerted different impacts on the water temperature. The TCs originating in the pre-monsoon season cooled the surface waters by ~3 °C. However, even the strongest TCs during the post-monsoon season did not show any such cooling effect (Sengupta et al. 2008). The cyclone Hudhud deepened the isothermal layer by favoring upwelling and thus bringing nutrient-rich cold waters from the bottom (Girishkumar et al. 2019). Similar observations prevailed during most of the TCs (13) that occurred between 1980 and 2015 (Qiu et al. 2019).

3.5.2 *Effect on Primary Productivity and Phytoplankton Bloom*

Several studies indicated that the monsoonal discharges could not stimulate algal bloom despite bringing a high amount of nutrients in the estuaries adjacent to BoB. A highly turbid water column that inhibits light penetration in photic layers hampers the photosynthetic potential of phytoplankton (Vinayachandran and Mathew 2003). The occurrence of phytoplankton blooms has been observed during the dry winter seasons in the absence or reduced freshwater discharge, as observed by Acharyya et al. (2012) in Godavari Estuary and Biswas et al. (2007) in Hooghly-Matla Estuary. Nutrient contribution from the adjoining mangrove patches, comparatively transparent water column due to decreased discharge, and cloud-free clear sky facilitated photosynthesis during the dry seasons and led to bloom formation. Though the occurrence of bloom formation is absent from the Mahanadi Estuary, the highest seasonal mean chlorophyll-a prevailed during the post-monsoon by Naik et al. (2009).

Thus, monsoonal discharge, in a way, delimited phytoplankton productivity. On the contrary, TCs lead to intense mixing of ocean waters, subsequently resulting in entrainment and upwelling of nutrients, which promotes phytoplankton blooms in the coastal BoB (Gomes et al. 2000). Based on the remotely sensed data, Maneesha et al. (2012) have reported that episodic atmospheric events enhanced phytoplankton biomass, thereby increasing the primary productivity, which remained sustained for up to 2–3 weeks. Estuarine and coastal ocean's biogeochemistry concerning the TCs such as Nargis, Laila, Phailin, and Hudhud showed that in all these cases, local Ekman pumping of nutrients into the photic layer enhanced the phytoplankton productivity (Lotliker et al. 2014; Baliarsingh et al. 2015). The degree of enhanced productivity largely depended upon the translation speed of the cyclone along with its prevailing period (Chacko 2017).

3.5.3 Variability in Dissolved Oxygen Levels Induced by Physical Forcing

DO, a crucial parameter that regulates the overall health of an aquatic ecosystem, was significantly controlled by tides, freshwater discharge, monsoonal rain, and TCs in the estuaries along the east coast of India. The estuaries that experience high freshwater discharge like Hooghly and Mahanadi had significantly higher DO all round the year, compared to those estuaries, which experience lower freshwater discharge, like Godavari and Krishna (Kumari and Rao 2013). Godavari and Krishna exhibit very low annual mean DO levels, which indicated that these estuaries are hypoxic in nature. Seasonally, most of the estuaries exhibit very high DO during the monsoon, when the estuaries are flooded with freshwater, compared to that observed in the dry seasons (Kumari and Rao 2013). Depending on the severity of the depletion in DO, the shallow continental shelf waters off the Tamil Nadu coast in the Bay of Bengal also suffer from hypoxia (Satpathy et al. 2013). This showed that physical forcing like a monsoon and freshwater discharge enhances the DO levels in these estuaries. Kumari and Rao (2013) reported that semidiurnal tides led to the fluctuation of DO in these estuaries, as the contribution of DO from freshwater and seawater continually changed depending on the tidal phases. TCs are also found to enhance DO levels in the estuaries and nearshore coastal waters because of drastic enhancement in productivity induced by the cyclones. Though the water surface DO levels increase substantially during the TCs, a reduction in DO levels concomitantly takes place in the bottom layers. However, once the effect of cyclone subsides, re-oxygenation of the bottom layers starts taking place (Sarma et al. 2013).

3.5.4 Nutrient Dynamics in Relation to Physical Forcing

The Indian Ocean receives about 0.22 ± 0.05, 0.11 ± 0.03, and 1.03 ± 0.26 Tg of dissolved inorganic nitrogen, phosphorus, and silicate per year, respectively, from the rivers and estuaries of India, out of which 76% falls into the BoB through the estuaries ending in the east coast (Krishna et al. 2016). The main reason behind such a disparity between the BoB and the Arabian Sea is the higher freshwater discharge covering a comparatively larger catchment area of the rivers ending on the east coast of India. A bulk-some part of the nutrients received by these estuaries throughout the annual cycle flows to the BoB during the monsoon season. Mukhopadhyay et al. (2006) reported 71% of inorganic nitrogen, 76% of inorganic phosphorus, and 76% of silicate of the annual total drains through the Hooghly Estuary during the monsoon season.

During the dry seasons, groundwater and exchanges at the sediment-water interface act as the principal source of nutrients to these estuaries (Sarma et al. 2010). In general, wind speed in the dry season is not enough to break the surface stratification in the BoB. Therefore, the nutrient supply through vertical mixing is not sufficient for algal growth. On the contrary, strong winds associated with cyclones provide nutrients into the photic layer by physical churning and eroding the density layer (Vinayachandran and Mathew 2003). Baliarsingh et al. (2015), in this regard, observed a two- to fourfold increase in nitrogenous and phosphorus nutrients in the coastal waters of BoB after the passage of cyclone Hudhud. Thus, all the physical forcing events that take place in these estuaries enhance the nutrient levels in the water.

3.6 Summary and Conclusion

Overall, we can summarize that the physical forcing and extreme atmospheric events significantly regulate the biogeochemistry of the estuaries along the east coast of India. The estuaries situated in the northern part of the coastline receive substantially higher river discharge than those situated down south, and there is a significant difference in the overall biogeochemical dynamics between these two sets of estuaries. Monsoonal rain favors thermal stratification in estuaries, which are micro-meso-tidal in nature. Macro-tidal estuaries did not show such monsoon induced stratification. Tropical cyclones favored phytoplankton bloom. However, monsoonal discharge delimited biological productivity. Monsoon-induced freshwater discharge and tropical cyclones both lead to significant enhancement in nutrient concentration. Many of the estuaries in the southern end of the coastline are experiencing hypoxia.

References

Acharyya T, Sarma VVSS, Sridevi B et al (2012) Reduced river discharge intensifies phytoplankton bloom in Godavari estuary, India. Mar Chem 132:15–22

Akhand A, Chanda A, Manna S et al (2016) A comparison of CO_2 dynamics and air-water fluxes in a river-dominated estuary and a mangrove-dominated marine estuary. Geophys Res Lett 43(22):11–726

Bahinipati CS (2014) Assessment of vulnerability to cyclones and floods in Odisha, India: a district-level analysis. Curr Sci 107(12):1997–2007

Baliarsingh SK, Chandanlal P, Lotliker AA et al (2015) Biological implications of cyclone Hudhud in the coastal waters of northwestern Bay of Bengal. Curr Sci 109(7):1243–1245

Bhalla R, Sekhon B (2010) Seasonal physico-chemical characteristics assessment and primary production in the planktonic community of Godavari River water, Nashik (Maharashtra). Environ Conserv J 11(1–2):41–45

Bharathi MD, Sarma VVSS (2019) Impact of monsoon-induced discharge on phytoplankton community structure in the tropical Indian estuaries. Reg Stud Mar Sci 31:100795

Biswas H, Mukhopadhyay SK, Sen S et al (2007) Spatial and temporal patterns of methane dynamics in the tropical mangrove dominated estuary, NE coast of Bay of Bengal, India. J Mar Syst 68(1–2):55–64

Bonneton P, Filippini AG, Arpaia L et al (2016) Conditions for tidal bore formation in convergent alluvial estuaries. Estuar Coast Shelf Sci 172:121–127

Chacko N (2017) Chlorophyll bloom in response to tropical cyclone Hudhud in the Bay of Bengal: bio-Argo subsurface observations. Deep Sea Res I Oceanogr Res Pap 124:66–72

Chatterjee M, Shankar D, Sen GK et al (2013) Tidal variations in the Sundarbans estuarine system, India. J Earth Syst Sci 122(4):899–933

Christopher PB, Luettich RA Jr, Powers SP et al (2002) Estimating the spatial extent of bottom-water hypoxia and habitat degradation in a shallow estuary. Mar Ecol Prog Ser 230:103–112
Das S, Giri S, Das I, Chanda A et al (2017) Nutrient dynamics of northern bay of Bengal (nBoB)—emphasizing the role of tides. Reg Stud Mar Sci 10:116–134
De TK, De M, Das S et al (2011) Phytoplankton abundance in relation to cultural eutrophication at the land-ocean boundary of Sunderbans, NE coast of Bay of Bengal, India. J Environ Stud Sci 1(3):169–180
Deepthy R, Balakrishnan S (2005) Climatic control on clay mineral formation: evidence from weathering profiles developed on either side of the Western Ghats. J Earth Syst Sci 114:545–556
Deshmukh JU, Ambore NE (2006) Seasonal variations in physical aspects of pollution in Godavari river at Nanded, Maharashtra. Indian J Aqua Biol 21(2):93–96
Ganguly D, Deya M, Chowdhurya C et al (2011) Coupled micrometeorological and biological processes on atmospheric CO_2 concentrations at the land-ocean boundary, NE coast of India. Atmos Environ 45:3903e3910
Girishkumar MS, Thangaprakash VP, UdayaBhaskar TVS et al (2019) Quantifying tropical cyclone's effect on the biogeochemical processes using profiling float observations in the Bay of Bengal. J Geophys Res Oceans 124(3):1945–1963
Gomes HR, Goes JI, Saino T (2000) Influence of physical processes and freshwater discharge on the seasonality of phytoplankton regime in the Bay of Bengal. Cont Shelf Res 20:313–330
Jithin AK, Unnikrishnan AS, Fernando V et al (2017) Observed tidal currents on the continental shelf off the east coast of India. Cont Shelf Res 141:51–67
Krishna H, Hosmani S, Jayashnakar M (2012) Physico-chemical and bacteriological parameters of Kaveri river at Talakaveri region: a comparative study. Nat Mon Ref J Res Sci Technol 1(6):1–15
Krishna MS, Prasad MHK, Rao DB et al (2016) Export of dissolved inorganic nutrients to the northern Indian Ocean from the Indian monsoonal rivers during discharge period. Geochim Cosmochim Acta 172:430–443
Kumar R, Singh RD, Sharma KD (2005) Water resources of India. Curr Sci 89:794–811
Kumari VR, Rao IM (2009) Estuarine characteristics of lower Krishna river. Indian J Mar Sci 38(2):215–223
Kumari VR, Rao IM (2013) Hypoxia in Indian estuaries—Krishna and Godavari estuarine systems—a case study. Clim Chang Impact Ecosyst:110
Lotliker AA, Srinivasakumar T, Reddem VS (2014) Cyclone Phailin enhanced the productivity following its passage: evidence from satellite data. Curr Sci 106(3):360–361
Maneesha K, Murty VSN, Ravichandran M et al (2012) Upper ocean variability in the Bay of Bengal during the tropical cyclones Nargis and Laila. Prog Oceanogr 106:49–61
Manna RK, Satpathy BB, Roshith CM et al (2013) Spatio-temporal changes of hydro-chemical parameters in the estuarine part of the river Ganges under altered hydrological regime and its impact on biotic communities. Aquat Ecosyst Health 16(4):433–444
Mohapatra M, Bandyopadhyay BK, Tyagi A (2012) Best track parameters of tropical cyclones over the North Indian Ocean: a review. Nat Hazards 63(3):1285–1317
Mukhopadhyay SK, Biswas HD, De TK et al (2006) Fluxes of nutrients from the tropical River Hooghly at the land–ocean boundary of Sundarbans, NE Coast of Bay of Bengal, India. J Mar Syst 62(1–2):9–21
Murty PLN, Bhaskaran PK, Gayathri R et al (2016) Numerical study of coastal hydrodynamics using a coupled model for Hudhud cyclone in the Bay of Bengal. Estuar Coast Shelf Sci 183:13–27
Muthusamy S, Sivakumar K, Durai AS et al (2018) Ockhi cyclone and its impact in the Kanyakumari District of Southern Tamilnadu, India: an aftermath analysis. Int J Recent Res Asp Special Issue:466–469
Naik S, Acharya BC, Mohapatra A (2009) Seasonal variations of phytoplankton in Mahanadi estuary, east coast of India. Indian J Mar Sci 38(2):184–190
Nandi G, Neogy S, Roy AK et al (2020) Immediate disturbances induced by tropical cyclone Fani on the coastal forest landscape of eastern India: a geospatial analysis. Remote Sens Appl Soc Environ:100407. https://doi.org/10.1016/j.rsase.2020.100407

Panda UC, Sundaray SK, Rath P et al (2006) Application of factor and cluster analysis for characterization of river and estuarine water systems–a case study: Mahanadi River (India). J Hydrol 331(3–4):434–445

Pattanaik S, Acharya D, Sahoo RK et al (2019) Short-term variability of physico-chemical properties and pCO_2 fluxes off Dhamra estuary from North-Eastern India. J Indian Soc Remote 47(7):1197–1208

Pattanaik S, Chanda A, Sahoo RK et al (2020a) Contrasting intra-annual inorganic carbon dynamics and air–water CO_2 exchange in Dhamra and Mahanadi Estuaries of northern Bay of Bengal, India. Limnology 21(1):129–138

Pattanaik S, Roy R, Sahoo RK et al (2020b) Air-Sea CO_2 dynamics from tropical estuarine system Mahanadi, India. Reg Stud Mar Sci 36:101284

Qiu Y, Han W, Lin X et al (2019) Upper-ocean response to the super tropical cyclone Phailin (2013) over the freshwater region of the Bay of Bengal. J Phys Oceanogr 49(5):1201–1228

Raha A, Das S, Banerjee K et al (2012) Climate change impacts on Indian Sunderbans: a time series analysis (1924–2008). Biodivers Conserv 21(5):1289–1307

Rao RS (1994) Role of remote sensing in flood management—the May 1990 cyclone. Int J Remote Sens 15(8):1557–1558

Reddy BSR, Rao VR (1994) Seasonal variation in temperature and salinity in the Gautami-Godavari estuary. Proc Indian Acad Sci Ear Planet Sci 103(1):47–55

Sahoo B, Bhaskaran PK (2016) Assessment on historical cyclone tracks in the Bay of Bengal, east coast of India. Int J Climatol 36:95–109

Sahoo B, Bhaskaran PK (2018) Multi-hazard risk assessment of coastal vulnerability from tropical cyclones–a GIS based approach for the Odisha coast. J Environ Manag 206:1166–1178

Sarma VVSS, Gupta SNM, Babu PVR (2009) Influence of river discharge on plankton metabolic rates in the tropical monsoon driven Godavari estuary, India. Estuar Coast Shelf Sci 85(4):515–524

Sarma VVSS, Prasad VR, Kumar BSK et al (2010) Intra-annual variability in nutrients in the Godavari estuary, India. Cont Shelf Res 30(19):2005–2014

Sarma VVSS, Kumar NA, Prasad VR et al (2011) High CO_2 emissions from the tropical Godavari estuary (India) associated with monsoon river discharges. Geophys Res Lett 38(8):L08601

Sarma VVSS, Sridevi B, Maneesha K et al (2013) Impact of atmospheric and physical forcings on biogeochemical cycling of dissolved oxygen and nutrients in the coastal Bay of Bengal. J Oceanogr 69(2):229–243

Sarma VVSS, Krishna MS, Prasad VR et al (2014) Distribution and sources of particulate organic matter in the Indian monsoonal estuaries during monsoon. J Geophys Res Biogeosci 119(11):2095–2111

Sarwade AB, Kamble NA (2014) Evaluation of physicochemical parameters of river Krishna, Sangli Maharashtra. Octa J Environ Res 2(4):329–337

Satpathy KK, Panigrahi S, Mohanty AK et al (2013) Severe oxygen depletion in the shallow regions of the Bay of Bengal off Tamil Nadu coast. Curr Sci 104(11):1467–1469

Sengupta D, Goddalehundi BR, Anitha DS (2008) Cyclone-induced mixing does not cool SST in the post-monsoon north Bay of Bengal. Atmos Sci Lett 9(1):1–6

Senthilkumar B, Purvaja R, Ramesh R (2008) Seasonal and tidal dynamics of nutrients and chlorophyll a in a tropical mangrove estuary, southeast coast of India. Indian J Mar Sci 37(2):132–140

Soman MK, Kumar KK (1990) Some aspects of daily rainfall distribution over India during the south-west monsoon season. Int J Climatol 10:299–311

Thomas J, Prasannakumar V (2016) Temporal analysis of rainfall (1871–2012) and drought characteristics over a tropical monsoon-dominated state (Kerala) of India. J Hydrol 534:266–280

Vijayan P, Senthilmurugan S, Pugazhendy K et al (2018) Analysis of physicochemical parameters water samples from Cauvery River in Thanjavur district, Tamil Nadu. Int J Biol Sci 3(1):223–227

Vijith V, Sundar D, Shetye SR (2009) Time-dependence of salinity in monsoonal estuaries. Estuar Coast Shelf Sci 85:601–608

Vinayachandran PN, Mathew S (2003) Phytoplankton bloom in the Bay of Bengal and its intensification by cyclones. Geophys Res Lett 30:1572

Chapter 4
Carbon Dynamics of the Estuaries Along the East Coast of India

Kunal Chakraborty, Jayashree Ghosh, Trishneeta Bhattacharya, Anirban Akhand, R. S. Mahendra, and Vinu Valsala

Abstract The estuaries are highly dynamic in terms of carbon cycling, considering the emission of CO_2 to the atmosphere, which is quantitatively significant for the global carbon cycle. The estuaries along the east coast of India, which flow into the northwestern coastal Bay of Bengal, are located in the tropics extending between latitudes 9°N and 22°N. The carbon dynamics of these estuaries is primarily influenced by monsoonal river discharge and seawater intrusion into the estuary. The solubility and biological pumps characterize the potential of an estuarine ecosystem as a source or sink of CO_2. In most cases, the inner estuaries act as useful sieves for terrestrial/riverine inputs and provide a way for the passage of carbon to the atmosphere. The transport of carbon from estuaries plays a significant role in determining whether the adjacent coastal ocean is a net source or sink of CO_2. Among all the estuaries on the east coast of India, the dissolved inorganic carbon concentration in the Hooghly estuary is the highest. However, in the Chilka lagoon, the dissolved organic carbon concentration is higher than the Godavari and Hooghly estuary. The Indian monsoonal estuaries along the east coast of India emit relatively less CO_2 to the atmosphere than elsewhere in the world. The estuaries supply a high amount of organic carbon flux, in addition to excess inorganic carbon, which is mineralized in

K. Chakraborty (✉) · T. Bhattacharya · R. S. Mahendra
Indian National Centre for Ocean Information Services, Ministry of Earth Sciences, Government of India, Hyderabad, India
e-mail: kunal.c@incois.gov.in

J. Ghosh
Indian National Centre for Ocean Information Services, Ministry of Earth Sciences, Government of India, Hyderabad, India

School of Ocean Science and Technology, Kerala University of Fisheries and Ocean Studies, Kochi, India

A. Akhand
Coastal and Estuarine Environment Research Group, Port and Airport Research Institute, Yokosuka, Japan

V. Valsala
Indian Institute of Tropical Meteorology, Ministry of Earth Sciences, Pune, India

S. Das, T. Ghosh (eds.), *Estuarine Biogeochemical Dynamics of the East Coast of India*, https://doi.org/10.1007/978-3-030-68980-3_4

the estuary and elevates the pCO_2 levels. The significant changes in the land-use pattern, along with human intervention, have influenced the transport of carbon from land to adjacent coastal ocean to a great extent over the last few decades.

Keywords Estuaries · Northwestern coastal Bay of Bengal · Organic carbon dynamics · Inorganic carbon dynamics · CO_2 flux

4.1 Introduction

An estuary is commonly known as the transition zone between the river and the sea where salinity variations, circulation, and tidal influence act together to define the physicochemical characteristics of the region. Perillo (1995) defined an estuary as "a semi-enclosed coastal body of water that extends to the effective limit of tidal influence, within which seawater entering from one or more free connections with the open sea, or any other saline coastal body of water, is significantly diluted with freshwater derived from land drainage and can sustain euryhaline biological species from either, part or the whole of their life cycle."

The structure of estuaries is essential because of their characteristic of being a transition zone between the anthropogenic input-laden rivers and the ocean deficit in inorganic elements. The anthropogenic inputs, which are primarily inorganic and derived from the weathering of rocks, are the primary source of nutrients to the estuary and hence one of the most critical drivers of the biogeochemical cycles within the estuary. The estuarine waters are, in general, understudied all over the world, and this is especially true for the Indian estuaries (Sarma et al. 2011, 2012). The biogeochemical cycling of materials in these tropical estuaries is different from those in estuaries in other parts of the world because of the high atmospheric temperatures, seasonal changes in the monsoon circulation, and many other reasons (Sarma et al. 2011). River discharges containing both particulate and dissolved matter in organic and inorganic forms have a profound effect on the carbon cycle of the ocean. The amount of carbon input from river discharge can be taken to be a measure of how much the anthropogenic activities play a part in controlling the carbon dynamics. Therefore, the study of estuarine ecosystems in carbon cycling is an essential step in that direction.

4.2 East Coast of India

More than 30 million people reside along the coastlines of India, and they are dependent on the ocean resources for survival. The Indian coastline is being subjected to a wide variety of anthropogenic pressures. Local factors such as changes in land use, particularly agricultural activities, urbanization and industrialization, construction of new ports all along the Indian coast, etc. make the Indian coastal waters especially vulnerable to anthropogenic perturbations. It will have far-reaching consequences on biogeochemistry and ecology, adversely affecting marine living resources. In this chapter, we have discussed the carbon dynamics of the estuaries along the east coast of India by making use of the reported results of various previously done studies on different estuaries located along the east coast of India.

4.2.1 Major Estuaries and Their Characteristics

The estuaries along the east coast of India, namely, Matla, Hooghly, Dhamra, Mahanadi, Godavari, Krishna, Vellar, Cauvery, and Chilka lagoon, are considered in this chapter based on the reports available in the literature (Fig. 4.1). The Hooghly estuary is a perennial estuary being fed by the glaciers, while the Matla estuary is a marine-dominated estuary with no connection with the rivers in their northern end. However, it receives some amount of riverine freshwater from the Hooghly River through different waterways through Hatania Doania Canal (Ray et al. 2018). The Sundarban and Pichavaram mangroves are located on the northeast and southeast coast of India, respectively. The Pichavaram is located between two estuaries, Vellar in the north and Coleroon in the south. The mangroves in the Gautami-Godavari estuary are an active site for the mineralization and subsequent efflux of CO_2 into the atmosphere (Bouillon et al. 2003a, b). The estuaries on the southeast coast of India are known as monsoonal estuaries due to their distinct runoff periods and non-steady-state behavior (Vijith et al. 2009). During the monsoon season, when the river discharge is high, these estuaries entirely assume riverine conditions (Sarma 2009; Sarma et al. 2010, 2011). These estuaries are different from one another concerning their source of water input, which in itself creates differences in their dynamics of element cycling.

4.2.2 Significance of the Study

The solubility and biological pumps characterize the potential of an estuarine ecosystem as a sink or source of CO_2. In most cases, the inner estuaries act as useful sieves for terrestrial/riverine inputs and provide a way for the passage of carbon to the atmosphere. The transport of carbon from estuaries plays a significant role in determining whether the adjacent coastal ocean is a net source or sink of CO_2. The CO_2 from the atmosphere fixed by the terrestrial plants is either stored as plant biomass or in soil. It can also be decomposed back to CO_2 through respiration. This carbon can then enter rivers where transformation to inorganic carbon can take place. Organic and inorganic carbon in the dissolved form may be transported to the deep sea where they may reside for many centuries. This overall transfer and loss are accounted for by their cycling in the estuarine waters. Therefore, the study of dynamics in these regions is of utmost importance.

Fig. 4.1 Map of the major estuaries located along the east coast of India. (Sources: Esri, HERE, Garmin, Intermap, increment P Corp., GEBCO, USGS, FAO, NPS, NRCAN, GeoBase, IGN, Kadaster NL, Ordnance Survey, Esri Japan, METI, Esri China (Hong Kong), (c) OpenStreetMap contributors, and the GIS User Community)

4.3 Spatial Variability of the Carbon Dynamics

4.3.1 Inorganic Carbon Dynamics

In the Hooghly estuary, the dissolved inorganic carbon (DIC) concentration varied in the range of 2073–2560μmol kg^{-1} throughout the year (Mukhopadhyay et al. 2006). In the Mahanadi estuary, the total alkalinity (TA) was reported to vary between 1102 and 2206μmol kg^{-1}. DIC was found with a range of 1121–1181μmol kg^{-1} (Pattanaik et al. 2017). The TA and DIC in the Dhamra estuary were found to be in the range of 990–2235μmol kg^{-1} and 903–2070μmol kg^{-1}, respectively (Pattanaik et al. 2020). In both the Mahanadi and Dhamra estuaries, the variation of TA and DIC was mainly controlled by the mixing of river and ocean water. The variation in TA and DIC range was more prevalent in Dhamra estuary than in Mahanadi estuary. In another important fresh- and marine-water mixing zone in Odisha, the Chilka lagoon, DIC concentration varied widely between 315 and 2567μmol kg^{-1} (Muduli et al. 2013). The annual mean concentration of DIC (1586μmol kg^{-1}) in the Chilka lagoon was reportedly lower than that of the Hooghly estuary (2291μmol kg^{-1}; Mukhopadhyay et al. 2006). However, the DIC concentration was relatively higher than the Godavari estuary (1216μmol kg^{-1}; Table 4.1). In the case of the Gautami-Godavari estuarine system, a relatively small decrease in TA is observed along the salinity gradient of the Godavari estuary, i.e., with lower values (approximately 2260μmol kg^{-1}) in the marine end of the estuary. The DIC concentration in Subarnarekha, Rushikulya, Vamsadhara, Nagavali, and Krishna estuaries is 1340, 862, 1809, 2262, 2023, and 2915μmol kg^{-1}, respectively (Sarma et al. 2012). The DIC concentration for the estuaries Penna, Vellar, Ponnayaar, Cauvery, Ambalayaar, and Vaigai, located along the southeast coast of India, is 3068, 4166, 3447, 3420, 2686, and 2033μmol kg^{-1}, respectively (Sarma et al. 2012).

Table 4.1 Comparison of the annual mean concentration of TA and DIC of the estuaries along the east coast of India

Study site	TA (μmol kg^{-1})	DIC (μmol kg^{-1})	Authors
Matla estuary	1734	1611	Akhand et al. (2016)
Hooghly estuary	2400	2560	Mukhopadhyay et al. (2006)
Mahanadi estuary	1706	1211	Pattanaik et al. (2017)
Dhamra estuary	1612	1486	Pattanaik et al. (2020)
Chilka lagoon	…	2567	Muduli et al. (2013)
Godavari estuary	2260	1216	Bouillon et al. (2003a, b) and Sarma et al. (2012)
Krishna	…	2915	Sarma et al. (2012)
Vellar	…	4166	Sarma et al. (2012)
Cauvery	…	3420	Sarma et al. (2012)

4.3.2 *Organic Carbon Dynamics*

The dissolved organic carbon (DOC) constituted the majority of the total organic carbon in the Hooghly estuary. The DOC concentration was in the range of 205–224 μmol kg^{-1} in the Hooghly estuary (Ray et al. 2018). In the Chilka lagoon, the DOC concentration was in the range of 669–2815μmol kg^{-1}. The values were relatively higher compared to other Indian estuaries like Hooghly and Godavari (100–300μmol kg^{-1}; Sarin et al. 2002; Bouillon et al. 2003a, b). In Pichavaram mangroves, the DOC concentration varied between 192 and 410μM (Senthilkumar et al. 2008). The concentrations of particulate organic carbon (POC) were in the range of 33.8–368μmol kg^{-1} for the estuaries that received high discharge (>200 m^3 s^{-1}) of freshwater (Sarma et al. 2014). The POC constitutes 2.42% of the total carbon concentration in the Cauvery estuary (Ramanathan et al. 1996). The POC concentration in the Godavari estuary is 166.66μmol kg^{-1} (Bouillon et al. 2003a, b).

4.4 Seasonal Variability of the Carbon Dynamics

4.4.1 *Inorganic Carbon Dynamics*

The TA and DIC concentrations follow a consistent trend of monsoon minima and pre- and post-monsoon high in the case of Hooghly estuary, whereas in the case of Matla, the TA and DIC concentrations show monsoon high in comparison to the other two seasons (Akhand et al. 2016). In the Mahanadi estuary, the concentration of TA and DIC was observed to be maximum in February (2055μmol kg^{-1}), followed by December (1854μmol kg^{-1}), whereas the lowest TA and DIC concentrations were observed during August (1201μmol kg^{-1} and 1210μmol kg^{-1}, respectively). The TA and DIC concentrations of the Dhamra estuary were found to be high in December (2235μmol kg^{-1} and 2070μmol kg^{-1}, respectively) and the low concentrations were observed in August (990μmol kg^{-1} and 903μmol kg^{-1}, respectively; Pattanaik et al. 2020).

4.4.2 *Organic Carbon Dynamics*

The reported DOC concentration in the Chilka lagoon was maximum during the premonsoon due to in situ production (Gupta et al. 2008). During monsoon, both the DOC and POC concentrations were inversely related to salinity, suggesting that river runoff plays a vital role in the seasonal variability of DOC and POC in the lagoon. In the Pichavaram mangroves, the DOC concentration varied between 192 and 421μmol kg^{-1} in the dry and wet seasons, respectively (Senthilkumar et al. 2008).

4.5 CO_2 Flux Across the Air-Water Interface

The solubility of CO_2 in estuarine water is controlled by temperature and salinity. The seasonal variation of the air-water CO_2 flux in the Hooghly estuary is controlled by the intrusion of seawater into the river water. In a comparative study of the Hooghly and Matla estuaries by Akhand et al. (2016), it was reported that the Hooghly estuary was supersaturated with CO_2 throughout the year (annual mean ~2200μatm). However, in the case of the Matla estuary, it was not as oversaturated (annual mean ~530μatm) as the Hooghly estuary. Moreover, during the post-monsoon season, the outer region of the Matla estuary was undersaturated concerning CO_2 and subsequently acted as a sink of CO_2. Akhand et al. (2016) reported that the annual mean CO_2 emission from the Hooghly estuary was significantly higher than that of the Matla estuary. The outer estuarine region of the Sundarban mangroves was found as a sink of CO_2 in the winter months (Biswas et al., 2004; Akhand et al. 2013a, b), while the inner estuarine region of the Sundarban was also reported as a lesser sink during the winter months (Biswas et al. 2004; Mukhopadhyay et al. 2006).

In contrast, the Mahanadi estuary located on the east coast of India acted as a net source of CO_2 (Ganguly et al. 2011). Other studies on Sundarban estuary concluded that the Sundarbans could play a simultaneous role as the CO_2 source as well as a sink (Akhand et al. 2013a, b). The annual average air-water CO_2 fluxes at the mouth of Mahanadi and the Dhamra estuaries were −3.9 ± 21.4 (mean ± standard deviation) μmol m^{-2} h^{-1} and −2.9 ± 11.6μmol m^{-2} h^{-1}, respectively (Pattanaik et al. 2020; Table 4.2). The Mahanadi estuary acted as a CO_2 source toward atmosphere during monsoon months, and the Dhamra estuary acted as a CO_2 source during premonsoon. However, both the Mahanadi and Dhamra estuaries are situated in close vicinity to each other and have a similar climate regime. They both exhibited significant intra-annual variability for air-water CO_2 fluxes. Annually both these estuaries acted as mild sinks for atmospheric CO_2. The Mahanadi estuary acted as a CO_2 source during summer and monsoon months. High water temperature and re-mineralization of monsoonal runoff-driven organic load from a vast catchment area of Mahanadi River were considered to be the main factors behind the CO_2 source character, despite showing an autotrophic nature on the water surface. It can be inferred that the degassing of CO_2 from the water column was substantially more than the surface

Table 4.2 Comparison of air-water CO_2 flux values for the estuaries along the east coast of India

Study site	Air-water CO_2 flux (μmol m^{-2} h^{-1})	Authors
Matla estuary	593.16	Akhand et al. (2016)
Hooghly estuary	6818.37	Akhand et al. (2016)
Mahanadi estuary	−3.9	Pattanaik et al. (2020)
Dhamra estuary	−2.9	Pattanaik et al. (2020)
Chilka lagoon	3.56	Muduli et al. (2012)
Godavari estuary	2527	Sarma et al. (2011)
Krishna	283	Sarma et al. (2012)
Vellar	708.75	Sarma et al. (2012)
Cauvery	92.91	Sarma et al. (2012)

water net primary productivity (NPP), which leads to the CO_2 source nature during August, October, and December. In the case of Dhamra estuary, it is presumed that comparatively lower freshwater discharge in comparison with Mahanadi, which is evident from high salinity values, especially in February and December, led to insufficient availability of nutrients, which in turn reduce productivity. Moreover, the turbidity in the Dhamra estuary was much high compared to Mahanadi, which also inhibited the penetration of light in the euphotic zone.

In another study of air-water CO_2 flux in Mahanadi estuary, it was found that the seasonal oscillations in the air-water CO_2 gradient were negative during winter and positive during summer, wherein the estuary acted as a weak source (Pattanaik et al. 2020a, b). The average air-water CO_2 flux values in the estuary varied between −5.65 and 13.11 μmol m^{-2} h^{-1}. The study also showed that the Mahanadi estuarine system has dual characteristics, varying between CO_2 sink and source. It was also observed that seasonal changes in the air-water CO_2 gradient were more pronounced during winter compared to summer when the estuary acted as a net sink. However, the magnitude of this seasonal gradient was lower. The low pCO_2 (water) was observed in the Mahanadi estuary. Chilka lagoon showed a net CO_2 source character, which is a net heterotrophic system. The CO_2 efflux during monsoon was strong, and excess CO_2 transported by rivers represented only 15–16% of total efflux from the lagoon. The net ecosystem production derived through the mass balancing of O_2 and total carbon revealed that the large flux of trapped organic carbon from rivers is biologically respired in the lake, and the converted CO_2 is pumped out to the atmosphere instead of exporting it to the sea. In this study, it was concluded that such evasion rates of CO_2 could be a characteristic feature for lagoons having large dimensions. Coastal lagoons, along with the estuaries, develop into "hot spots" influencing carbon cycle, thus contributing to regional carbon budgets (Gupta et al. 2008). The CO_2 efflux from the Krishna, Vellar, and Cauvery estuary is 283, 708.75, and 92.91 μmol m^{-2} h^{-1}, respectively (Sarma et al. 2012). The observed pCO_2 values for the estuaries Subarnarekha, Rushikulya, Vamsadhara, Nagavali, and Krishna are 866, 7247, 293, 669, 515, and 7473 μatm, respectively (Sarma et al. 2012). In contrast, the observed pCO_2 values for the estuaries like Penna, Vellar, Ponnayaar, Cauvery, Ambalayaar, and Vaigai, located along the southeast coast of India, are 1767, 2351, 11,153, 2989, 359 and 400 μatm, respectively. A record magnitude of $pCO_2 > 30{,}000$ μatm in the Godavari estuary was found during the discharge period of 2008 (Sarma et al. 2011). The Indian estuaries act as a source of atmospheric CO_2 during the wet season. The magnitude of river discharge in an estuary is proportional to the pCO_2 levels and organic matter concentrations. It indicates that aerobic microbial respiration due to the large amounts of organic matter brought to the estuaries is a primary mechanism leading to CO_2 efflux to the atmosphere. The mean CO_2 flux from the Indian estuaries is significantly low. The low CO_2 fluxes in the Indian estuaries were attributed to significantly low flushing time of water in the estuary during the wet period. As a result, the Indian monsoonal estuaries emit relatively less CO_2 to the atmosphere than elsewhere in the world. Although pCO_2 and the resulting fluxes in the mangrove creeks were high, a relatively low degree of saturation of CO_2 with respect to the atmospheric CO_2 concentration is observed in the case of the Godavari estuary during the pre-monsoon season.

4.6 Drivers of Organic and Inorganic Carbon Dynamics

The Hooghly estuary carries a sizeable organic load, which subsequently leads to oversaturation of CO_2 and hence acts as a substantial source of CO_2. In contrast, the marine-dominated estuary of Matla in the Sundarban is almost disconnected from any perennial river sources from the north and showed lower values of CO_2 fluxes (Akhand et al. 2016). The aquatic environment and sediments receive a considerable quantity of organic matter from their surroundings, which leads to the net heterotrophic nature of air-water CO_2 fluxes (Alongi et al. 1998). If an estuarine ecosystem is heterotrophic, then in most cases, it becomes a significant source of CO_2 to the atmosphere. In such estuaries, total respiration exceeds gross primary production (Frankignoulle et al. 1998; Gattuso et al. 1998; Bouillon et al. 2003a, b, 2007; Abril and Borges 2004; Biswas et al. 2004; Hopkinson and Smith 2005; Zhai et al. 2005). Also, the river water entering estuaries generally has higher pCO_2 than the atmosphere due to organic carbon mineralization in soils, river waters, and sediments (Cole and Caraco 2001; Sarma et al. 2001; Richey et al. 2002; Ferguson et al. 2003; Gazeau et al. 2004; Hopkinson and Smith 2005; Kortelainen et al. 2006; Cole et al. 2007). However, if the residence time of estuaries is long, then the contribution of CO_2 from rivers to its emission from estuaries is relatively less (Borges et al. 2006). The rivers supply high flux of DOC, apart from excess DIC, which is mineralized in the estuary and elevates the pCO_2 levels. During the windy periods, the production is masked by high sediment resuspension that reduces the photic depth. The additional CO_2 that is being added to the water column from the sediments leads to the super-saturation of CO_2. Mangroves are known for exchanging organic and inorganic carbon with estuaries and oceans. The organic matter generated from the mangrove plants, followed by its subsequent degradation, is the primary source of DIC and DOC in the Hooghly estuary, whereas POC is linked to soil erosion. Mangroves are identified as a significant source of carbon (POC, DOC, and DIC) transported from the Sundarbans into the Bay of Bengal (Ray et al., 2018). The river input, *Avicennia* leaves, and phytoplankton are the potential DOM sources of the Sundarban region and contribute 55%, 22%, and 23%, respectively, to the total input. Similarly, for POC sources, significant contributions are from litterfall, river discharge, and phytoplankton production (Ray and Shahraki 2016).

4.7 Conclusion and Future Direction of Research

Estuaries have the most significant influence on the carbon dynamics of any coastal marine ecosystem. The estuaries exhibit the highest transfer rates of CO_2 toward the atmosphere than any other coastal system. Carbon dynamics in estuaries are continually changing, reflecting geomorphic and ecological responses to long- and short-term disturbances like climate change, nutrient loading, and land-use change. Influences of these drivers are profound in coastal systems, often more than that of the inland wetlands, and thus require special attention. Estuarine systems along the east coast of India are highly dynamic in terms of carbon cycling and are considered contributors to the global carbon cycle. This emission of CO_2 to the atmosphere is strongly supported by the net heterotrophic nature of these estuaries. However, the sparseness of data is the major limitation in the quantification of the spatial and temporal variability of inorganic and organic carbon in estuaries along the east coast of India. The understanding of estuarine carbon dynamics and evaluation of the emission of CO_2 from estuarine ecosystems have increased during the past years due to increasing data availability.

References

Abril G, Borges AV (2004) Carbon dioxide and methane emissions from estuaries, Chapter 7. In: Tremblay A et al (eds) Greenhouse gases emissions from natural environments and hydroelectric reservoirs: fluxes and processes. Environmental science series. Springer, New York, pp 87–207

Akhand A, Chanda A, Dutta S, Manna S, Hazra S, Mitra D, Rao KH, Dadhwal VK (2013a) Characterizing air–sea CO2 exchange dynamics during winter in the coastal water off the Hugli-Matla estuarine system in the northern Bay of Bengal, India. J Oceanogr 69(6):687–697

Akhand A, Chanda A, Dutta S, Manna S, Sanyal P, Hazra S, Rao KH, Dadhwal VK (2013b) Dual character of Sundarban estuary as a source and sink of CO2 during summer: an investigation of spatial dynamics. Environ Monit Assess 185(8):6505–6515

Akhand A, Chanda A, Manna S, Das S, Hazra S, Roy R, Choudhury SB, Rao KH, Dadhwal VK, Chakraborty K, Mostofa KMG (2016) A comparison of CO_2 dynamics and air-water fluxes in a river-dominated estuary and a mangrove-dominated marine estuary. Geophys Res Lett 43(22):11–726

Alongi DM, Sasekumar A, Tirendi F, Dixon P (1998) The influence of stand age on benthic decomposition and recycling of organic matter in managed mangrove forests of Malaysia. J Exp Mar Biol Ecol 225:197–218

Biswas H, Mukhopadhyay SK, De TK, Sen S, Jana TK (2004) Biogenic controls on the air—water carbon dioxide exchange in the Sundarban mangrove environment, northeast coast of Bay of Bengal, India. Limnol Oceanogr 49(1):95–101

Borges AV, Schiettecatte LS, Abril G, Delille B, Gazeau F (2006) Carbon dioxide in European coastal waters. Estuar Coast Shelf Sci 70(3):375–387

Bouillon S, Dahdouh-Guebas F, Rao AVVS, Koedam N, Dehairs F (2003a) Sources of organic carbon in mangrove sediments: variability and possible ecological implications. Hydrobiologia 495(1–3):33–39

Bouillon S, Frankignoulle M, Dehairs F, Velimirov B, Eiler A, Abril G, Etcheber H, Borges AV (2003b) Inorganic and organic carbon biogeochemistry in the Gautami Godavari estuary (Andhra Pradesh, India) during premonsoon: the local impact of extensive mangrove forests. Glob Biogeochem Cycles 17(4):25-(1–12)

Bouillon S, Dehairs F, Schiettecatte L-S, Borges AV (2007) Biogeochemistry of the Tana estuary and delta (northern Kenya). Limnol Oceanogr 52:46–57

Cole JJ, Caraco NF (2001) Carbon in catchments: connecting terrestrial carbon losses with aquatic metabolism. Mar Freshw Res 52:101–110

Cole JJ, Praire YT, Caraco NF, McDowell WH, Tranvik LJ, Striegl RG, Duarte CM, Kortelainen P, Downing JA, Middelburg JJ, Melack J (2007) Plumbing the global carbon cycle: integrating inland waters into the terrestrial carbon budget. Ecosystems 10:171–184

Ferguson AJP, Eyre BD, Gay JM (2003) Organic matter and benthic metabolism in euphotic sediments along shallow sub-tropical estuaries, northern New South Wales, Australia. Aquat Microb Ecol 33(2):137–154

Frankignoulle M, Abril G, Borges AV, Bourge I, Canon C, Delille B, Libert E, Théate J-M (1998) Carbon dioxide emissions from European estuaries. Science 282:434–436

Ganguly D, Dey M, Chowdhury C, Pattnaik AA, Sahu BK, Jana TK (2011) Coupled micrometeorological and biological processes on atmospheric CO_2 concentrations at the land–ocean boundary, NE coast of India. Atmos Environ 45(23):3903–3910

Gattuso J-P, Frankignoulle M, Wollast R (1998) Carbon and carbonate metabolism in coastal aquatic ecosystems. Annu Rev Ecol Syst 29:405–434

Gazeau F, Gentili B, Smith SV, Frankignoulle M, Gattuso J-P (2004) The European coastal zone: characterization and first assessment of ecosystem metabolism. Estuar Coast Shelf Sci 60(4):673–694

Gupta GVM, Sarma VVSS, Robin RS, Raman AV, Kumar MJ, Rakesh M, Subramanian BR (2008) Influence of net ecosystem metabolism in transferring riverine organic carbon to atmospheric CO2 in a tropical coastal lagoon (Chilka Lake, India). Biogeochemistry 87(3):265–285

Hopkinson CSJ, Smith EM (2005) Estuarine respiration: an overview of benthic, pelagic and whole system respiration. In: del Giorgio PA, Williams PJL (eds) Respiration in aquatic ecosystems. Oxford University Press, Oxford, pp 123–157

Kortelainen P, Rantakari M, Huttunen JT, Mattsson T, Alm J, Juutinen S, Larmola T, Silvola J, Martikainen PJ (2006) Sediment respiration and lake trophic state are important predictors of large CO_2 evasion from small boreal lakes. Glob Chang Biol 12(8):1554–1567

Muduli PR, Kanuri VV, Robin RS, Kumar BC, Patra S, Raman AV, Rao GN, Subramanian BR (2012) Spatio-temporal variation of CO_2 emission from Chilika Lake, a tropical coastal lagoon, on the east coast of India. Estuar Coast Shelf Sci 113:305–313

Muduli PR, Kanuri VV, Robin RS, Kumar BC, Patra S, Raman AV, Rao GN, Subramanian BR (2013) Distribution of dissolved inorganic carbon and net ecosystem production in a tropical brackish water lagoon, India. Cont Shelf Res 64:75–87

Mukhopadhyay SK, Biswas HDTK, De TK, Jana TK (2006) Fluxes of nutrients from the tropical River Hooghly at the land–ocean boundary of Sundarbans, NE Coast of Bay of Bengal, India. J Mar Syst 62(1–2):9–21

Pattanaik S, Sahoo RK, Satapathy DR, Panda CR, Choudhury SB, Mohapatra P (2017) Intra-annual variability of CO_2 flux in the Mahanadi estuary–a tropical estuarine system, India. Ann Mar Sci 1:005–012

Pattanaik S, Chanda A, Sahoo RK, Swain S, Satapathy DR, Panda CR et al (2020) Contrasting intra-annual inorganic carbon dynamics and air–water CO_2 exchange in Dhamra and Mahanadi Estuaries of northern Bay of Bengal, India. Limnology 21(1):129–138

Pattanaik S, Chanda A, Sahoo RK, Swain S, Satapathy DR, Panda CR, Choudhury SB, Mohapatra PK (2020a) Contrasting intra-annual inorganic carbon dynamics and air–water CO_2 exchange in Dhamra and Mahanadi Estuaries of northern Bay of Bengal, India. Limnology 21(1):129–138

Pattanaik S, Roy R, Sahoo RK, Choudhury SB, Panda CR, Satapathy DR, Majhi A, D'Costa PM, Sai MS (2020b) Air-Sea CO_2 dynamics from tropical estuarine system Mahanadi, India. Reg Stud Mar Sci 36:101284

Perillo GME (ed) (1995) Definitions and geomorphic classifications of estuaries. In: Geomorphology and sedimentology of estuaries. Developments in sedimentology 53. Elsevier Science, New York, pp 17–47

Ramanathan AL, Subramanian V, Das BK (1996) Biogeochemical studies in the Cauvery estuary, east coast of India. http://nopr.niscair.res.in/handle/123456789/36415

Ray R, Shahraki M (2016) Multiple sources driving the organic matter dynamics in two contrasting tropical mangroves. Sci Total Environ 571:218–227

Ray R, Baum A, Rixen T, Gleixner G, Jana TK (2018) Exportation of dissolved (inorganic and organic) and particulate carbon from mangroves and its implication to the carbon budget in the Indian Sundarbans. Sci Total Environ 621:535–547

Richey JE, Melack JM, Aufdenkampe AK, Ballester VM, Hess LL (2002) Outgassing from Amazonian rivers and wetlands as a large tropical source of atmospheric CO_2. Nature 416(6881):617–620
Sarin MM, Sudheer AK, Balakrishna K (2002) Significance of riverine carbon transport: a case study of a large tropical river, Godavari (India). Sci China C Life Sci (English Edition) 45(Supp):97–108
Sarma VVSS (2009) Influence of river discharge on plankton metabolic rates in the tropical monsoon driven Godavari estuary, India. Estuarine Coastal Shelf Sci 85:515–524. https://doi.org/10.1016/j.ecss.2009.09.003
Sarma VVSS, Kumar MD, Manerikar M (2001) Emission of carbon dioxide from a tropical estuarine system, Goa, India. Geophys Res Lett 28:1239–1242
Sarma VVSS, Prasad VR, Kumar BSK, Rajeev K, Devi BMM, Reddy NPC, Sarma VV, Kumar MD (2010) Intraannual variability in nutrients in the Godavari estuary, India. Cont Shelf Res 30:2005–2014. https://doi.org/10.1016/j.csr.2010.10.001
Sarma VVSS, Kumar NA, Prasad VR, Venkataramana V, Appalanaidu S, Sridevi B, Kumar BSK, Bharati MD, Subbaiah CV, Acharyya T, Rao GD (2011) High CO_2 emissions from the tropical Godavari estuary (India) associated with monsoon river discharges. Geophys Res Lett 38(8):4
Sarma VVSS, Viswanadham R, Rao GD, Prasad VR, Kumar BSK, Naidu SA, Kumar NA, Rao DB, Sridevi T, Krishna MS, Reddy NPC (2012) Carbon dioxide emissions from Indian monsoonal estuaries. Geophys Res Lett 39(3):L03602
Sarma VVSS, Krishna MS, Prasad VR, Kumar BSK, Naidu SA, Rao GD, Viswanadham R, Sridevi T, Kumar PP, Reddy NPC (2014) Distribution and sources of particulate organic matter in the Indian monsoonal estuaries during monsoon. J Geophys Res Biogeo 119(11):2095–2111
Senthilkumar B, Purvaja R, Ramesh R (2008) Seasonal and tidal dynamics of nutrients and chlorophyll a in a tropical mangrove estuary, southeast coast of India. Indian J Mar Sci 37:132–140
Vijith V, Sundar D, Shetye SR (2009) Time-dependence of salinity in monsoonal estuaries. Estuar Coast Shelf Sci 85:601–608. https://doi.org/10.1016/j.ecss.2009.10.003
Zhai W, Dai M, Cai W-J, Wang Y, Wang Z (2005) High partial pressure of CO_2 and its maintaining mechanism in a subtropical estuary: the Pearl river estuary, China. Mar Chem 93:21–32

Chapter 5
A Systematic Review of Biogeochemistry of Mahanadi River Estuary: Insights and Future Research Direction

Tamoghna Acharyya, Bikram Prativa Sudatta, Susmita Raulo, Sambit Singh, Suchismita Srichandan, Sanjiba Kumar Baliarsingh, Alakes Samanta, and Aneesh Anandrao Lotliker

Abstract The Mahanadi estuary is located at the mouth of the perennial river Mahanadi, the third largest in peninsular India. It experiences semidiurnal tides and is characterized as a microtidal partially mixed coastal plain estuary. The objective of this chapter is to synthesise current knowledge in Mahanadi estuary from the published litearature. Range values of various parameters have been reported from the mouth of the main branch of Mahanadi river estuary. In most of the cases, the water column was well-oxygenated, with concentration of Dissolved oxygen (DO) remaining higher than 5 mg.l^{-1}, indicating good health of the estuary. Dissolved inorganic macronutrients behave non-conservatively with nitrite, nitrate, ammonia, phosphate, and silicate ranging between 0.01–1.61, 0.02–76.89, 0.1–7.93, 0.01–58.1, and 0.23–145.4µmol.l^{-1}, respectively. The optimum water column transparency and sufficient availability of nutrients leads to proliferation of phytoplankton biomass during post-monsoon season. However, the biomass never reached the bloom proportion as reported total chlorophyll-*a* concnetration remained below 4.57 mg. m^{-3}. Tide plays a vital role in short-term temporal variability of water quality parameters in Mahanadi estuary, though more indepth study covering full tidal cycle needs to be done. Metal pollution from the nearby industries and its possible transfer via the food chain has been documented in the Mahanadi estuary. Cadmium, nickel, cobalt, and lead show high lability in the exchangeable fraction of sediment, posing a possible health hazard. Mahanadi River is experiencing declined river runoff and sediment, which poses a concern for possible alteration in estuarine and adjacent

T. Acharyya (✉) · B. P. Sudatta · S. Raulo · S. Singh
School of Sustainability, Xavier University, Bhubaneswar, India
e-mail: acharyyat@xsos.edu.in

S. Srichandan
Centurion University of Technology and Management, Bhubaneswar, India

S. K. Baliarsingh · A. Samanta · A. A. Lotliker
Indian National Centre for Ocean Information Services, Ministry of Earth Sciences, Government of India, Hyderabad, India

S. Das, T. Ghosh (eds.), *Estuarine Biogeochemical Dynamics of the East Coast of India*, https://doi.org/10.1007/978-3-030-68980-3_5

coastal biogeochemistry. Therefore, future studies in the estuary should focus on the long-term high-frequency observation that would lead to nowcasting or forecasting the water quality parameter.

Keywords Anthropogenic impact · Biogeochemistry · Estuary · Mahanadi · Metal · Pollution · Runoff · Sediment · Water quality

5.1 Introduction

India houses several large and small estuaries along its 7500 km long coastline. The total area covered by Indian estuaries ranges about 2.14 × 10^6 ha. As many as 33 rives on the east coast and 34 rivers on the west coast have formed estuaries in their mouth before opening into the Bay of Bengal and the Arabian Sea, respectively (Khan and Murugesan 2005). Indian estuaries are characterized by high runoff during wet Indian summer monsoon (June–September) that eventually decreases with the successive seasons, leading to striking temporal changes in salinity and velocity. These fragile yet highly productive coastal ecosystems of India are under increasing pressure from high population concentration, damming and diversion of freshwater, exploitation of renewable and non-renewable natural resources, dumping of solid waste, and industrial discharge and municipal sewage, marine litter, and oil pollution. These stressors, either individually or synergistically getting coupled with climate change, alter the estuaries' biogeochemical characteristics with largely unknown consequences for the future. In this chapter, taking the Mahanadi River estuary as a case study, we aim to synthesize published literature on biogeochemistry with reference to increased anthropogenic interferences on the estuarine ecosystem.

5.2 Mahanadi River and Catchment Area

The Mahanadi River (Maha, mighty; Nadi, river) is the third-largest east-flowing peninsular river of India. The river basin lies between 80°30′E to 86°50′E and 19°20′N to 23°35′N and extends 1,41,589 km^2 area (WRIS 2011) of which a large portion lies in the states of Chhattisgarh (53%) and Odisha (45%) and a minimal area in the states of Jharkhand and Maharashtra. It originates at 442 m above mean sea level (MSL) in Chhattisgarh state. After traversing a total distance of 851 km along the northeast-southwest direction, Mahanadi opens into the Bay of Bengal, forming a large deltaic plain in Odisha. The average annual runoff of the Mahanadi River is 66.8 Billion Cubic Meter (Central Water Commission 2013) with a peak discharge of 44,740 m^{3}s^{-1} during the monsoon (Konhauser et al. 1997) . The Mahanadi basin is circular with maximum length and width of 587 km and 400 km, respectively.

The catchment area covers nearly 4.28% of the total geographical area of the country. The average water temperature in the basin varies between 24 and 27 °C. The basin receives an average annual rainfall of 1291 mm, of which about 84% is received from the southwest monsoon, i.e., from July to September (Singh and Das 2018). Chemical composition and the sediment yield of the river water are dominantly controlled by carbonate rock weathering (Chakrapani and Subramanian 1993a, b). The main soil types found in the basin are red and yellow soils, mixed red and black soils, laterite soils, and deltaic soils. A large part of the basin is covered with agricultural land, accounting for 55% of the total area (Behera et al. 2018). The basin has a total of 74 irrigation and 5 hydroelectric projects. There is a total of 253 dams in the basin that are used for irrigation purposes. The Ministry of Water Resources has reported that the Mahanadi basin's water quality parameters have exceeded the tolerance limit during recent years due to agricultural runoff, chemicals and hazardous effluents released from industries, raw domestic sewage from municipalities, and biomedical waste from the growing healthcare facilities. Water is getting increasingly reallocated from agriculture to industries. The basin's catchment area extends over the major parts of Chhattisgarh and Odisha and comparatively smaller portions of Jharkhand, Maharashtra, and Madhya Pradesh. Hence, there is a constant conflict of inter-state water sharing of Mahanadi waters.

5.3 Mahanadi River Estuary

5.3.1 Climate

Climate of Mahanadi estuary is represented by the nearby port city Paradeep (Paradeep Development Authority). Minimum and maximum atmospheric temperatures have been recorded at 8.9–41.4 °C. The average summer and winter wind speed ranges between 35 - 42 Km/h and 18–24 Km/h, respectively. The average annual rainfall in the region remains 1400–1480 mm. 80–84% of the total annual rainfall occurs from July to September during SW monsoon. The region is prone to frequent depression and cyclone landfalls.

5.3.2 Hydrography and Coastal Processes

The Mahanadi River opens into the Bay of Bengal at Paradeep in Jagatsinghpur district. The other distributary, Devi that carries lesser flow volume, opens into the bay near Astaranga in Puri district. Devi river mouth is known for annual turtle (*Lepidochelys olivacea*) nesting events. Mahanadi estuary's main branch has attracted more scientific investigations due to its larger importance for human enterprise, frequent reports of water quality deterioration, and fish mortality events (refer to the Sect. 5.3.5).

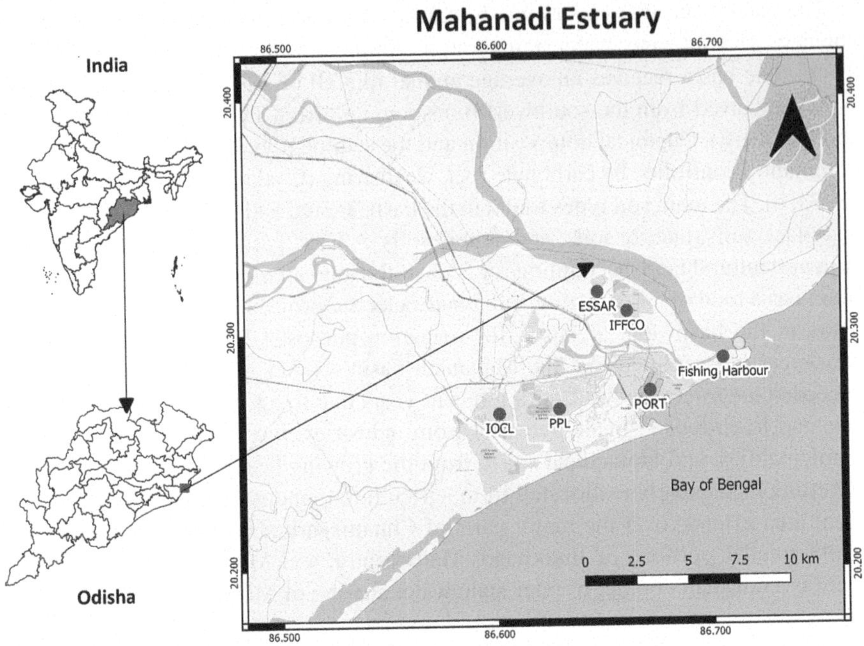

Fig. 5.1 Geographic location of the Mahanadi estuary. Red closed circle represents nearby industries. Yellow closed circle represents estuary mouth from where the water quality parameters have been synthesized. IOCL (Indian Oil Corporation Ltd.), PPL (Paradeep Phosphates Limited), PORT (Paradip Port), IFFCO (Indian Farmers Fertiliser Cooperative Limited), ESSAR (Essar steel plant)

In this chapter, we have confined our scope to the main branch, "Mahanadi estuary" (Fig. 5.1), which covers a length of 30–40 km and has a basin area of 9 km^2. The estuary's breadth ranges from 600 to 1200 m, whereas the depth varies from 6.0 to 14.0 m (Rao et al. 2007). Based on the salinity structure and mixing pattern, the Mahanadi estuary has been characterized as partially mixed during post-monsoon and well mixed during monsoon (Rao et al. 2004). It remains dominated by freshwater during southwest monsoon (July–September) as 75% of its annual freshwater flow is passed through the estuary during 3 months of a short time window (Das et al. 1997). During the peak flow, freshwater delivered through the estuary extends into the Bay of Bengal in the form of a plume which "first goes up in the northern direction, then it turns right and flows towards south" (Mishra 2004). Such a large volume of freshwater outflow during monsoon impacts coastal circulation as far as 90 km from the coastline (Rao et al. 2007). Freshwater gradually diminishes during the non-monsoon months to be majorly dominated by the tide and saline water. Tide in Mahanadi estuary is semidiurnal; the tidal range at the mouth varies from 1.45 to 2.20 m with a mean tidal range of 1.29 m (microtidal) (Dey et al. 2012; Naik et al. 2020). The tide can reach up to 42 km upstream from the mouth (Jhingran and Gopalakrishnan 1973). Representative current velocity in the estuary is estimated to be 0.8 ms^{-1} (Panigrahy et al. 1999). Estuarine flow completely reverses during flood and ebb tides. The flushing rate (the rate at which the total volume of the water in the estuary is exchanged) in the Mahanadi estuary has been modeled; the depth-integrated simulated value ranges from 910 m^3s^{-1} to 4190 m^3s^{-1} for post-monsoon (January) and monsoon (July), respectively (Rao et al. 2004). Irrespective of the season, sand dominates (>90%) in the estuarine sediment found at the mouth followed by silt and clay fraction (Raj et al. 2013). Mixing of fresh and saline water reduces the velocity of the transporting agent, which in turn leads to sand deposition (Satapathy et al. 2019). The Mahanadi has formed a long and narrow spit along with the coast due to intense littoral drift, which also causes the shifting of the mouth from south to north (Srinivasan et al. 1982).

5.3.3 *Estuarine Biodiversity*

5.3.3.1 Fauna

The Mahanadi's estuarine environment is subjected to the constant change in salinity - from fresh to brackish to marine at various locations from head to mouth, posing challenges to the physiology of animals to which few are only able to adopt. Zoological Survey of India (ZSI), Berhampur, surveyed 11 selected sites in the Mahanadi estuary from 1989 to 1992. As a result, a total of 231 species of vertebrates, viz., Aves (46 species), fish 180(species), snak 4(species) and testudine (1species), and 242 species of invertebrates belonging to echinoderms (2 species), shoreline insects (3 species), arachnid (1 species), crustaceans (44 species), hermit crabs (4 species), polychaetes (33 species), leech (1 species), mollusks (149 species), coelenterates (3 species), and sponges (2 species) were recorded (Alfred 1989). The diversity of fishes and crustaceans suggests that the Mahanadi estuarine system is one of India's most richest and productive estuaries. Fish populations are abundant with a large species diversity. Some marine fish move into the estuary to breed and use the estuary as a nursery ground to feed and grow before moving out to sea again. Fishes like salmon and eels use the estuary as a migratory route from rivers to the sea and vice versa. Only a few fish species live in the estuary throughout the year.

5.3.3.2 Flora

The estuarine mouth opposite to Paradeep port is dominated by anoxygenic mudflats rich in organic matter and crisscrossed by tidal creeks creating an ideal habitat for the mangrove vegetation. Mangrove flora are dominated by *Avicennia officinalis*, *A. alba*, *A. marina*, and *Rhizophora mucronata*, which are more abundant on the river mouth banks and are characterized by reddish and jointed pneumatophores. The other flora of mangrove swamps are *Bruguiera conjugata*, *Ceriops roxburghiana*, *Sonneratia apetala*, *Lumnitzera racemosa*, *Aegiceras corniculatum*, *Excoecaria agallocha*, *Arthrocnemum indicum*, *Salicornia brachiata*, *Suaeda maritima*, and *Suaeda mudijlora* (Alfred 1989).

5.3.4 Blue Economy

Being a hub of maritime transport and a productive fishery ground, the Mahanadi estuary plays a significant role in the region's blue economy by creating livelihood and promoting economic growth. Fisheries, both capture and culture, are important livelihood options for the local people. About 662 trawlers and 192 motorized boats are operational in the Paradeep fishing harbor, located at the estuarine mouth. The harbor's average annual fish landing is 25,000–30,000 MT; other crucial catch constitutes penaeid shrimps, crabs, lobsters, cephalopods, and mollusks (Paradeep Fishing Harbour). Export of shrimp from Paradeep Fishing Harbour generates foreign revenue of about Rs.150 crores, contributing to 40% of Odisha's seafood export. About 374 functional units of dry fish clusters in the area employ 3000 people in various capacities. The total annual turnover of dry fish is around 1900 MT, of which 10 MT per month is exported to Bangladesh (MSME 2016–2017).

Paradeep port, located near the confluence (20°15′–55.44″N 86°40′–34.62″E), is a major port on the east coast of India. In 2019–2020, it handled 112.67 million tons annual cargo majorly of crude oil, POL (petroleum, oil, and lubricants) products, iron ore, manganese ore, thermal and coking coal, finished steel, fertilizer, and fertilizer raw material (Paradip Port Trust). The port is administered by the Paradeep Port Trust (PPT), Govt. of India, with a total employee strength of 750.

As one of the six major petroleum, chemical, and petrochemical investment regions in India, the Paradeep area has witnessed rapid industrialization by attracting massive investment to the tune of 3.5 lakh crore (MSME 2016–2017). Several large-scale industries are currently operational: IFFCO Fertilizers, Paradeep Phosphates Ltd., Indian Oil Corporation-Paradeep Refinery, Skol Breweries Ltd., Paradeep Carbons Ltd., Cargill India (P) Ltd., and Essar steel plant (refer to Fig. 5.1 for location). These industries help spur the economic development of the region and employ many people directly or indirectly.

Natural creeks, mangrove forests, and scenic sea beaches make Paradeep a sought-after tourist destination in Odisha. In 2018, the total tourist footfall was about five lakhs (Statistical Bulletin 2018). This area's tourism sector employs many people in accommodation, food and beverage stalls, tour operation, and transportation services.

5.3.5 Anthropogenic Setting

The Mahanadi estuary receives back the untreated domestic wastewater and industrial effluents from port city Paradeep and industries such as fertilizer, paper, textile, distilleries, oil refineries, and others present along the course (refer to Table 5.1). The estuary also receives a large amount of agricultural runoff from the adjoining delta. Industry effluents and sewage from Paradeep township and Port Trust (PPT)

Table 5.1 Anthropogenic setup in and around Mahanadi estuary

Industry/establishment (year of establishment)	Manufactured product/anthropogenic setting	Major pollutants
Paradeep Port Trust (1962)	Import and export of goods mainly crude oil, petrochemicals, and ores	Trace metals, suspended solid, nutrients, BOD, petroleum hydrocarbon, pathogenic bacteria
East Coast Breweries and Distilleries Ltd., Paradeep (1969)	Breweries	Biochemical oxygen demand, suspended solid, pH, mineral, acids, sugars
Paradeep Township (2002 as a municipality)	Urban population	Nutrients, biochemical oxygen demand, oil, heavy metals, pathogenic bacteria
Paradeep Phosphates Ltd., Paradeep (PPL) (1981)	Diammonium phosphate, NPK, ammonium phosphate sulfate, Muriate of potash, ammonia and sulfuric acid, gypsum	pH, biochemical oxygen demand, nutrients, trace metals
Fishing Harbour (1996)	Harboring stations for fishing boats	Sewage, oil, nutrients, heavy metal, petroleum hydrocarbon, BOD, nutrients
IFFCO (previously OSWAL fertilizer plant at Paradeep) (2005)	DAP/NPK, sulfuric acid, phosphoric acid	pH, biochemical oxygen demand, nutrients, trace metals (like Zn, Cu, etc.)
Cargill India (P) Ltd (2011)	Edible oil, vegetable oil	Agricultural waste

are discharged into the Atharabanki river, which ultimately joins the Mahanadi River near the confluence point in the Bay of Bengal. Panigrahy et al. (2014) reported petroleum hydrocarbon ranging from 0.14 to 5.27μg.l^{-1} along the Mahanadi transact; they attributed it to the waste discharges from many fishing boats berthed in fishing harbor adjacent to the Mahanadi estuary. Mass fish kill events have been reported from small rivers and creeks, namely, Sahara, Atharabanki, and Kaudia, in and around Paradeep. Notably, all such events took place in July and August. Industrial effluents are thought to cause hypoxia along with high alkalinity, ammonia, and BOD (Times of India 07/06/2016).

5.3.6 *Biogeochemistry of Mahanadi Estuary*

Biogeochemically, estuaries are unique as they are juxtaposed at the land, ocean, and atmosphere interface. A constant flux of materials and energy takes place between these three reservoirs. Hence, a highly dynamic system like an estuary controlled by the tide and river runoff is influenced by various time scales from interannual, seasonal, fortnightly to diurnal scales.

The Mahanadi estuary has been an area of interest for investigators since the early 1970s. A combination of the keywords "Mahanadi" and "Estuary" on October 15, 2020, was used to extract bibliometric data from the SCOPUS database. A total of 19 and 11 research articles were obtained with these keywords appearing anywhere in the document and the title, respectively. The same keyword combinations delivered 31 research papers in the Google Scholar database, which has been used in this book chapter, including the SCOPUS output. While earlier studies were mainly focused on fisheries (Jhingran and Gopalakrishnan 1973) and conservative vs. non-conservative behavior of various cations and radiogenic elements (Borole et al. 1979), beginning from the latter part of the decade Mahanadi estuary started witnessing more diverse studies (Table 5.2). Frequency distribution of the publication pattern reveals that the majority of the studies are related to water quality (10/31) and phytoplankton (08/31); others are more specific and targeted, such as texture and heavy metal analysis of sediment, petroleum hydrocarbon, zooplankton, air-sea dynamics, etc. (Table 5.2). A wide variety of sampling strategies, both at space and time scales, was noticed. Investigators preferred seasonal sampling (pre-monsoon, monsoon, and post-monsoon, summer, monsoon, and winter) and single time observation over diurnal/tidal/seasonal/annual/interannual or their combinations. Spatial sampling also varied considerably across investigations; the majority of them covered the entire salinity gradient along the transect – a large number of studies also covered a single location which almost all the cases was at the confluence (20.29°N, 86.70°E) but few were along the river plume extended over the adjacent Bay of Bengal. For better inter-comparison among various measured parameters, we decided to stick to a single station (confluence station 20.29°N, 86.70°E, refer to Fig. 5.1) vis-à-vis to group all time scales to pre-monsoon (PRM), monsoon (MON), and post-monsoon (POM) based on the following criteria: June–September (monsoon), October–February (post-monsoon), and March–May (pre-monsoon), summer (pre-monsoon), and winter (post-monsoon). A range plot of various physico chemical parameters discussed below has been presented in Fig. 5.2.

5.3.6.1 Water Temperature

Temperature plays a vital role in an estuary as it determines the solubility of oxygen, the rate of photosynthesis, microbial activity, and degradation of organic matter. While optimal water temperature creates a conducive environment for aquatic plants and animals' growth in estuaries, rapid change (thermal pollution) can often be fatal. Surface water temperature in the Mahanadi estuary ranges between 22.08 and 31.4 °C (Pati et al. 2018; Pattanaik et al. 2020a, Srichandan et al. 2013) and follows air temperature pattern irrespective of the tide in all seasons (Das et al. 1997). Clear inter-seasonal variability exists as the highest temperature was recorded in the pre-monsoon, followed by post-monsoon and monsoon (Naik et al. 2009, 2020). Water temperature gradually increases from mouth to upstream in pre-monsoon and monsoon, but a reverse trend was noticed during the post-monsoon (Upadhyay 1988).

Table 5.2 Timeline of research themes and associated measured parameters in the Mahanadi estuary

Year	Research theme	Measured parameters
1979	Conservative vs. non-conservative behavior	Uranium 234 and 238, silicon, cations (Na^+, K^+, Ca^{2+}, Mg^{2+})
1984	Conservative vs. non-conservative behavior	Suspended matter, silicon, cations (Na^+, K^+, Ca^{2+}, Mg^{2+})
1988	Water quality	Water temperature, salinity, pH, dissolved oxygen, phosphate, nitrate and silicate
1995	Conservative vs. non-conservative behavior	Uranium 234 and 238, chlorinity
1997	Water quality	Water temperature, salinity, pH, dissolved oxygen, phosphate, nitrite, nitrate, ammonia and silicate
1999	Water quality	Salinity, pH, dissolved oxygen, biochemical oxygen demand, chlorophyll-*a*, phosphate, total phosphorus, nitrite, nitrate, ammonia and major cations (Ca^{2+}, Mg^{2+}) and anion (F^-)
2004	Satellite application	Salinity, Secchi disk depth, total suspended sediment concentration, chlorophyll-*a*
2006	Water quality	Conductivity, pH, dissolved oxygen, biochemical oxygen demand, total suspended sediment, total dissolved sediment, phosphate, total phosphorus, nitrite, nitrate, ammonia, total nitrogen, hardness, alkalinity, major cations (Na^+, K^+,Ca^{2+}, Mg^{2+}) and anions (Cl^- and SO^{4-})
2006	Water quality	Conductivity, pH, dissolved oxygen, biochemical oxygen demand, total suspended sediment, total dissolved sediment, phosphate, total phosphorus, nitrite, nitrate, ammonia, total nitrogen
2009	Water quality	Conductivity, pH, total dissolved sediment, nitrate, hardness, alkalinity, major cations (Na^+, K^+,Ca^{2+}, Mg^{2+}) and anions (Cl^- and SO^{4-})
2009	Phytoplankton taxonomy	Water temperature, salinity, pH, dissolved oxygen, total suspended solid, phytoplankton, chlorophyll-*a*
2009	Phytoplankton bloom in river plume	Salinity, nitrate, phosphate, chlorophyll-*a*, chlorophyll-*b*
2009	Water quality	pH, turbidity, dissolved oxygen, biochemical oxygen demand, turbidity, nitrate, fecal coliform
2011	Sediment heavy metals	Fe, Zn, Cu, Ni, Cd, Mn, Cr, Co, Pb
2012	Water quality	Temperature, salinity, dissolved oxygen, dissolved inorganic phosphate (DIP), dissolved inorganic nitrogen (DIN) (nitrate, nitrite, ammonium), dissolved silicate, pH, total alkalinity, pCO_2, chlorophyll-*a*, biochemical oxygen demand, primary productivity, phytoplankton taxonomy, phytoplankton identification, count and biovolume

(continued)

Table 5.2 (continued)

Year	Research theme	Measured parameters
2012	Water quality	Temperature, salinity, dissolved oxygen, biochemical oxygen demand, total suspended substances, nitrite, nitrate, ammonia, phosphate, silicate, total nitrogen, total phosphate, chlorophyll-*a*
2012	Surface water-dissolved metal	Fe, Mn, Zn, Cr, Cu, Co, Ni, Pb, Cd
2013	Sediment heavy metal and sediment texture	Cd, Pb, Hg, organic carbon, sand, silt and clay percentage
2013	Ocean color validation in the river plume	Chlorophyll-*a*
2013	Water quality	pH, salinity, dissolved oxygen, biochemicaloxygen demand, total suspended solid, nitrite, nitrate, ammonia, phosphate, silicate, total nitrogen, total phosphate, chlorophyll-*a*
2013	Zooplankton taxonomy	Temperature, depth, salinity, pH, dissolved oxygen, biochemical oxygen demand, nitrite, nitrate, ammonia, phosphate, silicate, total nitrogen, total phosphate, chlorophyll-*a*, zooplankton identification, count and biovolume
2014	Anthropogenic pollution	Petroleum hydrocarbon
2014	Sediment heavy metals	Fe, Mn, Zn, Cr, Cu, Co, Ni, Pb, Cd
2018	Phytoplankton taxonomy	Phytoplankton identification
2018	Phytoplankton taxonomy	Temperature, Secchi depth, salinity, pH, turbidity, dissolved oxygen, biochemical oxygen demand, alkalinity, nitrite, nitrate, ammonia, phosphate, silicate, chlorophyll-*a*, phytoplankton identification, count and biovolume
2019	Metal toxicity on human due to seafood consumption	Mn, Zn, Cd, Cu, Ni, Pb, Fe, organic carbon, sand, silt and clay percentage
2019	Phytoplankton dynamics	Temperature, Secchi depth, salinity, nitrite, nitrate, ammonia, phosphate, silicate, chlorophyll-*a*, size-fractionated chlorophyll-*a*, phytoplankton identification, count and biovolume
2020	Air-sea dynamics	Temperature, pH, salinity, dissolved oxygen, nitrate, ammonia, alkalinity, dissolved inorganic carbon, pCO_2, chlorophyll-*a*
2020	Phytoplankton pigment composition	Chlorophylls and carotenoids and xanthophyll, phytoplankton identification, count and biovolume
2020	Phytoplankton community dynamics	Temperature, pH, salinity, Secchi depth, total suspended solid, nitrite, nitrate, ammonia, phosphate, silicate, chlorophyll-*a*, phytoplankton identification, count and biovolume
2020	Heavy metal in adjacent shelf	Ti, Pb, Zn, Mn, Co, Ni, Cr, organic C

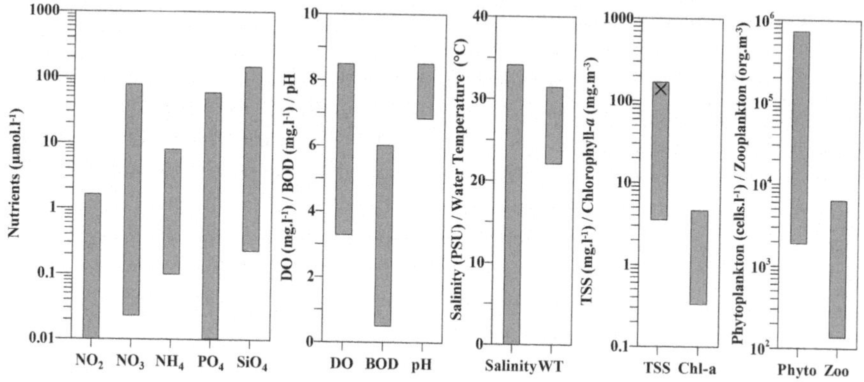

Fig. 5.2 Range of biogeochemical parameters (from published literature) in the Mahanadi estuary. Phyto: phytoplankton, Zoo: zooplankton, NO_2 nitrite, NO_3 nitrate, NH_4 ammonium, PO_4 phosphate, SiO_4 silicate, DO dissolved oxygen, BOD: biochemical oxygen demand, WT: water temperature, TSS: total suspended solid, Chl-*a* chlorophyll-*a*. The cross symbol in TSS bar represents the highest value ± standard deviation (167 ± 73.7 $mg.l^{-1}$)

5.3.6.2 Salinity

Salinity is the most dynamic entity in estuaries as it determines the spatial distribution of species; the less tolerant ones restrict them to the upper reaches, while the more tolerant and migratory species prefer brackish and higher salinities. Salinity has a strong seasonal trend in the Mahanadi estuary; it ranges from 0.01 PSU (during post-monsoon due to river runoff and dilution) to 34.1 PSU (during pre-monsoon due to tidal effect and high rate of evaporation) (Das et al. 1997; Dixit et al. 2013). The salinity gradient is from the head toward the sea, implying lower salinity values are recorded in the upper estuary/riverine end, whereas higher values are recorded from the lower estuary/oceanic end. During post-monsoon, the estuary is entirely dominated by saline water compared to monsoon when freshwater remains confined to the head with brackish or saline water in the mouth (Rao et al. 2004). Apart from the monsoon, the surface is always less saline than the bottom, indicating a partially mixed character of the estuary. Higher and lower salinity is associated with flood and ebb tide, respectively (Dixit et al. 2013).

5.3.6.3 Major Ions

Major cations such as Na^+, K^+, Ca^{2+}, and Mg^{2+} behave conservatively, implying their distribution across the salinity gradient follows the theoretical dilution line in the Mahanadi estuary (Borole et al. 1979). During lean periods, silicon behaves non-conservatively.

5.3.6.4 Total Suspended Solid (TSS) and Turbidity

Turbidity is a function of suspended solids in water. Highly turbid water full of suspended solids is not conducive to an estuary's health and its inhabiting life-forms since it deters photosynthesis and reduces DO. TSS varies from 3.53 to 240.7 mg.l^{-1}; usually lower and higher values are reported during post-monsoon and monsoon, respectively, in the Mahanadi estuary (Dixit et al. 2013; Pattanaik et al. 2017). Dixit et al. (2013) reported the highest TSS value during monsoon and lowest value during post-monsoon, which they attribute to contamination of estuaries with sewage and industrial effluents. During all the seasons, TSS is correlated negatively with the Secchi depth, which is a measure of water transparency (Naik et al. 2020).

5.3.6.5 Dissolved Oxygen

Dissolved oxygen (DO) is the most critical water quality parameter in estuaries since almost all life-forms depend on oxygen for respiration. Photosynthesis, turbulence, and wind mixing are known to influence DO concentration. Estuaries worldwide are receiving increasing volumes of organic matter and nutrients, leading to intense phytoplankton bloom and poor water quality. The Mahanadi estuary, in general, remains well-oxygenated (Sundaray et al. 2006) throughout the year (above the permissible limit for drinking water) as revealed by the reported range of values (3.28–8.48 mg.l^{-1}) (Dixit et al. 2013; Pati et al. 2018). DO concentration in the estuary goes up during post-monsoon but comes down during pre-monsoon/monsoon (Sarkar et al. 2019; Naik et al. 2009; Samantray et al. 2009; Upadhyay 1988). High DO concentration during post-monsoon along the estuary suggests abundant phytoplankton growth leading to high primary productivity, while low values during pre-monsoon may be due to reduced water volume and accelerated growth of microbes in higher temperature. Salinity also plays a crucial role in determining the DO of the estuary as the solubility of DO decreases with increased salinity (Mishra et al. 2018). Almost all the studies reported dissolved oxygen >5 mg/l, indicating the healthy state of the estuarine system concerning oxygen.

5.3.6.6 pH

pH maintains nutrient availability, acts as a buffering system hence is important for life-forms. The literature review suggests, pH in the Mahanadi estuary is seasonal (Pattanaik et al. 2020b) and varies between 6.84 and 8.5 (Mishra et al. 2018; Samantray et al. 2009). During post-monsoon, the estuary turns dominantly alkaline due to the influence of seawater and biological activity, which gradually decreases during pre-monsoon and monsoon due to the influence of freshwater influx and decomposition of organic matters (Upadhyay 1988). However, it is important to note that ocean water's influence on pH values in the Mahanadi estuary is not always well defined (Sundaray et al. 2006) as acidic effluent dumped in creeks from the nearby industries occasionally brings down the pH of the estuary. The tide controls pH; the higher and lower pH values are associated with flood and ebb tides, respectively (Das et al. 1997). Irrespective of the seasons, the pH gradient is always set up from the head to the estuary's mouth (Pattanaik et al. 2020b).

5.3.6.7 Nutrients

Many inorganic nutrients act as life-supporting components, but nitrogen (in the form of nitrite, nitrate and ammonia), phosphorus, and silicate are the most important macronutrients for phytoplankton growth in coastal waters. Nutrient distribution in estuaries is mainly dependent on season, tidal condition, and freshwater influx. The source of nutrients can be both autochthonous and allochthonous. The Mahanadi estuary is a nutrient-rich system where river discharge plays a vital role in nutrient dynamics (Mishra et al. 2018). Interestingly, removal of nutrients with increasing salinity is not well pronounced in the Mahanadi estuary due to the influence of effluent load via different point sources (Sundaray et al. 2006). In general, nutrient levels increase during low tide than high tide (Dixit et al. 2013).

Water column nitrite indicates fresh input of biodegradable organic load generated from sewage. Nitrite in Mahanadi ranges from 0.00 to 1.61 $\mu mol.l^{-1}$ (Das et al. 1997; Mishra et al. 2018). Seasonally it follows the pattern: monsoon > pre-monsoon > post-monsoon (Srichandan et al. 2013). Some studies also found high nitrite values during pre-monsoon which was attributed to point source or de-nitrification at the sediment-water interface (Naik et al. 2020). Nitrite level followed the tidal pattern as an increasing trend from flood to ebb tide was noticed (Das et al. 1997) during all three seasons (PRM, MON, and POM).

Nitrate in Mahanadi estuary ranges from 0.02 to 76.9 $\mu mol.l^{-1}$ (Mishra et al. 2009; Pati et al. 2018). Higher nitrate values are associated with the peak discharge period during monsoon (Pattanaik et al. 2020b; Srichandan et al. 2019). However, the progressive decrease during the post-monsoon and reaching the minimum can be attributed to biological utilization (Upadhyay 1988) and reduced microbial activity due to winter cooling (Pati et al. 2018). Like nitrite, nitrate also recorded some high values during post-monsoon, possibly due to anthropogenic addition (Naik et al. 2020). Nitrate exhibited an upward trend from flood tide to ebb tide during pre- and post-monsoon but little tidal variation was observed during the monsoon (Das et al. 1997). The lower value during flood tide of post-monsoon and pre-monsoon may be due to marine water incursion having less nitrate.

The primary source of ammonium in the estuary is local anthropogenic input rather than river runoff contributing to comparatively high concentration during pre-/post-monsoon (Dixit et al. 2013; Naik et al. 2020; Sundaray et al. 2006). Ammonia in the Mahanadi estuary ranges between 0.1 and 9.1 $\mu mol.l^{-1}$ (Dixit et al. 2013; Mishra et al. 2018). A couple of recent studies reported maximum ammonia concentration in monsoon and minimum in post-monsoon (Srichandan et al. 2013, 2019), attributed to high river runoff during monsoon as the likely source of ammonia in the estuarine system. Higher ammonia values were reported in the ebb tide during pre- and post-monsoon, indicating a possible influx of ammonium compounds from the adjacent fertilizer plant (Das et al. 1997).

Phosphate conentration varies between 0.01 and 58.1μmol.l^{-1} in Mahanadi estuary (Mishra et al. 2009, 2018). It exhibits relatively higher concentration during lean periods in the Mahanadi estuary (Naik et al. 2009), possibly due to local nutrient loading. Lower values of phosphate in the monsoon can be attributed to dilution and/or precipitation removal by silt-laden rainwater (Sundaray et al. 2006) favored by the acidic environment (pH 6) (Upadhyay 1988). However, a couple of studies done in different periods (Das et al. 1997; Srichandan et al. 2019) reported higher phosphate values during monsoon attributed to river runoff of fertilizer and weathering. In the middle and upper reaches of the estuary, the bottom layer's phosphate content was significantly higher than the surface waters except in monsoon (Upadhyay 1988), which recorded high phosphate values in surface water. Phosphate is also tidally modulated; high values of inorganic phosphate at ebb tide and low values at flood tides both during pre- and post-monsoon have been recorded, indicating limited inorganic phosphate input from the sea (Das et al. 1997).

Silicate concentration in the Mahanadi estuary ranged from 0.23 to 145.4μmol.l^{-1} (Mishra et al. 2018; Pati et al. 2018). It showed higher values during monsoon, possibly due to the input of weathered siliceous sediments from its catchments. Silicate concentration progressively reduces from post-monsoon to pre-monsoon due to higher biological uptake by diatoms and removal of dissolved silicate by adsorption onto suspended particles at higher salinity values (Naik et al. 2020). Silicate behaves near conservatively during the monsoon (Sundaray et al. 2006), possibly due to dilution and inorganic removal of silicate (Upadhyay 1988). The distribution of silicate reveals a higher concentration at the upper estuary, which gradually decreases downstream.

5.3.6.8 Metals

The presence of oil refineries and other heavy industries in nearby regions makes the Mahanadi estuary susceptible to metal pollution, both in water and in sediment (Sundaray et al. 2014). Relative abundance of various metals in estuarine water follows the trend Fe > Zn > Mn > Ni > Pb > Cu > Co > Cr > Cd (Sundaray et al. 2012). Zn, Co, and Cu behave conservatively, while Fe, Mn, Cr, Ni, and Pb behave non-conservatively in the estuary's mixing zone. Non-conservative mixing points to various removal processes such as flocculation, adsorption, and precipitation. The contribution of anthropogenic input of Cd is prominent in water.

Sediment metal concentration is relatively higher in estuaries than the freshwater due to the adsorption/accumulation process (Sundaray et al. 2014). The order of enrichment of metals in the sediment is Zn > Cu > Ni > Cr > Pb > Cd (Satapathy et al. 2019). Sediment metals such as Cd, Ni, Co, and Pb are highly available in their exchangeable fraction, posing a potential environmental risk for the benthic infauna and other aquatic biota (Sundaray et al. 2011). Fe–Mn oxides act as efficient scavengers for metals like Zn, Pb, Cu, Cr, Co, and Ni. Sediment organic carbon and grain size are important determinants of metal dynamics as sediments rich in organic content have higher cation exchange capacities hence can trap metal, while sandy and organically poor sediments have little ability to retain metal ions (Raj et al. 2013). However, binding competition among trace metals and type of organic carbon are important as well (Satapathy et al. 2019).

Cadmium is moderately polluted in the Mahanadi estuary sediment (Raj et al. 2013; Satapathy et al. 2019), possibly from the "fertilizer-based industrial effluents from IFFCO and PPL," which is a potential threat as Cd is not only toxic but also highly mobile (Sundaray et al. 2014). Recently, Satapathy et al. (2019) reported potential human health hazards due to edible oyster consumption (*Saccostrea cucullata*), which indicated the bioaccumulation of Cu and Zn. There lies substantial carcinogenic threat from Cr and Cd and non-carcinogenic health hazards from Mn, Ni, Cd, Fe, Pb, and Cu in the Mahanadi estuary due to seafood consumption, mostly edible oyster. Interestingly shelf sediment values of Ti, Pb, Zn, Mn, Co, Mo, and Ni indicate "no significant pollution by individual elements" on the local ecology (Nambiar et al. 2020). This observation demonstrates the Mahanadi estuary's role and its associated high sediment flux in removal or dilution of trace metals to the adjacent Bay of Bengal. This, in essence, indicates that the Mahanadi estuary acts as a metal-retaining and processing hotspot, keeping the adjoining shelf sediment healthy.

5.3.6.9 Chlorophyll-*a* and Other Carotenoids

Chlorophyll-*a* is the primary photosynthetic pigment in phytoplankton. Due to its ubiquitous presence, chlorophyll-*a* is used as a proxy for phytoplankton biomass. It ranged between 0.33 and 4.57 mg/m^3 in the Mahanadi estuary (Mishra et al. 2018; Pattanaik et al. 2017). Chlorophyll-*a* was higher during post-monsoon than pre-monsoon and monsoon season (Sarkar et al. 2019; Dixit et al. 2013); however, sporadic peaks in other seasons are not uncommon either (Pattanaik et al. 2020b). The upper estuary displayed relatively higher surface chlorophyll-*a* compared to the lower estuary. The Mahanadi outer estuary (estuary formed through river plume over the Bay of Bengal) shows a decrease in the chlorophyll-*a* concentration while moving away from the coast toward offshore waters (Nagamani et al. 2013). The plume chlorophyll is not well correlated with water transparency owing to the strong influence of suspended sediment concentration (Mishra 2004).

Accessory pigments in phytoplankton execute photoprotective and photosynthetic roles. Owing to their exclusive presence in some groups, they provide valuable insights into the size and class distribution of phytoplankton. In the Mahanadi estuary the relative prevalence of accessory pigments is as follows: zeaxanthin > chlorophyll-*b* > fucoxanthin > β-carotene > alloxanthin > violaxanthin (Srichandan et al. 2020).

5.3.6.10 Biochemical Oxygen Demand (BOD)

Biochemical oxygen demand (BOD) is a measure of biochemically oxidizable organic matter present in water. BOD is a good indicator of the bacterial load as well as the degree of organic pollution in aquatic bodies – higher BOD indicates less oxygen demand and low water quality, whereas low BOD signifies generally purer water. Higher BOD values are found in the Mahanadi estuary compared to the adjacent coastal Bay of Bengal, indicating the likelihood of contamination by industrial effluents and municipal sewage in the estuary (Khadanga et al. 2012). BOD in the Mahanadi estuary ranges between 0.5 and 6.0 $mg.l^{-1}$ (Mishra et al. 2018; Samantray et al. 2009). Some high spikes have been reportedly caused by "Atharbanki" creek that carries sewage and industrial effluents directly into the estuary from Paradeep port, Paradeep township, and adjoining industries (Dixit et al. 2013). River runoff plays a critical role in BOD values as it has been shown that sewage contamination, if any, gets diluted by a high flushing rate and massive volume of freshwater runoff in the Mahanadi estuary (Samantray et al. 2009). BOD is positively correlated with chlorophyll-*a*, indicating its dependence on phytoplankton biomass and its life cycle apart from anthropogenic organic loading.

5.3.6.11 Alkalinity, DIC, and pCO_2

Alkalinity is a measure of the capacity of a water body (estuary in our case) to resist acidification, whereas DIC (dissolved inorganic carbon) is made from the carbonate and bicarbonate pools. The average total alkalinity values ranged between 1067 and 2159$\mu mol.kg^{-1}$, whereas DIC values were in the order of 1057–1844$\mu mol.kg^{-1}$ in the Mahanadi estuary (Pattanaik et al. 2020b). The partial pressure of CO_2 (pCO_2) of the estuarine water varied between 100 and 1571 μatm. Net pCO_2 flux remained negative throughout the year as pCO_2 water was low compared to the atmosphere except in monsoon when the flux was positive across the estuary. The net annual CO_2 flux from the estuary (9.334 km^2) to the atmosphere was calculated to be 0.135 Gg (Dey et al. 2012).

5.3.6.12 Phytoplankton Community

Phytoplankton are autotrophic free-floating algae that live in the euphotic water column and ubiquitous in the aquatic milieu. Estuarine environments host wide variety ofphytoplankton communities. The esturaine phytoplankton growth is fueled by riverine nutrient discharge, anthropogenic input, and marine influx from adjoining sea (Gao and Song 2005; Ketchum 1967). In the context of the Mahanadi estuary, five studies (Dey et al. 2013; Mishra et al. 2018; Naik et al. 2009; Srichandan et al. 2019, 2020) have explored the phytoplankton taxonomic diversity and spatio-temporal distribution against various environmental variables. Based on the available literature (Dey et al. 2013; Mishra et al. 2018; Naik et al. 2009; Srichandan et al. 2019), a total of 282 phytoplankton species have been reported from the Mahanadi estuary. The total number of phytoplankton species reported from the Mahanadi estuary follows the rank order of Bacillariophyta > Chlorophyta > Cyanophyta > Dinophyta > Euglenophyta > Charophyta > Haptophyta and Ochrophyta (Fig. 5.3). Among the reported 169 species of Bacillariophyta, *Chaetoceros* sp., *Coscinodiscus* sp., *Skeletonema costatum, Coscinodiscus gigas, Thalassiothrix longissima, Synedra* sp., *Melosira* sp., *Thalassiosira* sp., *Cerataulus heteroceros, Asterionellopsis glacialis, Melosira sulcata, Rhizosolenia alata, Leptocylindrus danicus, Aulacoseira granulata, Odontella sinensis, Coscinodiscus granii, Rhizosolenia setigera, Thalassiosira subtilis*, and *Thalassionema nitzschioides* were observed to be dominant.

Regarding the phytoplankton abundance, the Mahanadi estuary has been observed with a range of 1.9×10^3 to 7.23×10^5 cells.l^{-1}. Naik et al. (2009) have reported higher phytoplankton abundance in post-monsoon, i.e., October–November with 4.09×10^4 to 5.11×10^4 cells.l^{-1}. They have also reported that the dominance of *Coscinodiscus gigas* and *Thalassiothrix longissima* was responsible for higher abundance during the post-monsoon season. The phytoplankton abundance was 3.16×10^4 to 4.31×10^4 cells.l^{-1} in pre-monsoon (May–June), while 3.00×10^4 to 4.07×10^4 cells.l^{-1} in summer (February–March). Similar to Naik et al. (2009), Dey et al. (2013) also reported higher phytoplankton abundance during post-monsoon with cell abundance of 7.23×10^5 cells.l^{-1}. During this period, the phytoplankton community was primarily dominated by *Pediastrum* sp., *Trichodesmium* sp., *Ceratium* sp., *Oscillatoria* sp., and *Nostoc* sp. During monsoon and pre-monsoon, phytoplankton densities were 0.68×10^5 cells. l^{-1} and 5.20×10^4 cells.l^{-1},

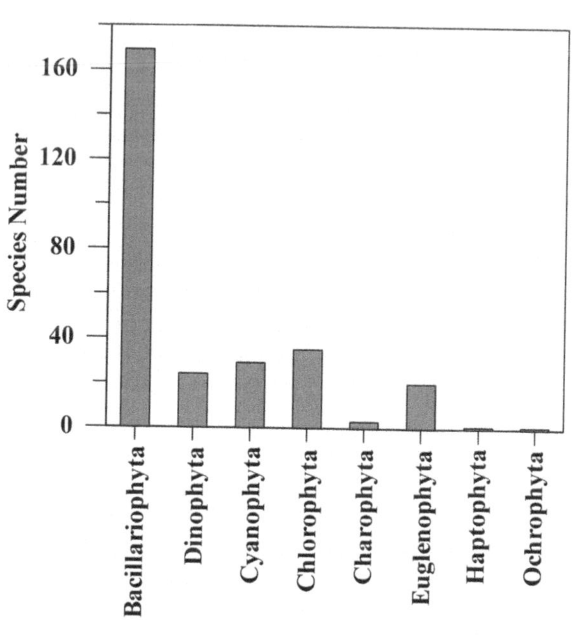

Fig. 5.3 Reported phytoplankton species number in the Mahanadi estuary

respectively. However, a contrasting observation was reported by Mishra et al. (2018) with the highest and lowest density during monsoon (22.40×10^4 cells.l^{-1}) and pre-monsoon (10.10×10^4 cells.l^{-1}), respectively. A recent study on seasonal dynamics of phytoplankton in the Mahanadi estuary has reported the highest abundance during pre-monsoon (16.53×10^4 cells.l^{-1}) followed by post-monsoon (1.46×10^4 cells.l^{-1}) and monsoon (1.9×10^3 cells.l^{-1}) (Srichandan et al. 2013). In the Mahanadi estuary, salinity and nitrogenous nutrients significantly influence phytoplankton communities such as Bacillariophyta and Cyanophyta (Naik et al. 2009). A set of environmental variables such as water column transparency, turbidity, salinity, and surface water temperature control phytoplankton distribution in this estuary (Mishra et al. 2018; Srichandan et al. 2019).

5.3.6.13 Zooplankton Community

Zooplankton, the planktonic animals, play an essential role in estuarine food webs by carrying forward bioenergy from autotrophic producers to higher trophic levels. Zooplankton community dynamics quickly respond to perturbation in ambient water quality and are thus widely used as an ecosystem health indicator. A total of 96 zooplankton species and 17 types of larval plankton have been documented so far from Mahanadi estuary (Fig. 5.4). Srichandan et al. (2013) have reported 86 holoplankton belonging to 53 genera represented by 12 diverse groups and 16 meroplankton. Among 12 groups of zooplankton, copepods were more diverse. Tidal oscillation influences the distribution of zooplankton groups in this estuary. The dominance of crustacean larvae demonstrated the Mahanadi estuary as a conducive site for breeding and spawning of shellfishes. Naik et al. (2013) have cataloged 19 groups with dominance of copepods, mysids, decapods, and chaetognaths. *Paracalanus parvus* and *Pontella andersoni* were found to be the most abundant species (Naik et al. 2013).

Zooplankton density has been observed within 137–6363 org.m^{-3} in the Mahanadi estuary (Naik et al. 2013; Srichandan et al. 2013) (Fig. 5.4). Zooplankton population density followed well-marked seasonal and tidal variation (Srichandan et al. 2013). Zooplankton density was highest during post-monsoon (532 org.m^{-3}) followed by monsoon (231 org.m^{-3}) and pre-monsoon (137 org.m^{-3}). In the context of tidal variation, zooplankton density was higher at high tide during post-monsoon and pre-monsoon while mid-tide during monsoon. Naik et al. (2013) studied the seasonal variation in zooplankton, reporting the highest density during post-monsoon (6363 org.m^{-3}) followed by monsoon (4702 org.m^{-3}) and pre-monsoon (1203 org.m^{-3}), respectively.

Copepods were the most dominant group in the Mahanadi estuary irrespective of the seasons, albeit varying densities (Srichandan et al. 2013). Copepods were more abundant during post-monsoon with a density of 487 org.m^{-3}, which subsequently decreased to 120 org.m^{-3} in pre-monsoon and further increased to 175 org.m^{-3} in monsoons. Naik et al. (2013) also reported higher copepod density (5131 org.m-3) during post-monsoon. Following post-monsoon, copepod density decreased to 3297 org.m^{-3} and 909 org.m^{-3} during monsoon and pre-monsoon, respectively.

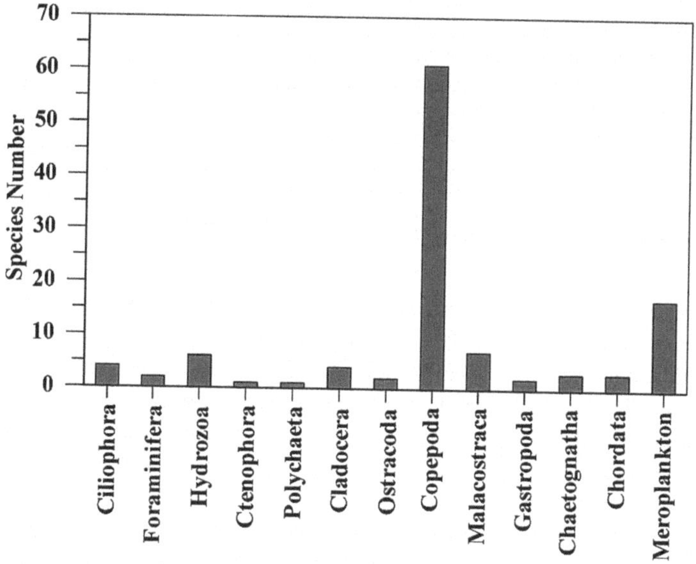

Fig. 5.4 Reported zooplankton species number in the Mahanadi estuary

5.3.7 *Large-Scale Impact Including Climate Change*

Environmental flows are important from the perspective of maintaining the natural ecosystem of the river and its functionality and are also crucial for the livelihood of local communities such as fishermen and river-bed cultivators who are dependent on the river. Bastia and Equeenuddin (2016) reported sediment load from the Mahanadi River was declining at the rate of 0.515×10^6 tons per year between 1980 and 2010; they attributed it to the increased number of dam construction (70–253 throughout 1980–2010) along the river. Water bodies (dam and aquaculture farms) in the Mahanadi River basin have witnessed upward trends, as revealed in the decadal time series analysis (Behera et al. 2018). Panda et al. (2011) reported a significant decline in sediment load (0.95×10^6 t/yr), attributed to streamflow diversion for irrigation, drinking, and industrial requirements. Due to the loss of protective vegetation along the coast, declining sediment has turned the lower Mahanadi basin prone to erosion. A recent finding suggests almost 65% of the coastline along the Mahanadi delta (including the port city Paradeep) is undergoing severe erosion, putting the inhabiting coastal communities under existential threat (Mukhopadhyay et al. 2018). A contrasting result derived from the model studies by Dunn et al. (2018) has shown sediment fluxes in Mahanadi are instead showing an increasing trend of sediment flux, but that can be compromised by anthropogenic pressure in the twenty-first century. It has undergone tremendous changes due to the development of aquaculture and agriculture activities during the last two decades. It was found that the delta was occupied by dense mangrove (12.6%), open mangrove

(3.3%), aquaculture (12.9%), and agriculture (30.9%) in 2006. A loss of 2606 ha mangrove area and an increase of 3657 ha aquaculture area was observed from 1973 to 2006, depicting the proliferation of aquaculture industry (Pattanaik and Prasad 2011).

5.4 Conclusion and Future Research Needs

The Mahanadi estuary is dynamic when it comes to its role in material transfer and transformation. This mighty river even extends its estuary beyond the land border in the Bay of Bengal as far as 90 km in the form of a plume. It contributes significantly to the region's blue economy being a productive fish landing site, harboring Paradeep port and heavy industries such as steel and fertilizer. However, it has come at the cost of receiving effluents from the fishing harbor, port, and industries, which are detrimental for the rich floral and faunal biodiversity and the coastal communities. The water quality of the Mahanadi estuary, in general, is warm, fresh to alkaline, turbid, fresh to brackish, well-oxygenated, and nutrient-rich. The reports on estuarine water quality have been designated as "good" though sporadic events of point and non-point pollution have been reported. Seasonal and spatial variabilities exist due to monsoon-driven river runoff and saltwater exchange with the Bay of Bengal through tide. Phytoplankton biomass becomes highest during post-monsoon with Bacillariophyta as the dominant group. Zooplankton also showed similar prevalence during post-monsoon dominated by copepods. Over the year, the Mahanadi River has been extensively dammed across its length, which raises concern for the altered biogeochemistry of the estuary and adjacent coastal waters.

However, the current knowledge is not complete in various fronts. For example, shoreline and bathymetry play an important role in estuarine hydrographic features; such data are grossly missing from the Mahanadi estuary, which calls for evaluating their trends over a year. Most of the Mahanadi estuary studies have been confined to routine water quality parameters that were also short term and site specific; none or very few looked at them from the biogeochemistry angle. The essence of biogeochemistry lies in studying the inherent coupling and interdependencies between various spheres of the system under study, i.e., hydrosphere (water), biosphere (life), lithosphere (sediment), and atmosphere (air). For instance, studies on sediment/water interaction and air/water interaction are very sparse in the Mahanadi estuary.

Estuaries are a highly dynamic system. There is also a need to shift focus from routine, discrete measurement of water quality parameter to more long-term, continuous monitoring keeping different variabilities in mind such as tidal (high/low and spring/neap), diurnal, annual, and interannual. That would help budget or develop a predictive model of the water quality that would benefit the coastal stakeholders through nowcasting or forecasting.

Similarly, studies on biotic diversities in the Mahanadi estuary have focused on phytoplankton and zooplankton taxonomy; the missing links are their interaction (predator, prey relationship), succession, community dynamics, etc. which can be studied further. Another exciting area of investigation would be studying invasive species of phytoplankton/zooplankton/larvae in the Mahanadi estuary, which is likely due to the Paradeep port's proximity. Food web linkages between higher (fish) and lower (plankton) trophic levels and its interaction with abiotic factors such as nutrient availability would help develop sustainable capture fishery in the region.

The Mahanadi estuary is situated in a highly cyclone-prone area of Odisha, where cyclones veer to make their landfall. Climate change would exacerbate such extreme weather mesoscale events, requiring climate-linked scenario modeling for biogeochemistry. Finally it can be said despite several studies in Mahanadi estuary a lot of uncharted areas and unanswered questions remain which needs to be addressed.

Acknowledgments The authors acknowledge the financial support received from the Coastal Monitoring Program by the Indian National Centre for Ocean Information Services (INCOIS), Ministry of Earth Sciences, Govt. of India. The first author (TA) acknowledges the Vice Chancellor's encouragement and support, Xavier University, Bhubaneswar. This is INCOIS contribution number 403.

References

Alfred JRB (1989) Fauna of Mahanadi estuary, Orissa, VIII+1-218 pp.

Bastia F, Equeenuddin SM (2016) Spatio-temporal variation of water flow and sediment discharge in the Mahanadi River, India. Glob Planet Chang 144:51–66. https://doi.org/10.1016/j.gloplacha.2016.07.004

Behera MD, Tripathi P, Das P, Srivastava SK, Roy PS, Joshi C, Behera PR, Deka J, Kumar P, Khan ML, Tripathi OP, Dash T, Krishnamurthy YVN (2018) Remote sensing based deforestation analysis in Mahanadi and Brahmaputra river basin in India since 1985. J Environ Manag 206:1192–1203. https://doi.org/10.1016/j.jenvman.2017.10.015

Borole DV, Mohanti M, Ray SB, Somayajulu BLK (1979) Preliminary investigations on dissolved uranium and silicon and major elements in the Mahanadi estuary. Indian Acad Sci Section A Phys Sci 88A:161–170

Central Water Commission (2013) https://doi.org/10.1152/physrev.1962.42.3.467

Chakrapani GJ, Subramanian V (1993a) Heavy metals distribution and fractionation in sediments of the Mahanadi River basin, India. Environ Geol 22:80–87. https://doi.org/10.1007/BF00775288

Chakrapani GJ, Subramanian V (1993b) Rates of erosion and sedimentation in the Mahanadi river basin, India. J Hydrol 149:39–48. https://doi.org/10.1016/0022-1694(93)90098-T

Das J, Das SN, Sahoo RK (1997) Semidiurnal variation of some physico-chemical parameters in the Mahanadi estuary, East Coast of India. Indian J Mar Sci 26:323–326

Sarkar SD, Sahoo AK, Gogoi P, Raman RK, Munivenkatappa MH, Kumari K, Mohanty BP, Das BK (2019) Phytoplankton biomass in relation to flow dynamics: the case of a tropical river Mahanadi, India. Trop Ecol. 60(4):485-94 https://doi.org/10.1007/s42965-019-00048-7

Dey M, Chowdhury C, Pattnaik AA, Ganguly D, Mukhopadhyay SK, De TK, Jana TK (2013) Comparison of Monsoonal change of water quality parameters between 1983 and 2008 in a tropical estuary in Northeastern India: role of phytoplankton and community metabolism. Mar Ecol 34:14–29. https://doi.org/10.1111/j.1439-0485.2012.00519.x

Dixit PR, Kar B, Chattopadhyay P, Panda CR (2013) Seasonal variation of the physicochemical properties of water samples in Mahanadi estuary, East Coast of India. J Environ Prot (Irvine, Calif) 4:843–848. https://doi.org/10.4236/jep.2013.48098

Dunn FE, Nicholls RJ, Darby SE, Cohen S, Zarfl C, Fekete BM (2018) Projections of historical and 21st century fluvial sediment delivery to the Ganges-Brahmaputra-Meghna, Mahanadi, and Volta deltas. Science of the total environment 642:105-116. https://doi.org/10.1016/j.scitotenv.2018.06.006

Gao X, Song J (2005) Phytoplankton distributions and their relationship with the environment in the Changjiang Estuary, China. Mar Pollut Bull 50:327–335. https://doi.org/10.1016/j.marpolbul.2004.11.004

Jhingran VG, Gopalakrishnan V (1973) Estuarine fisheries resources of India in relation to adjacent seas. J Mar Biol Assoc India 15:323–334

Ketchum BH (1967) Phytoplankton nutrients in estuaries. In: Lauff GH (ed) Estuaries. Am Assoc Adv Sci. Horn–Shafer Company, Washington, DC 83:329–335

Khadanga MK, Das S, Sahu BK (2012) Seasonal variation of the water quality parameters and its influences in the Mahanadi estuary and near coastal environment, East Coast of India National Institute of Ocean Science and Technology (NIOT). World Appl Sci J 17(6):797–801

Khan SA, Murugesan P (2005) Polychaete diversity in Indian estuaries. Indian J Mar Sci 34:114–119

Konhauser KO, Powell MA, Fife WS, Longstaffle FJ, Tripathy S. (1997) Trace element geochemistry of river sediment, Orissa state, India. J Hydrol 193, 258–269.

Mishra AK (2004) Retrieval of suspended sediment concentration in the estuarine waters using IRS-1C WiFS data. Int J Appl Earth Obs Geoinf 6:83–95. https://doi.org/10.1016/j.jag.2004.08.001

Mishra RK, Shaw BP, Sahu BK, Mishra S, Senga Y (2009) Seasonal appearance of Chlorophyceae phytoplankton bloom by river discharge off Paradeep at Orissa Coast in the Bay of Bengal. Environ Monit Assess 149:261–273. https://doi.org/10.1007/s10661-008-0200-2

Mishra S, Nayak S, Pati SS, Nanda SN, Mahanty S, Behera A (2018) Spatio temporal variation of phytoplankton in relation to physicochemical parameters along Mahanadi estuary & inshore area of Paradeep coast, north east coast of India in Bay of Bengal. Indian J Geo-Mar Sci 47:1502–1517

MSME (2016–2017) http://dcmsme.gov.in/dips/2016-17/DIPS%20of%20jagatsinghpur%202016-17.pdf. Accessed on 06/11/2020

Mukhopadhyay A, Ghosh P, Chanda A, Ghosh A, Ghosh S, Das S, Ghosh T, Hazra S (2018) Threats to coastal communities of Mahanadi delta due to imminent consequences of erosion – present and near future. Sci Total Environ 637–638:717–729. https://doi.org/10.1016/j.scitotenv.2018.05.076

Nagamani PV, Hussain MI, Choudhury SB, Panda CR, Sanghamitra P, Kar RN, Das A, Ramana IV, Rao KH (2013) Validation of chlorophyll-a algorithms in the coastal waters of Bay of Bengal initial validation results from OCM-2. J Indian Soc Remote Sens 41:117–125. https://doi.org/10.1007/s12524-012-0203-x

Naik S, Acharya BC, Mohapatra A (2009) Seasonal variations of phytoplankton in Mahanadi estuary, East Coast of India. Indian J Mar Sci 38:184–190

Naik S, Mahapatro D, Behera DP, Kumar M, Panda CR, Mishra RK (2013) Spatio-temporal study of zooplankton community in Mahanadi Estuary, Bay of Bengal. Int J Ecosyst 03:185–195. https://doi.org/10.5923/j.ije.20130306.04

Naik S, Mishra RK, Panda US, Mishra P, Panigrahy RC (2020) Phytoplankton community response to environmental changes in Mahanadi estuary and its adjoining coastal waters of Bay of Bengal: a multivariate and remote sensing approach. Remote Sens Earth Syst Sci. https://doi.org/10.1007/s41976-020-00036-9

Nambiar R, Shah C, Kumar J, Shrivastav PS, Bhushan R (2020) Assessment of contaminants in the northwestern Bay of Bengal. Environ Sci Pollut Res 27:34090–34098. https://doi.org/10.1007/s11356-020-09576-5

Panda DK, Kumar A, Mohanty S (2011) Recent trends in sediment load of the tropical (Peninsular) river basins of India. Glob Planet Chang 75:108–118. https://doi.org/10.1016/j.gloplacha.2010.10.012

Panigrahy PK, Das J, Das SN, Sahoo RK (1999) Evaluation of the influence of various physico-chemical parameters on coastal water quality, around Orissa, by factor analysis. Indian J Mar Sci 28:360–364

Panigrahy PK, Satapathy DR, Panda CR, Kar RN (2014) Dispersion pattern of petroleum hydrocarbon in coastal water of Bay of Bengal along Odisha and West Bengal, India using geospatial approach. Environ Monit Assess 186:8303–8315. https://doi.org/10.1007/s10661-014-4004-2

Paradeep Development Authority, Paradeep. http://www.pdaparadeep.in/index.php?option=com_content&view=article&id=46&Itemid=53. Accessed on 06/11/2020

Paradeep Fishing Harbour, Paradeep. http://www.msfhparadeep.com/infrastructure.html. Accessed on 06/11/2020

Paradip Port Trust, Paradeep. https://www.paradipport.gov.in/. Accessed on 06/11/2020

Pati A, Behera RK, Dash S, Mohapatra PK, Sarangi RK, Raut D, Mohanty B, Patnaik L (2018) First and new record of Ceratium vulture v. sumatranum and Pediastrum species from coastal waters of Paradip, Bay of Bengal, East Coast of India. Indian J. Geo-Mar Sci 47:1169–1171

Pattanaik C, Prasad SN (2011) Assessment of aquaculture impact on mangroves of Mahanadi delta (Orissa), East coast of India using remote sensing and GIS. Ocean Coast Manag 54:789–795. https://doi.org/10.1016/j.ocecoaman.2011.07.013

Pattanaik S, Sahoo RK, Satapathy DR, Panda CR, Choudhury SB (2017) Variation in carbonate system and air-water CO2 flux during summer in the Mahanadi estuary, India. Int J Adv Agric Environ Eng 3:387–390. https://doi.org/10.15242/ijaaee.c1216010

Pattanaik S, Chanda A, Sahoo RK, Swain S, Satapathy DR, Panda CR, Choudhury SB, Mohapatra PK (2020a) Contrasting intra-annual inorganic carbon dynamics and air–water CO2 exchange in Dhamra and Mahanadi Estuaries of northern Bay of Bengal, India. Limnology 21:129–138. https://doi.org/10.1007/s10201-019-00592-0

Pattanaik S, Roy R, Sahoo RK, Choudhury SB, Panda CR, Satapathy DR, Majhi A, D'Costa PM, Sai MVRS (2020b) Air–Sea CO2 dynamics from tropical estuarine system Mahanadi, India. Reg Stud Mar Sci 36:101284. https://doi.org/10.1016/j.rsma.2020.101284

Raj S, Jee PK, Panda CR (2013) Textural and heavy metal distribution in sediments of Mahanadi estuary, East coast of India. Indian J Mar Sci 42:370–374

Rao AD, Dash S, Babu SV (2004) Numerical study of the circulation and sediment transport in the Mahanadi estuary. Nat Hazards 32:219–237. https://doi.org/10.1023/B:NHAZ.0000031315.73712.18

Rao AD, Dash S, Jain I, Dube SK (2007) Effect of estuarine flow on ocean circulation using a coupled coastal-bay estuarine model: an application to the 1999 Orissa cyclone. Nat Hazards 41:549–562. https://doi.org/10.1007/s11069-006-9049-2

Samantray P, Mishra BK, Panda CR, Rout SP (2009) Assessment of water quality index in Mahanadi and Atharabanki Rivers and Taldanda Canal in Paradip area, India. J Hum Ecol 26:153–161. https://doi.org/10.1080/09709274.2009.11906177

Satapathy S, Panda CR, Jena BS (2019) Risk-based prediction of metal toxicity in sediment and impact on human health due to consumption of seafood (Saccostrea cucullata) found in two highly industrialised coastal estuarine regions of Eastern India: a food safety issue. Environ Geochem Health 41:1967–1985. https://doi.org/10.1007/s10653-019-00251-4

Singh RK, Das M (2018) Mahanadi: the great river. In: The Indian rivers. Springer Hydrogeology, pp 309–318. https://doi.org/10.1007/978-981-10-2984-4_25

Srichandan S, Panda CR, Rout NC (2013) Seasonal distribution of zooplankton in Mahanadi estuary (Odisha), East Coast of India: a taxonomical approach. Int J Zool Res 9:17–31

Srichandan S, Baliarsingh SK, Prakash S, Lotliker AA, Parida C, Sahu KC (2019) Seasonal dynamics of phytoplankton in response to environmental variables in contrasting coastal ecosystems. Environ Sci Pollut Res 26:12025–12041. https://doi.org/10.1007/s11356-019-04569-5

Srichandan S, Baliarsingh SK, Lotliker AA, Prakash S, Samanta A, Sahu KC (2020) A baseline investigation of phytoplankton pigment composition in contrasting coastal ecosystems of n orth-western Bay of Bengal. Mar Pollut Bull 160:111708. https://doi.org/10.1016/j.marpolbul.2020.111708

Srinivasan R, Rao KS, Kapileswar PS (1982) Studies on the morphological changes in the Mahanadi estuary and Hukitola barrier island with the aid of photo-interpretation technique. J Indian Soc Photo-Interpretation Remote Sens 10:39–44. https://doi.org/10.1007/BF02990612
Statistical Bulletin (2018) Department of Tourism, Govt. of Odisha
Sundaray SK, Panda UC, Nayak BB, Bhatta D (2006) Multivariate statistical techniques for the evaluation of spatial and temporal variations in water quality of the Mahanadi river-estuarine system (India) – a case study. Environ Geochem Health 28:317–330. https://doi.org/10.1007/s10653-005-9001-5
Sundaray SK, Nayak BB, Bhatta D (2009) Environmental studies on river water quality with reference to suitability for agricultural purposes: Mahanadi river estuarine system, India – a case study. Environ Monit Assess 155:227–243. https://doi.org/10.1007/s10661-008-0431-2
Sundaray SK, Nayak BB, Lin S, Bhatta D (2011) Geochemical speciation and risk assessment of heavy metals in the river estuarine sediments – a case study: Mahanadi basin, India. J Hazard Mater 186:1837–1846. https://doi.org/10.1016/j.jhazmat.2010.12.081
Sundaray SK, Nayak BB, Kanungo TK, Bhatta D (2012) Dynamics and quantification of dissolved heavy metals in the Mahanadi River estuarine system, India. Environ Monit Assess 184:1157–1179. https://doi.org/10.1007/s10661-011-2030-x
Sundaray SK, Nayak BB, Lee BG, Bhatta D (2014) Spatio-temporal dynamics of heavy metals in sediments of the river estuarine system: Mahanadi basin (India). Environ Earth Sci 71:1893–1909. https://doi.org/10.1007/s12665-013-2594-6
TOI 07/06/2016. https://timesofindia.indiatimes.com/city/bhubaneswar/OSPCB-starts-probe-into-mass-fish-death/articleshow/52633321.cms. Accessed on 06/11/2020
Upadhyay S (1988) Physico-chemical characteristics of the Mahanadi estuarine ecosystem, East Coast of India. Indian J Mar Sci 17:19–23
WRIS. (2011). River basins. Water Resource Information System of India http://india- ris.nrsc.gov.in/wrpinfo/index.php?title=Basins. Last accessed on 20 February 2021

Chapter 6
Mercury-Resistant Marine Bacterial Population in Relation to Abiotic Variables at the Bay of Bengal, India

Hirak R. Dash and Surajit Das

Abstract Sediment and water samples were collected during monsoon and summer seasons for 2 years from four study sites, viz. Paradeep, Chilika, Gopalpur, and Rushikulya, along the Bay of Bengal, Odisha, India. Samples were analysed for abiotic parameters such as pH, salinity, temperature and total mercury content. Similarly, total heterotrophic bacterial load and mercury-resistant marine bacterial load were analysed in the collected samples. The pH of water samples in all the study sites varied from 6.41 ± 0.014 to 8.4 ± 0.141 during the summer season, whereas during the monsoon season it varied from 6.535 ± 0.077 to 8.55 ± 0.070. Similarly, the salinity in the study sites ranged from 9 ± 1.414 psu to 36.5 ± 0.707 psu during the summer season and 4.5 ± 0.707 psu to 34.5 ± 0.707 psu during the monsoon season. The temperature of the water also followed the same trend that varied from 26.55 ± 1.343 °C to 38.5 ± 0.141 °C in the summer season, and during the monsoon season the range was 21.85 ± 0.494 °C to 28.95 ± 0.636 °C. Sediment samples showed to contain a higher amount of mercury at Rushikulya (2.48 and 2.26 ppb) among the study sites in both seasons followed by Paradeep, Chilika and Gopalpur. Huge fractions of marine bacteria were found to be mercury resistant in nature, and a strong positive correlation ($r = 0.94$ and 0.93 in water and sediment samples) was obtained between mercury content and the mercury-resistant marine bacterial populations in the Bay of Bengal, India.

Keywords Bay of Bengal · pH · Salinity · Temperature · Mercury · Total heterotrophic bacteria · Mercury-resistant marine bacteria

---

H. R. Dash · S. Das (✉)
Laboratory of Environmental Microbiology and Ecology (LEnME), Department of Life Science, National Institute of Technology, Rourkela, Odisha, India
e-mail: surajit@nitrkl.ac.in

S. Das, T. Ghosh (eds.), *Estuarine Biogeochemical Dynamics of the East Coast of India*, https://doi.org/10.1007/978-3-030-68980-3_6

6.1 Introduction

Marine ecosystems are maintained by the transfer of energy from the primary producers through the intermediate consumers to the top predators including humans (Chan and Ruckelshaus 2010). The aggregate effect of these interactions constitutes ecosystem function for natural benefits such as fisheries and aquaculture production, water purification, shoreline protection as well as recreation. However, the growing anthropogenic pressure and climate changes have profound consequences on the marine ecosystem. The primary direct consequences include an increase in ocean temperature, acidity, sea level rise, increased ocean stratification, decreased sea ice extent, altered pattern of ocean circulation, precipitation and fresh water input (Doney et al. 2012; Das and Mangwani 2015).

Changes in ocean temperature and other physico-chemical parameters have a direct impact on the physiological functioning, behaviour and productivity of the inhabitant organisms. This ultimately leads to the change in size, structure, spatial range and seasonal abundance of the populations. These changes in turn are reflected in altered species interaction, change of cascades from primary producers to other trophic levels, changes in community structure and ecosystem functioning resulting in the disruption in biological interactions. A healthy aquatic ecosystem is dependent on its physico-chemical and biological parameters (Venkatesharaju et al. 2010). Thus, the quality of an ecosystem provides sufficient information about the available resources for supporting life in the ecosystem. Therefore, assessment of water quality parameters is essential to identify the magnitude and source of any pollution load.

Mercury occurs naturally in the environment which is referred to be the background mercury concentrations. This metal is present usually in small concentrations in combustion materials, and in certain instances it is introduced to the atmosphere through volatilization from the natural sources of soils and lakes. However, it is difficult to trace the source of mercury in the environmental media and biota (USEPA 1976). Mercury contamination in the aquatic environment is a result of atmospheric deposition and direct discharge of mercury from the point and non-point sources, i.e. waste disposal and soil lixiviation by agriculture (Schroeder and Munthe 1998). In the marine environment, mercury occurs in any one of the following forms, i.e. Hg^0, Hg^{2+}, methylmercury, dimethylmercury and particulate and colloidal mercury (Morel et al. 1998). The average residence time of mercury in the marine environment has been estimated to be 20–30 years in comparison to 0.8–2 years in the atmosphere (Gworek et al. 2016). Such higher residence time worsens the situation as mercury is removed much slowly from the marine environment in comparison to the atmosphere. The total concentration of mercury differs from one oceanic environment to another. The average concentration of mercury in ocean was reported in the order of 1.5 picomolar (pM) (Lamborg et al. 2002). However, a higher concentration has been found in the Mediterranean (2.5 pM) (Cossa et al. 1997; Cinnirella et al. 2019) and North Atlantic (2.4 pM) (Mason et al.

1998) and the lower concentrations observed in the Pacific Ocean (1.2 pM) (Laurier et al. 2004). Besides, an increase in mercury concentration in sea water of Hardangerfjord ecosystem, Norway, is witnessed in terms of European Union Hg maximum level in demersal fish species (Azad et al. 2019). Modelling analyses have revealed that the concentration of mercury is not in a steady state in ocean basins, and with respect to the atmospheric inputs, it is likely to increase over the decades (Sunderland and Mason 2007).

Though mercury resistance and bacteria have been studied widely, few reports are available till date regarding mercury-resistant bacteria from the marine environment. As ocean is an open system, mercury vapours released by resistant bacteria due to volatilization process can become a part of the local mercury cycle and thus re-pollute the environment which has been evidenced in the Amazon River basin (Lacerda et al. 2012). Though Jaysankar and Ramaiah (2006) studied extensively the distribution of mercury-resistant bacteria along the Indian coast, the Bay of Bengal of Odisha region is untapped. Thus, a detailed study has been carried out with time scale investigation of mercury-resistant marine bacterial (MRMB) population in the Bay of Bengal along the Odisha coast, and the effect of the physico-chemical parameters on the population dynamics has been investigated.

6.2 Materials and Methods

6.2.1 Study Sites and Sampling

Four study sites were selected along the Odisha coast of the Bay of Bengal, viz. Paradeep, Chilika, Rushikulya and Gopalpur (Fig. 6.1). Water and sediment samples from ten stations of each study site were collected with a minimum distance of 100 m between the stations. During the study period (2010–2012), sampling was conducted in the monsoon and summer seasons.

6.2.2 Analysis of Physico-Chemical Parameters

Both water and sediment samples were analysed for various physico-chemical parameters such as pH, salinity, temperature and total mercury content following the standard procedure (APHA 1992). Analysis of mercury content was performed as per the protocol of the Environmental Protection Agency (USEPA 1976) and analysed after acid digestion by cold vapour atomic absorption spectrophotometer (Perkin Elmer, AAnalyst™ 200). All the samples were analysed in triplicates and represented as mean ± SD.

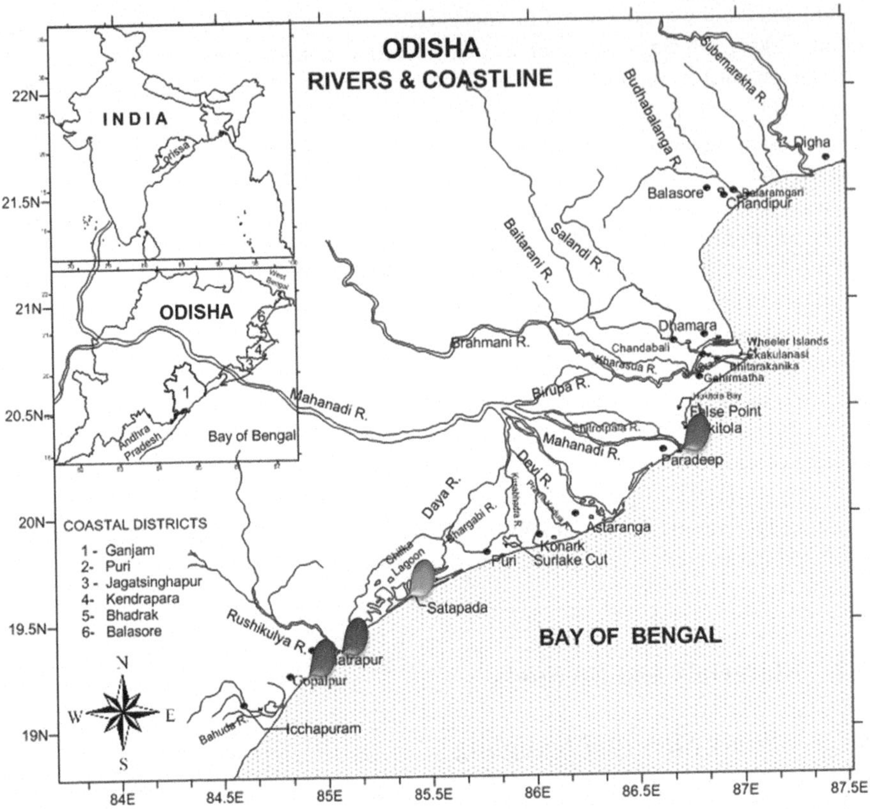

Fig. 6.1 Rivers and coastline map of the Bay of Bengal along the Odisha coast, India. The bookmarked places are the study sites

6.2.3 Isolation and Enumeration of Total Heterotrophic Bacteria (THB) and Mercury-Resistant Marine Bacteria (MRMB)

Enumeration of THB was conducted on Zobell's Marine agar medium (Hi-Media, India) by diluting 1 g or 1 ml of sediment and water samples and plating aliquots of 100μl of the diluents. Similarly, for enumeration of MRMB, the same procedure was followed with sea water nutrient (SWN) agar medium (peptone 5 g, yeast extract 3 g, agar 15 g, aged sea water 500 ml, de-ionized water 500 ml, pH 7.5 ± 0.1) supplemented with 10 ppm of mercury as $HgCl_2$ following Jaysankar and Ramaiah (2006). Plates were incubated at 25 °C for 48 h and the number of colony-forming units was calculated. THB and MRMB were calculated and expressed as CFU/g and CFU/ml for sediment and water samples, respectively. The percentage of MRMB was calculated using the formula: % of MRMB = (MRMB/THB) × 100.

6.2.4 Statistical Analysis

The F-value (two-way ANOVA, $p < 0.05$) was calculated. The result of the F-value was compared with the tabulated values to determine the significant variation in physico-chemical parameters as well as the biological parameters between the seasons of sampling and the study sites. The correlation coefficient between the various physico-chemical parameters and the THB and MRMB populations was calculated. Multiple regression analysis (Statistica 9.1) was performed to determine the influence of different physico-chemical parameters on MRMB and THB populations. Multiple regression equation was also determined for both MRMB and THB populations along the study sites.

6.3 Results

6.3.1 Physico-Chemical Parameters

The variation of pH along the stations during the two collection periods has been presented in Fig. 6.2. In Paradeep, the pH value ranged from 7.6 (Station 7) to 8.52 (Station 1) with the mean (±SD) value of 8.1875 ± 0.373. pH of water during monsoon at Chilika averaged at 7.1525 ± 0.279 and the values ranged from 6.535 (Station 4) to 7.495 (Station 2). Study site Rushikulya showed a pH range of 6.6 (Station 4) to 8.55 (Station 6) with the mean (±SD) of 7.66 ± 0.637181293. At Gopalpur, the pH also followed the same trend with the average value of 7.7215 ± 0.480 and the value between the stations ranged from 6.925 (Station 4) to 8.245 (Station 2). At Paradeep, the average pH value was 7.405 ± 0.294, while the values ranged from 6.95 (Station 7) to 7.75 (Station 9). Similarly, the pH value of Chilika water during summer varied from 6.41 (Station 5) to 7.01 (Station 10),

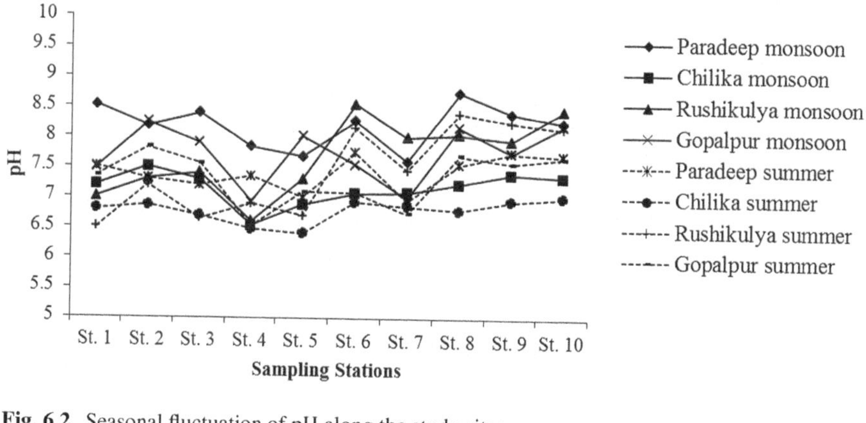

Fig. 6.2 Seasonal fluctuation of pH along the study sites

while the mean (±SD) was at 6.778 ± 0.198. At Gopalpur, the mean value of pH in water samples was 7.3115 ± 0.436138676, while the range of pH varied from 6.53 (Station 4) to 7.82 (Station 2). Rushikulya also showed similar range of pH during the study period from 6.5 (Station 1) to 8.4 (Station 8) with the mean (±SD) of 7.435 ± 0.744.

Seasonal variation of salinity along the study sites during the study period has been presented in Fig. 6.3. The average value of salinity at Paradeep during monsoon was found to be 19.2 ± 6.481 and the value ranged from 9.5 psu (Station 1) to 27 psu (Station 3). Chilika water during the monsoon season witnessed the lowest salinity level with an average value of 30.1 ± 3.51 psu and the value ranged from 4.5 psu (Station 7) to 9 psu (Station 3). Similarly, at Rushikulya the salinity value ranged from 6.5 to 33.5 psu. Monsoon sampling of water sample at Gopalpur showed a salinity range of 23.5 psu (Station 4) to 34.5 psu (Station 8) with the mean (±SD) of 30.1 ± 3.518 psu. During summer, at the Paradeep study site the salinity value was found to be at an average of 26.6 ± 6.367 psu, whereas the salinity level ranged from 17.5 psu (Station 2) to 34 psu (Stations 5 and 6). The salinity value during the summer season at the Chilika study site also witnessed the lowest level of salinity with the average of 11.15 ± 1.292 psu. The range of salinity at this study site was 9 psu (Station 9) to 13.5 psu (Station 4). However, the salinity value at Rushikulya varied at a wide range from 10 psu (Station 2) to 33.5 psu (Station 7). The average salinity at this study site was found to be 19.5 ± 6.824 psu. At Gopalpur, the salinity level was found to be in the range of 27.5 psu (Stations 4 and 8) to 36.5 (Station 2) with the average value of salinity at this study site being 32.75 ± 3.513 psu.

The temperature of water samples from the study has been presented in Fig. 6.4. The temperature of Paradeep water during the monsoon season ranged from 23 °C (Station 6) to 28 °C (Station 10). During this time, the temperature of Chilika water was found to be in the range of 24 °C (Station 9) to 27 °C (Station 5) with the mean (±SD) of 25.55 ± 0.955 °C. The water temperature of the Rushikulya study site showed the lowest temperature during the monsoon season with an average value of

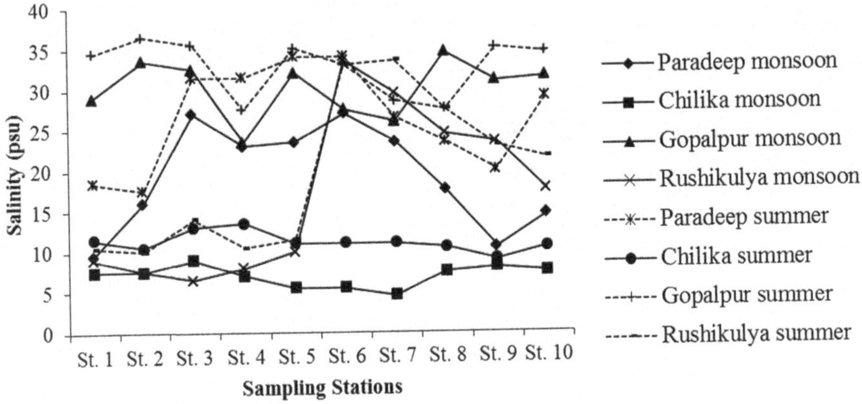

Fig. 6.3 Seasonal fluctuation of salinity along the study sites

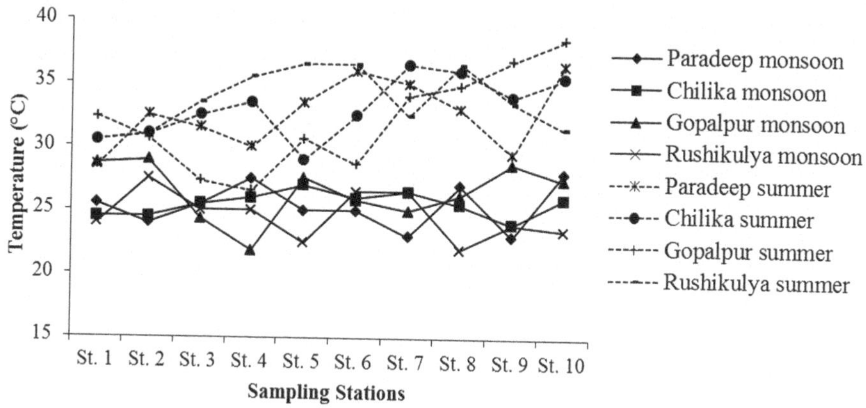

Fig. 6.4 Seasonal fluctuation of temperature along the study sites

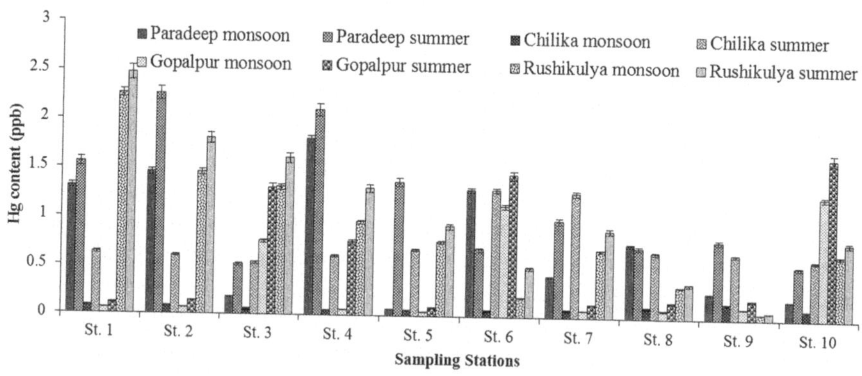

Fig. 6.5 Seasonal fluctuation of Hg content along the study sites in collected water samples

24.65 ± 1.79 °C, while the temperature ranged from 22 °C (Station 8) to 27.5 °C (Station 2). Water samples of Gopalpur during the monsoon season ranged from 21.85 °C (Station 4) to 28.95 °C (Station 2) with an average value of 26.46 ± 2.27 °C. Water samples from Paradeep during the summer season ranged from 28.5 °C (Station 1) to 36.5 °C (Station 10) with the average value of 32.6 ± 2.746 °C. The average temperature of water samples from Chilika during the summer season was found to be 33.1 ± 2.74 °C, while the sample temperatures ranged from 29 °C (Station 5) to 36.5 °C (Station 7). The Rushikulya estuary also witnessed the same range of temperature in the water samples during the monsoon season and the observed temperature range was found to be 30.5 °C (Station 1) to 36.5 °C (Stations 5 and 6) with an average temperature of 33.75 ± 2.371 °C. At Gopalpur, the samples showed a temperature range of 26.55 °C (Station 4) to 38.5 °C (Station 10) with an average value of 32.035 ± 4.01 °C.

The total mercury content was determined in both water (Fig. 6.5) and sediment samples (Fig. 6.6) collected during the study period throughout the study sites.

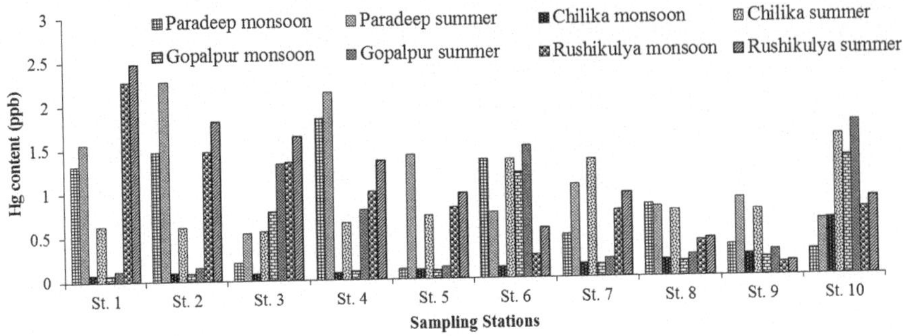

Fig. 6.6 Seasonal fluctuation of Hg content along the study sites in collected sediment samples

During the monsoon season, Paradeep water samples showed an average of 0.769 ± 0.63 ppb of mercury content, while the value ranged from 0.065 ppb (Station 5) to 1.45 ppb (Station 2). Chilika water also contained a trace amount of mercury at a level of 0.045 ppb (Station 4) to 0.155 ppb (Station 9) during the monsoon season with the average value of 0.769 ± 0.637 ppb. The average value of mercury content in water samples of Rushikulya during the monsoon season was found to be the highest among the study sites, which was 0.866 ± 0.664 ppb and the values ranged between 0.055 ppb (Station 9) and 2.26 ppb (Station 1). An average of 0.362 ± 0.28 ppb of mercury was found in the water samples of Gopalpur and the mercury content ranged from 0.045 ppb (Station 5) to 1.25 ppb (Station 10). Water samples from the study sites during the summer season were analysed for the total mercury content. During the summer season, water samples from Paradeep showed mercury content in the range of 0.51 ppb (Station 3) to 2.1 ppb (Station 4) with the average value of 1.15 ± 0.639 ppb. The average mercury content in Chilika water samples was found to be 0.751 ± 0.283 ppb, while the mercury content was in the range of 0.53–1.295 ppb. Rushikulya samples showed an average mercury content of 1.068 ± 0.731 ppb, while the variation was from 0.07 ppb (Station 9) to 2.47 ppb (Station 1). In Gopalpur water samples, the range of mercury content was found to be 0.11 ppb (Station 1) to 1.65 ppb (Station 10) with the mean (±SD) of 0.59 ± 0.34 ppb.

Sediment samples showed higher mercury content irrespective of the study sites and the season of sample collection. During the monsoon season, the average mercury content of the sediment samples in Paradeep was found to be in the range of 0.115 ppb (Station 5) to 1.84 ppb (Station 4) with the mean (±SD) of 0.824 ± 0.221 ppb. Similarly, the average value of total Hg content in the Chilika sediment samples was found to be 0.18 ± 0.021 ppb, while the values ranged from 0.080 (Station 3) to 0.64 ppb (Station 10). Sediment samples from Rushikulya varied from 0.145 ppb (Station 9) to 2.27 ppb (Station 1) with the mean (±SD) of 0.921 ± 0.390 ppb. An average of 0.417 ± 0.284 ppb of Hg content was found in the sediment samples from Gopalpur with the range of 0.065 ppb (Station 1) to 1.35 ppb (Station 10). In the summer season, Paradeep sediment samples showed a higher level of mercury content at a range of 0.54 ppb (Station 3) to 2.275 ppb (Station 2) with the average of 1.205 ± 0.209 ppb. Similarly, in Chilika sediment samples, the total mercury content varied from 0.56 ppb (Station 3) to 1.595 (Station 10) with the mean (±SD) of 0.896 ± 0.216 ppb. The average value of the total mercury content in Rushikulya sediment during the summer season was found to be 1.123 ± 0.050 ppb and the values of which varied between 0.16 ppb (Station 9) and 2.48 ppb (Station 1). Gopalpur sediment samples contained an average of 0.652 ± 0.273 ppb, while the total mercury content varied from 0.12 ppb (Station 1) to 1.75 ppb (Station 10).

6.3.2 Enumeration of THB

Total heterotrophic bacteria (THB) enumerated along various study sites have been given in Tables 6.1, 6.2, 6.3, and 6.4. Water samples from Paradeep were found to contain THB at the range of 1.33 ± 0.45 × 10^5 CFU/ml (Station 9) to 8.7 ± 0.2 × 10^5 CFU/ml (Station 1) and 1.265 ± 0.15 × 10^5 CFU/ml (Station 9) to 8.40 ± 3.00 × 10^5 CFU/ml (Station 7), respectively, during monsoon and summer seasons. The number of THB found to be present in the sediment samples from Paradeep during the monsoon season was 0.75 ± 0.15 × 105 CFU/g (Station 5) to 3.50 ± 1.0 × 105 CFU/g (Station 4). However, the THB population varied from 7.40 ± 1.00 × 10^4 CFU/g (Station 8) to 15.60 ± 2.00 × 10^4 (Station 5) CFU/g during the summer season. In Chilika, the THB population varied slightly during the two seasons, and the bacterial population was found to vary from 2.205 ± 1.33 × 10^5 CFU/ml (Station 8) to 8.70 ± 0.1 × 10^5 CFU/ml (Station 1) and 1.555 ± 0.25 × 10^5 CFU/ml (Station 9) to 7.70 ± 1.10 × 10^5 CFU/ml (Station 1) during monsoon and summer seasons, respectively, in the water samples. Similarly, THB in the sediment samples varied from 5.40 ± 1.27 × 10^6 CFU/g (Stations 4 and 7) to 8.95 ± 1.20 × 10^6 CFU/g (Station 8) and 3.35 ± 0.35 × 10^6 CFU/g (Station 1) to 7.75 ± 1.34 × 10^6 CFU/g (Station 3), respectively, in monsoon and summer seasons. Rushikulya water samples contained THB at a range of 1.42 ± 1.40 × 10^5 CFU/ml (Station 5) to 8.50 ± 2.00 × 10^5 CFU/ml (Station 1) and 1.27 ± 0.30 × 10^4 CFU/ml (Station 8) to 8.65 ± 2.50 × 10^4 CFU/ml (Station 5), respectively, during monsoon and summer seasons. Similarly, during the monsoon season, the sediment samples contained THB at a range of 2.06 ± 0.81 × 10^6 CFU/g (Station 10) to 9.60 ± 2.00 × 10^6 CFU/g (Station 2), whereas during the summer season the population varied from 1.26 ± 0.20 × 10^5 CFU/g (Station 7) to 5.60 ± 2.0 × 10^5 CFU/g (Station 2). Similarly, the THB population in the sediment samples of Gopalpur varied from 0.19 ± 0.04 × 10^5 CFU/g (Station 8) to 9.05 ± 5.50 × 10^5 CFU/g (Station 6) and 0.13 ± 0.04 × 10^4 CFU/g (Station 10) to 12.65 ± 1.50 × 10^4 CFU/g (Station 8), respectively, in monsoon and summer seasons. The water sample of the same study site harboured THB at a range of 1.27 ± 0.20 × 10^5 CFU/ml (Station 5) to 8.50 ± 2.00 × 10^5 CFU/ml (Station 1) and 1.56 ± 0.05 × 10^5 CFU/ml (Station 4) to 8.65 ± 2.50 × 10^5 CFU/ml (Station 1) during monsoon and summer seasons, respectively.

Table 6.1 Heterotrophic bacterial population and mercury-resistant marine bacterial population from the water and sediment samples collected from Paradeep during 2010–2012

St. no.	Geographical position	Total heterotrophic bacterial population				Mercury-resistant marine bacterial population				% of mercury-resistant bacteria			
		Sediment		Water		Sediment		Water		Sediment		Water	
		Monsoon (×10^5 CFU/g) ±SD	Summer (×10^4 CFU/g) ±SD	Monsoon (×10^5 CFU/ml) ±SD	Summer (×10^5 CFU/ml) ±SD	Monsoon (×10^4 CFU/g) ±SD	Summer (×10^4 CFU/g) ±SD	Monsoon (×10^3 CFU/ml) ±SD	Summer (×10^4 CFU/ml) ±SD	Monsoon	Summer	Monsoon	Summer
01	20°17.542′N and 086°42.996′E	0.87 ± 0.20	9.55 ± 0.25	8.7 ± 0.2	6.45 ± 1.15	7.65 ± 1.50	7.60 ± 2.00	6.40 ± 0.10	7.70 ± 0.20	87.931	79.581	0.735	11.937
02	20°17.508′N and 086°42.966′E	6.55 ± 2.55	15.40 ± 2.90	5.8 ± 1.0	4.50 ± 3.00	8.90 ± 2.00	8.70 ± 2.00	7.70 ± 0.10	9.70 ± 0.10	13.587	56.493	1.327	21.555
03	20°17.082′N and 086°41.169′E	0.87 ± 0.15	11.40 ± 2.00	6.9 ± 2.0	6.35 ± 1.50	3.15 ± 0.50	2.80 ± 0.45	1.14 ± 0.10	1.87 ± 1.50	36	24.605	0.165	2.952
04	20°17.632′N and 086°43.009′E	3.50 ± 1.00	15.15 ± 1.55	5.2 ± 1.30	5.55 ± 1.50	9.70 ± 1.00	8.25 ± 0.50	8.80 ± 0.10	8.20 ± 1.00	27.7142	54.455	1.692	14.774
05	20°17.001′N and 086°41.233′E	0.75 ± 0.15	15.60 ± 2.00	5.3 ± 0.60	4.90 ± 0.60	1.33 ± 1.00	7.00 ± 1.00	4.40 ± 0.15	7.05 ± 1.50	17.682	44.871	0.830	14.387
06	20°17.082′N and 086°40.168′E	3.50 ± 1.00	15.85 ± 10.50	4.65 ± 1.50	3.75 ± 0.50	8.05 ± 5.50	3.75 ± 0.50	5.75 ± 1.50	4.15 ± 0.50	23	23.659	1.236	11.066
07	20°17.092′N and 086°41.269′E	1.05 ± 1.00	8.80 ± 1.00	8.85 ± 3.50	8.40 ± 3.00	6.30 ± 1.50	6.30 ± 1.00	3.80 ± 1.50	6.30 ± 1.00	59.715	71.590	0.429	7.5
08	20°17.082′N and 086°42.299′E	1.54 ± 1.24	7.40 ± 1.00	7.70 ± 1.00	5.55 ± 1.50	7.15 ± 2.00	4.15 ± 0.50	4.75 ± 2.0	5.15 ± 2.50	46.278	56.081	0.616	9.279
09	20°17.500′N and 086°42.905′E	1.35 ± 1.25	8.75 ± 1.50	1.33 ± 0.45	1.265 ± 0.15	5.60 ± 2.50	4.70 ± 1.00	3.30 ± 1.32	5.60 ± 2.00	41.328	53.714	2.471	44.268
10	20°17.541′N and 86° 43.008′E	0.95 ± 0.25	12.60 ± 2.00	3.65 ± 1.50	3.15 ± 0.50	3.65 ± 2.00	3.40 ± 2.00	1.29 ± 0.55	3.50 ± 3.00	38.219	26.984	0.354	11.111

Table 6.2 Heterotrophic bacterial population and mercury-resistant marine bacterial population from the water and sediment samples collected from Chilika during 2010–2012

St. no.	Geographical position	Total heterotrophic bacterial population				Mercury-resistant marine bacterial population				% of mercury-resistant bacteria			
		Sediment		Water		Sediment		Water		Sediment		Water	
		Monsoon ($\times10^6$ CFU/g) ±SD	Summer ($\times10^6$ CFU/g) ±SD	Monsoon ($\times10^5$ CFU/ml) ±SD	Summer ($\times10^5$ CFU/ml) ±SD	Monsoon ($\times10^4$ CFU/g) ±SD	Summer ($\times10^4$ CFU/g) ±SD	Monsoon ($\times10^3$ CFU/ml) ±SD	Summer ($\times10^4$ CFU/ml) ±SD	Monsoon	Summer	Monsoon	Summer
01	19°44.582′N and 85°12.768′E	8.1 ± 1.13	3.35 ± 0.35	8.70 ± 0.1	7.70 ± 1.10	3.5 ± 2.57	2.57 ± 0.35	5.35 ± 1.15	5.00 ± 1.50	0.317	1.044	0.614	6.493
02	19°44.468′N and 85°12.783′E	8.05 ± 2.19	4.55 ± 1.20	7.15 ± 1.0	6.30 ± 1.00	4.9 ± 4.20	4.20 ± 0.49	4.85 ± 1.05	4.14 ± 1.55	0.521	1.076	0.678	6.579
03	19°44.322′N and 85°12.698′E	6.05 ± 1.06	7.75 ± 1.34	6.65 ± 0.5	5.90 ± 5.00	1.92 ± 1.23	1.23 ± 0.92	2.18 ± 0.92	1.25 ± 0.35	0.204	0.248	0.327	2.118
04	19°44.283′N and 85°11.732′E	5.40 ± 1.27	5.17 ± 5.05	7.00 ± 0.5	6.40 ± 5.00	3.15 ± 2.57	2.57 ± 0.5	3.95 ± 0.35	1.97 ± 0.58	0.476	0.609	0.564	3.085
05	19°44.183′N and 85°12.754′E	7.05 ± 3.18	3.77 ± 2.30	5.65 ± 3.1	4.10 ± 0.89	6.95 ± 4.90	4.90 ± 1.5	4.75 ± 1.05	4.05 ± 1.45	0.695	1.843	0.840	9.878
06	19°44.123′N and 85°12.803′E	6.95 ± 2.33	5.40 ± 4.24	7.05 ± 2.72	4.08 ± 2.72	7.45 ± 5.20	5.20 ± 1.5	8.20 ± 0.30	10.20 ± 4.20	0.748	1.379	1.163	25
07	19°43.488′N and 85°10.638′E	5.40 ± 4.24	7.50 ± 1.69	2.62 ± 0.5	2.50 ± 0.50	8.40 ± 6.65	6.65 ± 1.0	7.80 ± 0.50	8.05 ± 1.55	1.231	1.12	2.977	32.135
08	19°43.482′N and 85°11.632′E	8.95 ± 1.20	6.12 ± 5.19	2.205 ± 1.33	2.565 ± 1.33	9.65 ± 7.85	7.85 ± 1.5	7.35 ± 4.40	6.90 ± 2.00	0.877	1.575	3.333	26.900
09	19°45.492′N and 85°12.686′E	7.40 ± 1.83	4.12 ± 3.78	5.40 ± 2.5	1.555 ± 0.25	2.07 ± 0.86	8.60 ± 4.90	6.95 ± 5.50	6.50 ± 2.00	1.162	5.018	1.287	41.800
10	19°44.482′N and 85°11.771′E	6.05 ± 4.94	3.47 ± 3.14	4.05 ± 0.68	3.515 ± 0.68	1.26 ± 0.74	7.40 ± 1.00	5.90 ± 1.00	5.80 ± 2.00	1.223	3.625	1.456	16.500

Table 6.3 Heterotrophic bacterial population and mercury-resistant marine bacterial population from the water and sediment samples collected from Rushikulya during 2010–2012

St. no.	Geographical position	Total heterotrophic bacterial population				Mercury-resistant marine bacterial population				% of mercury-resistant bacteria			
		Sediment		Water		Sediment		Water		Sediment		Water	
		Monsoon ($\times 10^6$ CFU/g) ±SD	Summer ($\times 10^5$ CFU/g) ±SD	Monsoon ($\times 10^5$ CFU/ml) ±SD	Summer ($\times 10^4$ CFU/ml) ±SD	Monsoon ($\times 10^5$ CFU/g) ±SD	Summer ($\times 10^4$ CFU/g) ±SD	Monsoon ($\times 10^4$ CFU/ml) ±SD	Summer ($\times 10^3$ CFU/ml) ±SD	Monsoon	Summer	Monsoon	Summer
01	19°22.647′N and 85°03.165′E	5.60 ± 2.00	4.70 ± 2.00	8.50 ± 2.00	4.65 ± 1.50	9.70 ± 1.00	9.55 ± 0.50	8.55 ± 5.00	9.55 ± 1.5	17.321	20.319	10.058	20.537
02	19°22.701′N and 85°03.168′E	9.60 ± 2.00	5.60 ± 2.00	7.60 ± 2.00	4.20 ± 3.00	8.45 ± 5.00	8.65 ± 1.50	7.55 ± 2.50	8.75 ± 0.50	8.802	15.446	9.934	20.833
03	19°22.824′N and 85°03.162′E	4.25 ± 0.65	3.80 ± 1.00	3.65 ± 1.50	7.60 ± 2.00	7.75 ± 5.00	7.50 ± 2.00	6.40 ± 1.00	7.55 ± 2.50	18.235	19.736	17.534	9.934
04	19°22.872′N and 85°03.166′E	8.55 ± 5.00	4.25 ± 1.50	8.25 ± 1.50	6.60 ± 2.00	6.65 ± 1.50	6.60 ± 2.00	5.60 ± 2.00	6.65 ± 1.50	7.777	15.529	6.787	10.075
05	19°22.907′N and 85°03.173′E	4 30 ± 2.00	3.50 ± 3.00	1.42 ± 1.40	8.65 ± 2.50	6.15 ± 5.00	5.95 ± 1.50	4.45 ± 2.50	5.40 ± 1.00	14.302	17.000	31.338	6.242
06	19°23.142′N and 85°03.170′E	4.70 ± 1.00	2.65 ± 1.50	4.25 ± 4.05	1.36 ± 1.10	4.50 ± 3.00	4.45 ± 2.50	2.58 ± 0.40	3.95 ± 1.50	9.574	16.792	6.070	29.044
07	19°23.247′N and 85°03.165′E	3.55 ± 3.50	1.26 ± 0.20	2.52 ± 0.45	6.55 ± 2.50	5.65 ± 5.00	5.40 ± 1.00	3.75 ± 1.50	5.50 ± 1.00	15.915	42.857	14.880	7.633
08	19°23.334′N and 85°03.171′E	5.20 ± 1.00	4.65 ± 1.50	5.85 ± 5.00	1.27 ± 0.30	5.30 ± 2.00	3.55 ± 0.50	2.82 ± 0.20	3.35 ± 1.50	10.192	7.634	4.820	26.377
09	19°23.401′N and 85°03.162′E	5.75 ± 5.00	3.65 ± 1.50	3.15 ± 0.70	8.50 ± 1.00	3.55 ± 1.50	1.82 ± 0.54	1.26 ± 0.20	1.31 ± 0.65	6.173	4.986	4.000	1.547
10	19°23.644′N and 85°03.158′E	2.06 ± 0.81	2.40 ± 0.45	2.21 ± 0.50	4.00 ± 2.00	5.75 ± 5.00	5.20 ± 1.00	3.20 ± 1.00	4.40 ± 1.00	27.912	21.621	14.479	11.000

Table 6.4 Heterotrophic bacterial population and mercury-resistant marine bacterial population from the water and sediment samples collected from Gopalpur during 2010–2012

St. no.	Geographical position	Total heterotrophic bacterial population				Mercury-resistant marine bacterial population				% of mercury-resistant bacteria			
		Sediment		Water		Sediment		Water		Sediment		Water	
		Monsoon ($\times10^5$ CFU/g) ±SD	Summer ($\times10^4$ CFU/g) ±SD	Monsoon ($\times10^5$ CFU/ml) ±SD	Summer ($\times10^5$ CFU/ml) ±SD	Monsoon ($\times10^3$ CFU/g) ±SD	Summer ($\times10^3$ CFU/g) ±SD	Monsoon ($\times10^3$ CFU/ml) ±SD	Summer ($\times10^4$ CFU/ml) ±SD	Monsoon	Summer	Monsoon	Summer
01	19°19.218′N and 084° 57.730′E	7.05 ± 0.75	6.65 ± 1.50	8.50 ± 2.00	8.65 ± 2.50	4.70 ± 1.00	3.15 ± 0.50	3.35 ± 1.50	3.60 ± 2.00	0.666	4.736	0.394	4.161
02	19°20.124′N and 084° 57.730′E	5.10 ± 2.00	3.60 ± 2.00	5.35 ± 5.00	5.75 ± 1.50	5.15 ± 0.50	4.75 ± 0.50	3.70 ± 1.00	5.20 ± 1.00	1.009	13.194	0.691	9.043
03	19°20.210′N and 084° 57.732′E	7.10 ± 0.80	3.55 ± 0.50	4.75 ± 0.50	2.75 ± 0.50	7.85 ± 0.50	7.65 ± 1.50	7.85 ± 0.50	8.20 ± 1.00	1.105	21.549	1.652	29.818
04	19°20.218′N and 084° 57.698′E	2.71 ± 2.15	8.65 ± 8.50	3.85 ± 0.50	1.56 ± 0.05	5.65 ± 0.50	6.80 ± 1.00	2.36 ± 0.15	7.85 ± 0.50	2.081	7.861	0.614	50.159
05	19°20.321′N and 084° 57.804′E	8.65 ± 5.00	6.25 ± 4.50	1.27 ± 0.20	2.41 ± 0.90	5.85 ± 0.50	4.20 ± 1.00	1.27 ± 0.10	3.15 ± 0.50	0.676	6.72	1.00	13.070
06	19°20.398′N and 084° 57.854′E	9.05 ± 5.50	9.15 ± 5.50	2.70 ± 1.00	3.70 ± 1.00	8.80 ± 1.00	8.35 ± 1.50	8.35 ± 1.50	8.70 ± 2.00	0.972	9.125	3.092	23.513
07	19°20.458′N and 084° 57.704′E	1.31 ± 0.65	6.90 ± 6.00	3.65 ± 1.50	2.85 ± 0.50	6.65 ± 0.50	5.15 ± 0.50	4.75 ± 1.50	5.20 ± 1.00	5.057	7.463	1.301	18.245
08	19°20.602′N and 084° 57.870′E	0.19 ± 0.04	12.65 ± 1.50	2.85 ± 0.50	3.70 ± 1.00	7.15 ± 0.50	5.65 ± 1.50	5.75 ± 0.50	6.00 ± 2.00	37.631	4.466	2.017	16.216
09	19°20.653′N and 084° 57.670′E	4.65 ± 1.50	3.65 ± 1.50	3.75 ± 1.50	5.80 ± 1.00	7.45 ± 0.50	6.25 ± 0.50	6.40 ± 1.00	6.65 ± 1.50	1.602	17.123	1.706	11.465
10	19°20.765′N and 084° 57.708′E	1.30 ± 0.55	0.13 ± 0.04	2.70 ± 1.00	6.50 ± 2.00	8.30 ± 2.00	8.75 ± 1.50	8.80 ± 1.00	9.40 ± 1.00	6.360	67.049	3.259	14.461

6.3.3 Enumeration of MRMB

MRMB enumerated along various study sites have been given in Tables 6.1, 6.2, 6.3, and 6.4. The MRMB population in Paradeep water samples varied from $1.14 \pm 0.10 \times 10^3$ CFU/ml (Station 3) to $8.80 \pm 0.10 \times 10^3$ CFU/ml (Station 4) and $1.87 \pm 1.50 \times 10^4$ CFU/ml (Station 3) to $9.70 \pm 0.10 \times 10^4$ CFU/ml (Station 2), respectively, during monsoon and summer seasons. Similarly, the population of MRMB found in the water samples collected during the monsoon and the summer seasons was $1.33 \pm 1.00 \times 10$ CFU/g (Station 5) to $9.70 \pm 1.00 \times 10$ CFU/g (Station 4) and $2.80 \pm 0.45 \times 10$ CFU/g (Station 3) to $8.70 \pm 2.00 \times 10$ CFU/g (Station 2), respectively. In Chilika water samples, the MRMB population was found to vary from $2.18 \pm 0.92 \times 10^3$ CFU/ml (Station 3) to $8.20 \pm 0.30 \times 10^3$ CFU/ml (Station 6) and $1.25 \pm 0.35 \times 10^4$ CFU/ml (Station 3) to $10.20 \pm 4.20 \times 10^4$ CFU/ml (Station 6), respectively, during monsoon and summer seasons. A similar result was also witnessed in the sediment samples that varied from $1.26 \pm 0.74 \times 10^4$ CFU/g (Station 10) to $9.65 \pm 7.85 \times 10^4$ CFU/g (Station 8) and $1.23 \pm 0.92 \times 10^4$ CFU/g (Station 3) to $8.60 \pm 4.90 \times 10^4$ CFU/g (Station 9) during monsoon and summer seasons, respectively. In Rushikulya sediments, the MRMB population varied from $3.55 \pm 1.50 \times 10^5$ CFU/g (Station 9) to $9.70 \pm 1.00 \times 10^5$ CFU/g (Station 1) and $1.82 \pm 0.54 \times 10^4$ CFU/g (Station 9) to $9.55 \pm 0.50 \times 10^4$ CFU/g (Station 1) during monsoon and summer seasons, respectively. Similarly, the MRMB population varied from $1.26 \pm 0.20 \times 10^4$ CFU/ml (Station 9) to $8.55 \pm 5.00 \times 10^4$ CFU/ml (Station 1) and $1.31 \pm 0.65 \times 10^3$ CFU/ml (Station 9) to $9.55 \pm 1.5 \times 10^3$ CFU/ml (Station 1), respectively, during monsoon and summer seasons. The range of MRMB found in the water samples of Gopalpur during the monsoon and the summer seasons was $1.27 \pm 0.10 \times 10$ CFU/ml (Station 5) to $8.80 \pm 1.00 \times 10$ CFU/ml (Station 10) and $3.15 \pm 0.50 \times 10$ CFU/ml (Station 5) to $9.40 \pm 1.00 \times 10$ CFU/ml (Station 10), respectively. Similarly, the sediment samples of this study site contained $4.70 \pm 1.00 \times 10^3$ CFU/g (Station 1) to $8.80 \pm 1.00 \times 10^3$ CFU/g (Station 6) and $3.15 \pm 0.50 \times 10^3$ CFU/g (Station 1) to $8.75 \pm 1.50 \times 10^3$ CFU/g (Station 10) MRMB, during monsoon and summer seasons, respectively.

6.3.4 Percentage (%) of Mercury-Resistant Bacteria

The percentage of mercury-resistant marine bacteria calculated along various study sites has been given in Tables 6.1, 6.2, 6.3, and 6.4. The percentage of mercury-resistant bacteria varied from 13.587% (Station 2) to 87.931% (Station 1) and 23.659% (Station 6) to 79.581% (Station 1) in sediment samples of Paradeep during monsoon and water samples, respectively. 0.165% (Station 3) to 2.471% (Station 9) of bacteria was resistant to mercury in water samples during the monsoon season, whereas 2.952% (Station 3) to 44.268% (Station 9) mercury-resistant bacteria were found in water samples during monsoon and summer seasons, respectively. In the Chilika water sample, the percentage of MRMB varied from 0.564% (Station 4) to 3.333% (Station 8) and 2.118% (Station 3) to 41.800% (Station 9), respectively, during monsoon and summer seasons. However, 0.476% (Station 4) to 1.223% (Station 10) and 0.248% (Station 3) to 5.018% (Station 9) of bacteria were found to be resistant to mercury in sediment samples of Chilika collected during monsoon and summer seasons, respectively. The percentage of mercury-resistant bacteria in Rushikulya sediment samples varied from 6.173% (Station 9) to 18.235% (Station 3) during the monsoon season, whereas a variation of 4.986% (Station 9) to 42.857% (Station 7) was observed during the summer season. Similarly, in water samples, 4.000% (Station 9) to 31.338% (Station 5) and 1.547% (Station 9) to 29.044% (Station 6) of bacteria were found to be resistant to mercury during monsoon and summer seasons, respectively. In Gopalpur sediment samples, 0.666% (Station 1) to 37.631% (Station 8) of bacteria and 6.72% (Station 5) to 67.049% (Station 10) of bacteria were mercury resistant during monsoon and summer seasons, respectively. In the monsoon season, the percentage of the mercury-resistant bacterial population varied from 0.394% (Station 1) to 3.259% (Station 10) in water samples and 4.161% (Station 1) to 50.159% (Station 4) during the summer season.

6.3.5 *Statistical Analysis*

Multiple regression analysis showed the correlation between various physico-chemical parameters and the biological parameters with the respective study sites and the sampling seasons. The results of multiple regression analysis and the multiple regression equation were obtained for THB and MRMB populations and are given in Tables 6.5 and 6.6.

6.4 Discussion

pH is the most critical factor of a marine ecosystem as it has the potential to change the distribution of carbon dioxide and carbon availability. Thus, it may alter the availability of trace metals and other essential nutrients. In addition, extreme pH can cause direct physiological effects on the inhabitant flora and fauna. A number of studies have also reported the significant change of pH in marine systems despite their strong buffering capacity due to the carbonate system in sea water (Guinotte and Fabry 2008; Fabry et al. 2008; Wootton et al. 2008; Hofmann et al. 2011). In the same line with that, the lowest pH observed among the study sites was found at Chilika (6.41) during the summer season and at Rushikulya (6.6) during the monsoon season. However, the upper level of pH was found at Rushikulya (8.4) during the summer season and at Paradeep (8.75) during the monsoon season. pH has been reported to be 8.50 ± 0.18 at coastal areas of Paradeep (Khadanga et al. 2012), which is of the same range obtained during the present study.

Table 6.5 Multiple regression equation obtained for mercury-resistant marine bacteria (MRMB) against physico-chemical parameters in the study sites in both summer and monsoon seasons

Study sites	Sampling season	R and R^2	Multiple regression equation
Bhitarkanika	Summer	$R = 0.8021$ $R^2 = 0.6433$	Y = −3979.73 − 753.667 pH + 814.635 salinity − 14.7226 temperature − 1167.01 total mercury
	Monsoon	$R = 0.8916$ $R^2 = 0.7949$	Y = −20751.7 + 5481.16 pH − 1060.23 salinity + 187.584 temperature − 90139.9 total mercury
Chilika	Summer	$R = 0.9493$ $R^2 = 0.9012$	Y = −39774.9 + 18877.3 pH − 8990.48 salinity + 528.212 temperature + 64520.5 total mercury
	Monsoon	$R = 0.8950$ $R^2 = 0.8011$	Y = 10171.9 + 667.689 pH − 1112.74 salinity − 170.665 temperature + 35930.2 total mercury
Paradeep	Summer	$R = 0.9537$ $R^2 = 0.9096$	Y = 35646 − 3297.36 pH − 989.02 salinity + 1125.16 temperature + 32716.2 total mercury
	Monsoon	$R = 0.9431$ $R^2 = 0.8896$	Y = 20296.4 − 1763.85 pH − 52.337 salinity − 116.189 temperature + 3676.23 total mercury
Gopalpur	Summer	$R = 0.9268$ $R^2 = 0.8590$	Y = 35611.7 + 9368.15 pH − 1968.78 salinity + 176.31 temperature + 31251.3 total mercury
	Monsoon	$R = 0.9212$ $R^2 = 0.8486$	Y = 29088.7 − 8135.25 pH + 1119.65 salinity + 116.926 temperature + 6051.89 total mercury
Rushikulya	Summer	$R = 0.9856$ $R^2 = 0.9715$	Y = −4841.92 + 177.622 pH + 1.21576 salinity + 147.317 temperature + 3853.21 total mercury
	Monsoon	$R = 0.9872$ $R^2 = 0.9746$	Y = 6002.96 − 3602.58 pH − 356.899 salinity + 2045.05 temperature + 27011.4 total mercury

Table 6.6 Multiple regression equation obtained for total heterotrophic bacteria (THB) against physico-chemical parameters in the study sites in both summer and monsoon seasons

Study sites	Sampling season	R and R^2	Multiple regression equation
Bhitarkanika	Summer	$R = 0.7920$ $R^2 = 0.6272$	Y = 4380500 − 475209 pH − 34430.4 salinity − 4353.48 temperature − 318843 total mercury
	Monsoon	$R = 0.5438$ $R^2 = 0.2957$	Y = 469609000 − 8623870 pH + 3505300 salinity − 16614600 temperature + 319657000 total mercury
Chilika	Summer	$R = 0.9338$ $R^2 = 0.8719$	Y = −3033160 + 575407 pH + 136305 salinity − 55181.1 temperature − 151873 total mercury
	Monsoon	$R = 0.8713$ $R^2 = 0.7592$	Y = 8704920 − 168900 pH − 25630.2 salinity − 246394 temperature − 5732670 total mercury
Paradeep	Summer	$R = 0.7346$ $R^2 = 0.5397$	Y = 4600840 − 527212 pH − 2955.92 salinity − 2313.91 temperature − 38366.9 total mercury
	Monsoon	$R = 0.2750$ $R^2 = 0.0756$	Y = 124951 + 46748.7 pH + 8395.83 salinity − 5595.28 temperature + 69895 total mercury
Gopalpur	Summer	$R = 0.7312$ $R^2 = 0.5347$	Y = −170959 + 7168.26 pH + 2626.73 salinity + 2441.73 temperature − 3385.28 total mercury
	Monsoon	$R = 0.5331$ $R^2 = 0.2842$	Y = 2000840 − 563921 pH + 50149.6 salinity + 46353.4 temperature + 30777.6 total mercury
Rushikulya	Summer	$R = 0.9792$ $R^2 = 0.9588$	Y = 964633 − 66999.6 pH − 795.929 salinity − 9308.83 temperature − 78069.5 total mercury
	Monsoon	$R = 0.7116$ $R^2 = 0.5064$	Y = 1998700 − 292749 pH + 13540.7 salinity + 13386.4 temperature + 182752 total mercury

Chilika lagoon is the most studied marine ecosystem along the Bay of Bengal due to its high fish productivity and tourist attraction. Many studies also found the same range of pH as obtained during this study, which was reported to be 6.4 ± 0.41 to 9.4 ± 0.46 (Nayak and Behera 2004) and 7.936 (Pradhan et al. 2012). Similarly, the pH range of Gopalpur water was also reported to be of 7.46–7.95 (Sahu et al. 2013) and 7.46–7.93 (Baliarsingh et al. 2013), which is well within the result obtained in the present study. However, the pH range of the Rushikulya study site varied widely from 6.6 to 8.4. The wide variation of pH along this study site has also been reported in many other studies such as 6.09–8.29 (Das et al. 2001) and 7.14–8.47 (Bramha et al. 2011). The lowering of pH of the marine ecosystems is a major source of concern nowadays as natural uptake of atmospheric CO_2 by ocean water has decreased the pH of surface sea water by 0.1 units in the last 200 years. Modelling experiments estimate the pH reduction in the range of 0.3–0.4 units over the next 300 years (Caldeira and Wickett 2005). Though at a minute level, but when the anthropogenic activities come into the picture, the local scenario of pH change becomes significant enough to harm the normal functioning of the marine ecosystem as observed in this study. pH reduction may affect the marine species in many ways such as it has impact on oxygen transport and respiration system, reduces sperm motility and fertilization success and increases the volume of the ocean, which is unsaturated with respect to aragonite and calcite (Pörtner et al. 2005; Barange and Perry 2009).

A significant change in temperature among the study sites was observed during the two seasons of sampling, i.e. summer and monsoon. The water temperature witnessed a highest high temperature at Gopalpur (36.85 °C) and a lowest at the same study site during the summer season. However, the temperature was found to change significantly during the monsoon season with a highest high temperature of 28.95 °C and the lowest low temperature of 21.85 °C at Gopalpur among the study sites. A similar trend of temperature was observed at many other marine ecosystems of the Bay of Bengal (Vinayachandran and Murty 2002; Bhat et al. 2004) and the North Indian Ocean (Shenoi et al. 2009). Sea surface temperature plays a major role in the sustainability of a marine ecosystem by influencing the development of eggs and other propagules, release of propagules, survival of larval stage of animals, survival of post-settlement juveniles as well as survival of adults (Hiscock et al. 2004). It has been estimated that there exists an increase in mean global sea surface temperature by 0.13 °C per decade since 1979 and the oceanic interior temperature by >0.1 °C since 1961 (Brierley and Kingsford 2009).

Salinity variation becomes sufficiently large to support temperature inversions (Shetye and Gouveia 1998). It has been reported that the surface waters of the Bay of Bengal have low salinity level and the circulation in the upper layer is dominated by the salinity gradients (Shetye et al. 1991). Thus, the salinity range as observed during the present study is well within the other reports from the other parts of the Bay of Bengal. The salinity range of Chilika among the study sites showed a lowest level in the range of 4.5–9 psu during the monsoon season and 9–13.5 during the summer season. This is due to the gradual chocking of the lagoon mouth, subsequently disturbing the tidal flux and altering the natural ecosystem. The salinity level of the lake has been reported to decline to 1.4–6.3 psu from 22.3 psu during 1957–1958 to 1995 (Gopikrishna et al. 2014). Other study sites show the salinity variation as per the other marine environments along the Bay of Bengal (Shankar and Sherye 2001; MoES, GoI 2010) and other coastal and marine environments (Chamarthi and Ram 2009; Kudale 2010).

Among the four study sites, the highest mercury content was found at Rushikulya in the range of 2.26–2.48 ppb irrespective of the sediment or water sample or sampling seasons. The order of mercury content among the study sites was Rushikulya > Paradeep > Chilika > Gopalpur. Though in most of the occasions, the level of mercury detected was well within the permissible limit of 0.001 ppm as prescribed by WHO (1993), the level may increase in the near future if not monitored and regulated very soon. Many studies have also confirmed the presence of high level of mercury along the Rushikulya region of the Bay of Bengal (Panda et al. 1992; Das et al. 2001; Bramha et al. 2011). Mercury pollution at Rushikulya estuary may be attributed to the Chloralkali plant at Ganjam, India. Paradeep also witnessed higher mercury content, and at four stations the mercury content was recorded to be on the upper level than the permissible limit. There exist many huge industries and port activities that contribute to the mercury load of this ecosystem. Other study sites showed a trace amount of mercury and gained mercury from the non-point sources. Mercury is also found to be present in many natural marine ecosystems contaminated from both point and non-point sources. Many wetlands such as mangrove wetlands of French Guiana (0.41 nmol/g) (Marchand et al. 2006), Sunderban mangrove wetland (17.2–43.32 ng/g) (Kwokal et al. 2008) and Gulf of Mexico (3.38 ng/l) (Hall et al. 2008) have been reported to contain trace amount of mercury which may be from the natural sources and the pollution from the non-point sources.

Marine microorganisms are of critical importance for the health of the marine environment as they are the integral part of the major geochemical cycles, fluxes and processes occurring in marine systems. They provide essential goods and services to the society in terms of production of oxygen, supporting sustainable supply of food, regulating the heath of the marine environment and providing largely untapped source of genetic information and biomolecules for industrial and medical applications and products. A healthy marine ecosystem is associated with the number of bacterial population inhabiting in that environment. Microbes are the first to respond to the changing environmental conditions, and hence, they should be monitored continuously for their population as well as diversity. Heavy metal-resistant bacteria in the marine environment are proposed to be an indicator of metal pollution in the marine environment (Das et al. 2007; Dash and Das 2014), and hence, mercury-resistant marine bacterial population has been monitored during the present study to assess mercury pollution level along the study sites.

The THB population in all the study sites was at a higher level in the monsoon season than that of the summer season. This may be due to the large fresh water inflow into the oceanic system during the monsoon season, thus increasing the bacterial load of the ecosystem. Contrary to that, the MRMB population was at a higher level during the summer season as compared to the monsoon season. Similarly, the percentage of mercury-resistant bacteria is also in an upper level during the summer season than that of the monsoon season as observed during the study. A strong positive correlation between the mercury-resistant marine bacterial population and the total mercury content in both water and sediment sample is obtained during this study. Total heterotrophic bacteria in the coastal environments of Andaman islands of India were in the range of 10^3–10^4 CFU/ml in water and 10^4–10^5 CFU/g in sediment samples (Swarnakumar et al. 2008), which is less than the number of population found in the present study, suggesting the higher productivity of the Odisha coast of the Bay of Bengal, India.

Mercury-resistant bacterial population is quite ubiquitous in nature and accounts for 1–10% of aerobic total heterotrophic bacteria (Muller et al. 2001). During this study, mercury-resistant marine bacteria have also been isolated from the sites without any direct source of mercury pollution. However, mercury-resistant bacterial population is correlated with the level of mercury contamination of a particular environment (Ramaiah and De 2003). In context to that, an unusual rise of mercury-resistant marine bacterial population in the coastal environments has been predicted by Ramaiah and De (2003). 96% of the CFU in water and 71.4% of sediment samples of the Bay of Bengal were reported to tolerate mercury at a level of 10 ppm (50μM), and the present study follows the same pattern of the percentage of MRMB along the Odisha coast of the Bay of Bengal. Many other reports also found large fractions of mercury-resistant marine bacteria that emerged due to anthropogenic activities (Reyes et al. 1999).

A strong positive correlation between mercury-resistant marine bacterial population and the total mercury content in the samples has been observed during this study. Barkay et al. (2003) also predicted the positive correlation of mercury in the environment with the mercury-resistant bacteria. Thus, this study will be helpful and can be applied to other marine ecosystems for estimating the load of mercury through mercury-resistant marine bacterial population.

6.5 Conclusion

The coastal and marine ecosystem is the most endemic ecosystem on the planet despite its many important services and functions. Odisha is of utmost importance due to the effect of the industrial revolution along this region as well as the presence of many internationally renowned ecosystems such as Chilika Lake and Bhitarkanika mangrove ecosystem. A decrease of pH, decrease of salinity mostly at Chilika and increased sea surface temperature were the highlights of this study, which are not healthy for the sustenance of the coastal ecosystem. The most horrifying facts are the gradual increase of mercury pollution and the subsequent rise of mercury-resistant marine bacterial population along with the study sites. The increased pollution levels are due to the anthropogenic activities both from the point and non-point sources. Thus, the ecosystem along the Odisha coast should be monitored continuously, and necessary regulations should be implemented to restore and regain its economic as well as biological benefits.

References

APHA, American Public Health Association (1992) Standard methods for the examination of water and waste water. Washington, DC. www.mwa.co.th/download/file_upload/SMWW_1000-3000.pdf

Azad AM, Frantzena S, Bank MS, Johnsen IA, Tessier E, Amouroux D, Madsen L, Maage A (2019) Spatial distribution of mercury in seawater, sediment, and seafood from the Hardangerfjord ecosystem, Norway. Sci Total Environ 667:622–637. https://doi.org/10.1016/j.scitotenv.2019.02.352

Baliarsingh SK, Sahu BK, Srichandan S, Sahu KC, Lotliker AA, Kumar TS (2013) Seasonal variation of phytoplankton community in Gopalpur creek: a tropical tidal backwater ecosystem, East Coast of India. Indian J GeoMar Sci 42:622–634. ISSN: 0975-1033

Barange M, Perry RI (2009) Physical and ecological impacts of climate change relevant to marine and inland capture fisheries and aquaculture. In: Cochrane K, De Young C, Soto D, Bahri T (eds) Climate change implications for fisheries and aquaculture: overview of current scientific knowledge. FAO Fisheries and Aquaculture Technical Paper. No. 530. FAO, Rome, pp 7–106. ISBN 978-92-5-106347-7

Barkay T, Miller SM, Summers AO (2003) Bacterial mercury resistance from atoms to ecosystems. FEMS Microbiol Rev 27:355–384. https://doi.org/10.1016/S0168-6445(03)00046-9

Bhat GS, Vecchi GA, Gadgil S (2004) Sea surface temperature of the Bay of Bengal derived from the TRMM microwave imager. J Atmos Ocean Technol 21:1283–1290. https://doi.org/10.1175/15200426(2004)021<1283:SSTOTB>2.0.CO;2

Bramha SN, Panda UC, Rath P, Mohanty PK, Satpathy KK (2011) Anthropological influence in coastal water and its impact on olive ridley turtle: a case study at Rushikulya mass nesting site. J Ecol Nat Environ 3:268–272. ISSN 2006-9847 ©2011

Brierley AS, Kingsford MJ (2009) Impacts of climate change on marine organisms and ecosystems. Curr Biol 19:R602–R614. https://doi.org/10.1016/j.cub.2009.05.046

Caldeira K, Wickett ME (2005) Ocean model predictions of chemistry changes from carbon dioxide emissions to the atmosphere and ocean. J Geophys Res 110:C09S04. https://doi.org/10.1029/2004JC002671

Chamarthi S, Ram PS (2009) Role of fresh water discharge from rivers on oceanic features in the northwestern Bay of Bengal. Mar Geod 32:64–76. https://doi.org/10.1080/01490410802662219

Chan KMA, Ruckelshaus M (2010) Characterizing changes in marine ecosystem services. Biol Reprod 54:1–6. https://doi.org/10.3410/B2-54

Cinnirella S, Bruno DE, Pirrone N, Horvat M, Živković I, Evers DC, Johnson S, Sunderland EM (2019) Mercury concentrations in biota in the Mediterranean Sea, a compilation of 40 years of surveys. Sci Data 6:205. https://doi.org/10.1038/s41597-019-0219-y

Cossa D, Martin JM, Takayanagi K, Sanjuan J (1997) The distribution and cycling of mercury species in the western Mediterranean. Deep-Sea Res II Top Stud Oceanogr 44:721–740. https://doi.org/10.1016/S0967-0645(96)00097-5

Das S, Patro SK, Sahu BK (2001) Variation of residual mercury in penaeid prawns from Rushikulya estuary, east coast of India. Indian J Mar Sci 30:33–37. ISSN: 0975-1033

Das S, Shanmugapriya R, Lyla PS, Khan SA (2007) Heavy metal tolerance of marine bacteria- an index of marine pollution. Natl Acad Sci Lett (India) 30:279–284

Dash HR, Das S (2014) Assessment of mercury pollution through mercury resistant marine bacteria in Bhitarkanika mangrove ecosystem, Odisha, India. Indian J GeoMar Sci 43:1103–1115. ISSN: 0975-1033

Das S, Mangwani N (2015) Ocean acidification and marine microorganisms: responses and consequences. Oceanologia 57:349–361. https://doi.org/10.1016/j.oceano.2015.07.003

Doney SC, Ruckelshaus M, Duffy JE, Barry JP, Chan F, English CA, Galindo HM, Grebmeier JM, Hollowed AB, Knowlton N, Polovina J, Rabalais NN, Sydeman WJ, Talley LD (2012) Climate change impacts on marine ecosystems. Annu Rev Mar Sci 4:11–37. https://doi.org/10.1146/annurev-marine-041911-111611

Fabry VJ, Seibel BA, Feely RA, Orr JC (2008) Impacts of ocean acidification on marine fauna and ecosystem processes. ICES J Mar Sci 65:414–432. https://doi.org/10.1093/icesjms/fsn048
Gopikrishna B, Sinha J, Kudale MD (2014) Impact on salinity of Chilika lake due to changes in the inlet system. Indian J Mar Sci 43:1–6. www.niscair.res.in/jinfo/ijms/ijms...articles/.../MS%20 21%20Edited.pdf
Guinotte JM, Fabry VJ (2008) Ocean acidification and its potential effects on marine ecosystems. Ann N Y Acad Sci 1134:320–342. https://doi.org/10.1196/annals.1439.013
Gworek B, Bemowska-Kałabun O, Kijeńska M, Wrzosek-Jakubowska J (2016) Mercury in marine and oceanic waters-a review. Water Air Soil Pollut 227:371. https://doi.org/10.1007/s11270-016-3060-3
Hall BD, Aiken GR, Krabbenhoft DF, Marvin-DiPasquale M, Swarzenski CM (2008) Wetlands as principal zones of methylmercury production in southern Louisiana and the Gulf of Mexico region. Environ Pollut 154:124–134. https://doi.org/10.1016/j.envpol.2007.12.017
Hiscock K, Southward A, Tittley I, Hawkins S (2004) Effects of changing temperature on benthic marine life in Britain and Ireland. Aquat Conserv Mar Freshwat Ecosyst 14:333–362. https://doi.org/10.1002/aqc.628
Hofmann GE, Smith JE, Johnson KS, Send U, Levin LA, Micheli F, Paytan A, Price NN, Peterson B, Takeshita Y, Matson PG, Crook ED, Kroeker KJ, Gambi MC, Rivest EB, Frieder CA, Yu PC, Martz TR (2011) High frequency dynamics of ocean pH: a multi-ecosystem comparison. PLoS One 6(12):e28983. https://doi.org/10.1371/journal.pone.0028983
Jaysankar D, Ramaiah N (2006) Occurrence of large fractions of mercury resistant bacteria in the Bay of Bengal. Curr Sci 91:368–372. http://drs.nio.org/drs/handle/2264/348
Khadanga MK, Das S, Sahu BK (2012) Seasonal variation of the water quality parameters and its influences in the Mahanadi estuary and near coastal environment, east coast of India. World Appl Sci J 17:797–801. ISSN: 1818-4952
Kudale MD (2010) Impact of port development on the coastline and the need for protection. Indian J GeoMar Sci 39:597–604. nopr.niscair.res.in/bitstream/.../10808/1/IJMS%2039(4)%20 597-604.pdf
Kwokal Z, Sarkar SK, Chatterjee M, Franciskovis-Bilinski S, Bilinski H, Bhattacharya A, Bhattacharya BD, Alam MA (2008) An assessment of mercury loading in core sediments of Sunderban Mangrove Wetland, India (a preliminary report). Bull Environ Contam Toxicol 81:105–112. https://doi.org/10.1007/s00128-008-9443-4
Lacerda LD, Bastos WR, Almeida MD (2012) The impacts of land use changes in the mercury flux in the Madeira River, Western Amazon. Ann Braz Acad Sci 84:73–82. https://doi.org/10.1590/S0001-37652012000100007
Lamborg CH, Fitzgerald WF, O'Donnell J, Torgersen T (2002) A non-steady state compartmental model of global scale mercury biogeochemistry with interhemispheric atmospheric gradients. Geochim Cosmochim Acta 66:1105–1118. https://doi.org/10.1016/S0016-7037(01)00841-9
Laurier FJG, Mason RP, Gill GA, Whalin L (2004) Mercury distributions in the North Pacific Ocean: 20 years of observations. Mar Chem 90:3–19. https://doi.org/10.1016/j.marchem.2004.02.025
Marchand C, Lallier-Verges E, Baltzer F, Alberic P, Cossa D, Baillif P (2006) Heavy metals distribution in mangrove sediments along the mobile coastline of French Guiana. Mar Chem 98:1–17. https://doi.org/10.1016/j.marchem.2005.06.001
Mason RP, Rolfhus KR, Fitzgerald WF (1998) Mercury in the North Atlantic. Mar Chem 61:37–53. https://doi.org/10.1016/S0304-4203(98)00006-1
Ministry of Earth Sciences, ICMAM PD, Chennai (2010) Monitoring of marine pollution through coastal ocean monitoring and prediction system (COMAPS) programme. http://www.icmam.gov.in/comaps/par.pdf
Morel FM, Kraepiel AM, Amyot M (1998) The chemical cycle and bioaccumulation of mercury. Annu Rev Ecol Syst 29:543–566. https://doi.org/10.1146/annurev.ecolsys.29.1.543
Muller A, Rasmussen L, Sorenson S (2001) Adaptation of the bacterial community to mercury contamination. FEMS Microbiol Lett 16:49–53. https://doi.org/10.1016/S0378-1097(01)00376-7

Nayak L, Behera DP (2004) Seasonal variation of some physicochemical parameters of Chilika lagoon (east coast of India) after opening of the new mouth near Sipakuda. Indian J Mar Sci 33:206–208. ISSN: 0379-5136
Panda KK, Lenka M, Panda BB (1992) Monitoring and assessment of mercury pollution in the vicinity of a chloralkali plant. III. Concentration and genotoxicity of mercury in the industrial effluent and contaminated water of Rushikulya estuary, India. Mutat Res Genet Toxicol 280:149–160. https://doi.org/10.1016/0165-1218(92)90043-Y
Pörtner HO, Langenbuch M, Michaelidis B (2005) Synergistic effects of temperature extremes, hypoxia, and increases in CO_2 on marine animals: from earth history to global change. J Geophys Res Oceans 110. https://doi.org/10.1029/2004JC002561
Pradhan V, Mohsin M, Gaikwad BH (2012) Assessment of physico chemical parameters of Chilika Lake water. Int J Res Environ Sci Technol 2:101–103. ISSN 2249–9695
Ramaiah N, De J (2003) Unusual rise in mercury resistant bacteria in coastal environments. Microb Ecol 45:444–454. drs.nio.org/drs/bitstream/handle/2264/.../Microb_Ecol_45_444o.pdf?...4
Reyes S, Frischer E, Sobecky A (1999) Characterization of mercury resistance mechanism in marine sediment microbial communities. FEMS Microbiol Ecol 30:273–284. https://doi.org/10.1111/j.1574-6941.1999.tb00655.x
Sahu BK, Baliarsingh SK, Srichandan S, Sahu KC (2013) Seasonal variation of zooplankton abundance and composition in Gopalpur creek: a tropical tidal backwater, east coast of India. J Mar Biol Assoc India 55:59–64. https://doi.org/10.6024/jmbai.2013.55.1.01715-10
Schroeder W, Munthe J (1998) Atmospheric mercury -an overview. Atmos Environ 32:809–822. www.quantracmoitruong.gov.vn/.../1_Schmeltz_Introduction%20Mercury
Shetye SR, Gouveia AD (1998) Coastal circulation in the Northern Indian Ocean. In The Sea, Vol. 11, A. R. Robinson and K. H. Brink (eds.). New York: John Wiley & Sons, Inc.
Shetye SR, Gouveia AD, Shenoi SSC, Michael GS, Sundar D, Almeida AM, Santanam K (1991) The coastal current off western India during the northeast monsoon. Deep-Sea Res 38:1517–1529
Shankar D, Sherye SR (2001) Why is mean sea level along the Indian coast higher in the Bay of Bengal than in the Arabian Sea? Geophys Res Lett 28:563–565. https://doi.org/10.1029/2000GL012001
Shenoi SSC, Nasnodkar N, Rajesh G, Joseph KJ, Suresh I, Almeida AM (2009) On the diurnal ranges of sea surface temperature (SST) in the North Indian Ocean. J Earth Syst Sci 118:483–496. www.ias.ac.in/jess/oct2009/d8je-104.pdf
Sunderland EM, Mason RP (2007) Human impacts on open ocean mercury concentrations. Global Biogeochem Cycles 21. https://doi.org/10.1029/2006GB002876
Swarnakumar N, Sahu M, Sivakumar K, Thangaradjou T (2008) Assessment of microbial population in the coastal environs of the Little Andaman Islands, India. Indian J Mar Sci 37:146–152. nopr.niscair.res.in/bitstream/.../1878/1/IJMS%2037(2)%20146-152.pdf
U.S. Environmental Protection Agency (1976) National recommended water quality criteria. Washington DC. http://water.epa.gov/
Venkatesharaju K, Ravikumar P, Somashekhar RK, Prakash KL (2010) Physico-chemical and bacteriological investigation on the river Cauvery of Kollegal stretch in Karnataka. Kathmanadu Univ J Sci Eng Technol 6:10–15. www.nepjol.info/index.php/KUSET/article/download/3310/2849
Vinayachandran PN, Murty VSN (2002) Babu VR (2002) observations of barrier layer formation in the Bay of Bengal during summer monsoon. J Geophys Res 107:8018. https://doi.org/10.1029/2001JC000831
WHO (1993) Guidelines for drinking water quality. Revision of the 1984 guidelines. Final task group meeting. Geneva, 21–25 September, 1992. www.who.int/water_sanitation_health/dwq/2edaddvol2a.pdf
Wootton JT, Pfister CA, Forester JD (2008) Dynamic patterns and ecological impacts of declining ocean pH in a high-resolution multi-year dataset. PNAS 105:18848–18853. https://doi.org/10.1073/pnas.0810079105

Chapter 7
Persistent Organic Pollutants in the Coastal and Estuarine Regions Adjoining the Indian Periphery of the Bay of Bengal

Sanghamitra Basu, Abhra Chanda, Sourav Das, and Subarna Bhattacharyya

Abstract Persistent organic pollutants (POPs) comprise an array of organic pollutants, renowned for their toxicity and bioaccumulation potential. These POPs exist in abundance in the coastal and estuarine sediments, water column, and biotas, which are impacted by city sewage, agricultural runoffs, aquaculture practices, and others. This chapter summarized the findings from existing literature about the distribution and range of the principal classes of POPs. We considered dichlorodiphenyltrichloroethane and its metabolites (DDTs) and the isomers of hexachlorocyclohexanes (HCHs) as the main organochlorine pesticides (OCPs). We also discussed the congeners of polychlorinated biphenyls (PCBs), polycyclic aromatic hydrocarbons (PAHs), and polybrominated diphenyl ethers (PBDEs). The present chapter focused on the POP accumulation in the sediments, water column, and selected biotas in the estuaries along the east coast of India. Overall, these estuaries exhibited an abundance of HCHs and DDTs with negligible to moderate pollution levels of PCBs, PAHs, and PBDEs. The pollution led by HCHs and DDTs was found to be significantly high in northern estuaries like that of Hooghly and Sundarban estuaries, compared to the southern estuaries like Vellar, Cauvery, and Coleroon estuaries of the east coast of India.

Keywords POP · DDT · HCH · PAH · PCB · PBDE · OCP · Congeners · Isomers · Estuaries

---

S. Basu (✉) · S. Bhattacharyya
School of Environmental Studies, Jadavpur University, Kolkata, West Bengal, India

A. Chanda · S. Das
School of Oceanographic Studies, Jadavpur University, Kolkata, West Bengal, India

S. Das, T. Ghosh (eds.), *Estuarine Biogeochemical Dynamics of the East Coast of India*, https://doi.org/10.1007/978-3-030-68980-3_7

7.1 Introduction

Persistent organic pollutants (POPs) are xenobiotic lipid-soluble compounds having low water solubility. These compounds exhibit toxic character owing to their tendency to bioaccumulate in several smaller life forms and eventually biomagnify through the ecological food chain (Sarkar et al. 2012). Dichlorodiphenyltrichloroethane and its metabolites (DDTs) and all the isomers of hexachlorocyclohexanes (HCHs) principally constitute the class of organochlorine pesticides (OCPs). The congeners of polychlorinated biphenyls (PCBs), polycyclic aromatic hydrocarbons (PAHs), and polybrominated diphenyl ethers (PBDEs) are some of the other polluting POPs. Several countries have banned most of these compounds following the restrictions imposed by the Stockholm Convention (Lallas 2001). Few POPs find their way to the open environment due to natural causes like smaller amounts of PAH due to forest fire. All the other POPs are released into the environment exclusively due to anthropogenic activities. Manufacturing capacitors and transformers require PCBs. PAHs generate as a by-product of waste incineration and fossil fuel combustion. PBDEs find their use in the textile, plastics, and automobile industries. Though most of the developed countries have adopted draconian regulations against banning these products, India is still in its infancy and continues to be one of the largest producers of some POPs like DDT, which is mostly used in the public health sector mainly to combat disease like malaria (Zanardi-Lamardo et al. 2019). Indiscriminate disposal of materials containing POPs and leaching and weathering of such materials along with runoffs from agricultural fields and industrial effluents augment the POP levels in the rivers, which eventually reach the estuaries and contaminate the sediments and biota of the Indian coastal region (Sarkar et al. 2012). Therefore, it has become imperative to investigate and comprehend the present POP pollution scenario in the estuarine and coastal sectors of India. Though we recognize the adverse impacts of these POPs throughout the world, very few studies have been carried out in the estuaries along the east coast of India, even though this coastline has some of the crucial megacities and industrial hubs of India. The present chapter collated all the findings and observations made to date in the coastal region along the east coast of India.

7.2 POPs in Coastal and Estuarine Sediments

Coastal and estuarine sediments act as both short-term and long-term repositories of various POPs, which make these POPs bioavailable to several bottom-dwelling marine organisms (Sarkar et al. 2012) and hence mark the beginning of trophic chain transfer of these pollutants. Among the significant estuaries along the Indian east coast, Hooghly estuary, followed by the estuaries within the Sundarban mangroves, has perhaps received the most attention in this regard (Table 7.1). Guzzella et al. (2005) recorded the presence of hexachlorocyclohexane isomers (ΣHCHs), ΣDDTs,

Table 7.1 Range of observed persistent organic pollutants in the sediments and water column of the major estuaries along the east coast of India

Study site	Sampling time	HCH	DDT	PCB	PAH	PBDE	Authors
Sediments (in ng g^{-1})							
Hooghly est.	April 2003	0.11–0.40	0.18–1.93	0.18–2.33	2.5–1081		Guzzella et al. (2005)
Hooghly est.	NM			0.23–31.4	9.39–4249	0.08–29.0	Binelli et al. (2007, 2008)
Hooghly est.	March 2015		ND–8.97	ND–13.5	15.4–1731		Zanardi-Lamardo et al. (2019)
Hooghly est.	NM	0.10–0.60	0.14–18.6	0.28–7.7	3.3–630		Mitra et al. (2019)
Sundarban est.	Nov–Dec 2005	0.05–12	0.05–11.5				Sarkar et al. (2008)
Sundarban est.	NM				4880–20,000		Balu et al. (2020)
Vellar est.	NM	1.8–25.8	0.4–7.1				Venugopalan and Rajendran (1984)
Cauvery est.	NM	4.35–158.4	0.69–4.85				Rajendran and Subramanian (1999)
Cooum river, Ennore estuary, Pulicat Lake					13–31,425		Goswami et al. (2016)
Water column (in ng l^{-1})							
Hooghly est.	2005	3.41–19.70	6.01–44.10				Purkait et al. (2009)
Sundarban est.	NM				ND–125,000		Balu et al. (2020)
Chennai coast	Jan 1998		5.6–12.5	1.9–4.4			Rajendran et al. (2005)
Cauvery est.	1990–1992	3.2–182.0	0.8–4.2				Rajendran and Subramanian (1997)

ND not detected, *NM* not mentioned

and ΣPCBs in the Hooghly estuary in the range of 0.11–0.40, 0.18–1.93, and 0.18–2.33 ng g^{-1} dry weight, respectively. They attributed the widespread use of HCH and DDTs in the agricultural sector and malaria-prevention sanitary activities behind such recorded magnitudes. Among the four important isomers of HCH (α, β, γ, and δ), α-HCH was the most abundant, followed by γ-HCH, and the other two were below detection level in most of the samples. Sarkar et al. (2008), while analyzing the estuarine sediments of Sundarban mangroves, observed ΣHCHs and ΣDDTs ranged between 0.05 and 12 ng g^{-1}, and 0.05 and 11.5 ng g^{-1}, respectively. Such an observation indicated that both HCHs and DDTs were substantially higher in the Sundarban mangroves than Hooghly estuary. Moreover, contrary to Guzzella et al. (2005), Sarkar et al. (2008) observed the highest dominance of γ-HCH followed by β-HCH and found all the four isomers to be present in significant quantities. Sarkar et al. (2008) argued that α-HCH and γ-HCH potentially transformed to β-HCH, which was sustained in the sediments due to its recalcitrant character and low vapor pressure. Like HCHs, DDTs occurred in the sediment samples of Sundarban and the lower stretch of Hooghly estuary with varying percentage contribution of their metabolites (Guzzella et al. 2005; Sarkar et al. 2008). pp′-DDT and op′-DDT were at much higher levels compared to DDE and DDD following an overall hierarchy of pp′-DDT > pp′-DDD > pp′-DDE > op′-DDT > op′-DDD > op′-DDE (Sarkar et al. 2012). pp′-DDT:pp′-DDE ratio in the Hooghly and Sundarbans was higher than 0.33, which indicated that DDT is still in use in the agricultural sector (whereas a ratio lower than 0.33 indicates the transformation of existing pp′-DDT to pp′-DDE) (Stranberg et al. 1998). Zanardi-Lamardo et al. (2019), in a very recent study on the Hooghly River estuary and Sundarbans, observed the highest concentrations of pp′-DDT among all the other congeners of DDT, which prompted them to conclude that fresh input of DDT is continuing in this estuary.

Besides Hooghly and Sundarban, sediments from Vellar estuary, Cauvery River estuary, and its distributary, Coleroon estuary have recorded substantial concentrations of POPs. Venugopalan and Rajendran (1984) recorded ΣHCH range of 1.8–25.8 ng g^{-1} wet weight and ΣDDT range of 0.4–7.1 ng g^{-1} dry weight in the sediments of Vellar estuary. Rajendran and Subramanian (1999) observed HCH and DDT ranging from 4.35 to 158.4 ng g^{-1} dry weight and 0.69 to 4.85 ng g^{-1} dry weight, respectively, in the Cauvery and Coleroon estuaries.

In these southern estuaries, they observed that pp′-DDE, which happens to be a breakdown product of DDT, comprised the majority of the total DDTs, which implies that the fresh input of DDT is not taking place to that extent as observed in the case of Hooghly. However, among the HCH isomers, α-HCH was the most abundant, followed by β-HCH and γ-HCH, indicating its continued usage in the agricultural sector. However, these studies took place about two decades back, and there are no recent observations.

Besides HCH and DDT, reports of substantial concentrations of PCBs, PAHs, and PBDEs are available from the east coast estuaries. However, most of these studies took place in the Hooghly estuary and the Sundarbans part. PCBs in Sundarban sediments ranged between 0.23 and 31.4 ng g^{-1} with significantly varying concentrations of individual congeners having no specific trend in terms of spatial variability (Binelli et al. 2008; Sarkar et al. 2012). On the whole, tri-chlorinated biphenyls to octa-chlorinated biphenyls were the principal PCB congeners found in the sediments, and the high molecular weight PCBs were more prevalent than the low molecular weight PCBs, owing to the lesser persistency and higher volatility of the latter compared to the former (Binelli et al. 2008). PAHs exhibited low to moderate concentrations (2.5–1081 ng g^{-1}) in the sediments adjoining the Hooghly estuary with a prominent declining trend from the inner estuarine stations to the river mouth (Guzzella et al. 2005). However, Balu et al. (2020) reported an exceedingly high range of PAH (4880–20,000 ng g^{-1}) covering the sediments of five locations through the estuaries of Indian Sundarban. Goswami et al. (2016) observed an even bigger range of PAH (13–31,425 ng g^{-1}) in the Cooum river, Ennore estuary, and

Pulicat Lake of Tamil Nadu. Binelli et al. (2007) observed PBDEs ranging from 0.08–29.0 ng g^{-1} in the mangrove sediments of Sundarban with the highest concentrations in the sewage outfall zone of the Kolkata metropolis, i.e., Ghusighata. The authors wrangled that the burning of electronic wastes and sewage effluents from textile industries could be the fundamental source of PBDEs in these estuarine regions.

7.3 POPs in Estuarine Water Column

Compared to sediments, very few studies characterized POP levels in the estuarine water column of the east coast of India. In the Coleroon and Cauvery estuaries, Rajendran and Subramanian (1997) observed ΣHCH and ΣDDT concentrations of 3.2–182.0 and 0.8–4.2 ng l^{-1}, respectively. They observed a predominance of α-HCH among the isomers of HCH. However, there was no significant difference between the DDT compounds and their metabolites. Kumarasamy et al. (2012) analyzed an array of organochlorine pesticides (OCPs) in the Tamiraparani River estuary, which meets the Bay of Bengal through the Gulf of Mannar. ΣOCP ranged from 0.1 to 79.9 ng l^{-1} with heptachlor, o,p′-DDE, dieldrin, o,p′-DDD, and mirex being the dominant contaminants. The upper reaches of the Tamiraparani River estuary had higher OCP levels in the water, and the authors observed that the construction of dams reduced the OCP load in the seaward end. Though the sediments of Hooghly and Sundarban estuaries have received substantial attention, only a single study so far (Purkait et al. 2009) reported the OCP levels in the sewage waters near Kolkata. Purkait et al. (2009) observed alarming levels of ΣHCH and ΣDDT varying from 3.41 to 19.70 and 6.01 to 44.10 mg l^{-1}, respectively, in the sewage waters, which potentially polluted the upper reaches of Hooghly estuary. On the contrary, Rajendran et al. (2005) observed much less ΣPCB (1.9–4.4 ng l^{-1}) and ΣDDT (5.6–12.5 ng l^{-1}) levels in the coastal waters near Chennai, Mahabalipuram, Pondicherry, Cuddalore, Nagapattinam, and Mandapam. Recently, Balu et al. (2020) measured PAH in the aquatic column of the Sundarban estuaries. They reported a PAH range varying from non-detectable to 125,000 ng l^{-1}.

7.4 POPs in Marine Biota

Marine mammals, particularly cetaceans owing to their long life span and position in the higher trophic chain, are highly susceptible to POP bioaccumulation. Karuppiah et al. (2005) reported substantially high concentrations of ΣHCH (95–765 ng g^{-1}), ΣDDT (3330–23,330 ng g^{-1}), and ΣPCB (210–1220 ng g^{-1}) in the blubber of three dolphin species, namely, *Tursiops truncatus*, *Sousa chinensis*, and *Stenella longirostris*, accidentally caught in the southeast coast of India. Similarly, Kannan et al. (2005) reported very high concentrations of ΣHCH (maximum,

1200 ng g^{-1}) and ΣDDT (maximum, 10,000 ng g^{-1}), whereas low to moderate concentrations of ΣPBDE (range, 0.98–18.0 ng g^{-1}) and ΣPCB (28–390 ng g^{-1}) in the Irrawaddy dolphin (*Orcaella brevirostris*) accidentally caught from the Chilika Lake, Mahanadi estuary. Zuloaga et al. (2009) reported substantial accumulation of PAH in different body parts of three intertidal bivalve mollusks, namely, *Sanguilonaria acuminata*, *Macoma birmanica*, and *Meretrix meretrix*, from the Sundarbans. The total PAH followed the hierarchy: visceral mass (1433 ng g^{-1}) > adductor muscle (454 ng g^{-1}) > gill (203 ng g^{-1}) > mantle (139 ng g^{-1}) > podium (128 ng g^{-1}) > siphon (93 ng g^{-1}) > shell (<25.5 ng g^{-1}).

7.5 Serious Health Effects of POP Pollution

The POPs have a long residence time in the environment. Being hydrophobic in nature, they prefer to accumulate in the lipids of living organisms and persist there for a long time (Scheringer 2009). Medical research has associated a suite of lethal human diseases with POP pollution ranging from reproductive disorder, cardiovascular malfunction, endocrine disturbance, obesity, diabetes, and various forms of cancer (Alharbi et al. 2018). Research shows that POPs are capable of damaging the DNA structure of several marine biotas as well (Sarker et al. 2018). As human beings often get exposed to a wide variety of POPs, it becomes difficult to attribute any particular health effect to a specific compound (Walker 2008). Exposure to mixed POP compounds often leads to additive and synergistic effects. Almost all the OCPs are endocrine disruptors that give rise to several neurological disorders affecting mostly the neonatal human population exposed to such toxic substances (Crinnion 2011). Various POPs lead to reproductive disorders and are capable of transmutation from the mothers to their fetus through the placenta (Vizcaino et al. 2014). The lipophilic nature of the POPs enables these compounds to bind with the proteins, and it leads to cardiovascular diseases (Ljunggren et al. 2014). Many of these POPs are active carcinogens. Though the actual reason behind causing cancer still needs further elaborate study, several studies associated various types of cancer with the ingestion of POPs through food (Yu et al. 2010). Mitra et al. (2019), while working on the sediments of Hooghly estuary, observed that the accumulated PAH could lead to significantly high cancer risk to the local people who dwell around these sediments. Accidental ingestion and dermal contact are the principal pathways that make this population vulnerable.

7.6 Summary and Conclusion

Compared to many other pollutants, there is a paucity of information on persistent organic pollutants in the estuarine waters, sediments, and biotas of the Indian east coast. From the existing literature survey, we can infer that HCH isomers and parent

DDT congeners exist at substantial levels, which evince their ongoing usage in the agricultural and health sectors. The concentration of these POPs was significantly high in the sediments and water column of the Hooghly estuary as compared to the southern estuaries of the east coast like Vellar, Cauvery, and Coleroon estuaries. Such an increased level of pollution in the Hooghly estuary can be due to the higher freshwater discharge and the effect of sewage water. Although there are studies, which reported alarming levels of POPs in the bodies of some marine organisms, the number of such studies is scarce. Dolphins, so far, have received much attention. The collation of data made in this chapter warrants the need for more exhaustive and extensive studies in many major estuaries of this region, viz., the Mahanadi, Godavari, and Krishna. This study also shows that India needs some stringent regulations at the central government level to curb the usage of these dangerous pollutants. Though various POPs find its applicability in the industrial and health sector, controlled use and a ban on some of the products are the only options left to fight out the evil posed by these substances. Bioremediation techniques by deploying genetically engineered microbial community that feeds on the POPs are coming up as an excellent option to remove the existing POP load from the open environment. Chemical absorbents are also an option, which needs further study and advancement that can effectively remove these pollutants from our ambiance.

References

Alharbi OM, Khattab RA, Ali I (2018) Health and environmental effects of persistent organic pollutants. J Mol Liq 263:442–453

Balu S, Bhunia S, Gachhui R, Mukherjee J (2020) Assessment of polycyclic aromatic hydrocarbon contamination in the Sundarbans, the world's largest tidal mangrove forest and indigenous microbial mixed biofilm-based removal of the contaminants. Environ Pollut 266:115270. https://doi.org/10.1016/j.envpol.2020.115270

Binelli A, Sarkar SK, Chatterjee M et al (2007) Concentration of polybrominated diphenyl ethers (PBDEs) in Sundarban mangrove wetland, northeastern part of bay of Bengal (India). Mar Pollut Bull 54:1220–1229

Binelli A, Sarkar SK, Chatterjee M et al (2008) A comparison of sediment quality guidelines for toxicity assessment in the Sunderban wetlands (bay of Bengal, India). Chemosphere 73(7):1129–1137

Crinnion WJ (2011) Polychlorinated biphenyls: persistent pollutants with immunological, neurological, and endocrinological consequences. Altern Med Rev 16(1):5–13

Goswami P, Ohura T, Guruge KS et al (2016) Spatio-temporal distribution, source, and genotoxic potential of polycyclic aromatic hydrocarbons in estuarine and riverine sediments from southern India. Ecotoxicol Environ Safety 130:113–123

Guzzella L, Roscioli C, Vigano L et al (2005) Evaluation of the concentration of HCH, DDT, HCB, PCB and PAH in the sediments along the lower stretch of Hugli estuary, West Bengal, Northeast India. Environ Int 31:523–534

Kannan K, Ramu K, Kajiwara N et al (2005) Organochlorine pesticides, polychlorinated biphenyls and polybrominated diphenyl ethers in Irrawaddy dolphins from India. Arch Environ Contam Toxicol 49:415–420

Karuppiah S, Subramanium A, Obbard JP (2005) Organochlorine residues in odontocete species from the southeast coast of India. Chemosphere 60:891–897

Kumarasamy P, Govindaraj S, Vignesh S et al (2012) Anthropogenic nexus on organochlorine pesticide pollution: a case study with Tamiraparani river basin, South India. Environ Monit Assess 184(6):3861–3873
Lallas PL (2001) The Stockholm convention on persistent organic pollutants. Am J Int Law 95(3):692–708
Ljunggren SA, Helmfrid I, Salihovic S et al (2014) Persistent organic pollutants distribution in lipoprotein fractions in relation to cardiovascular disease and cancer. Environ Int 65:93–99
Mitra S, Corsolini S, Pozo K et al (2019) Characterization, source identification and risk associated with polyaromatic and chlorinated organic contaminants (PAHs, PCBs, PCBzs and OCPs) in the surface sediments of Hooghly estuary, India. Chemosphere 221:154–165
Purkait S, Ganguly M, Aktar MW et al (2009) Impact assessment of various parameters polluting ganga water in Kolkata region: a study for quality evaluation and environmental implication. Environ Monit Assess 155(1–4):443–454
Rajendran RB, Subramanian A (1997) Pesticide residues in water from river Kaveri, South India. Chem Ecol 13(4):223–236
Rajendran RB, Subramanian AN (1999) Chlorinated pesticide residues in surface sediments from the river Kaveri, South India. J Environ Sci Health B 34(2):269–288
Rajendran RB, Imagawa T, Tao H et al (2005) Distribution of PCBs, HCHs and DDTs, and their ecotoxicological implications in bay of Bengal, India. Environ Int 31(4):503–512
Sarkar SK, Binelli A, Riva C et al (2008) Organochlorine pesticide residues in sediment cores of Sunderban wetland, north eastern part of bay of Bengal, India and their ecotoxicological significance. Arch Environ Contam Toxicol 55:358–371
Sarkar SK, Satpathy KK, Jonathan MP et al (2012) Persistent organic pollutants (POPs) in sediments and biota in coastal environments of India. In: Lichtfouse E, Schwarzbauer J, Robert D (eds) Environmental chemistry for a sustainable world. Springer, Dordrecht, pp 375–406
Sarker S, Vashistha D, Sarker MS, Sarkar A (2018) DNA damage in marine rock oyster (Saccostrea Cucullata) exposed to environmentally available PAHs and heavy metals along the Arabian Sea coast. Ecotoxicol Environ Safety 151:132–143
Scheringer M (2009) Long-range transport of organic chemicals in the environment. Environ Toxicol Chem Int J 28(4):677–690
Stranberg B, Van Bavel B, Bergavist PA et al (1998) Occurrence, sedimentation and spatial variations of organochlorine contaminants in settling particulate matter and sediments in the northern part of the Baltic Sea. Environ Sci Technol 32:1754–1759
Venugopalan VK, Rajendran N (1984) Pesticide pollution effects on marine and estuarine resources. DAE research project report, Centre for Advanced Study in Marine Biology, Annamalai University, Parangipettai, pp 1–316
Vizcaino E, Grimalt JO, Fernández-Somoano A, Tardon A (2014) Transport of persistent organic pollutants across the human placenta. Environ Int 65:107–115
Walker CH (2008) Organic pollutants: an ecotoxicological perspective. CRC Press, Boca Raton, USA
Yu HY, Guo Y, Zeng EY (2010) Dietary intake of persistent organic pollutants and potential health risks via consumption of global aquatic products. Environ Toxicol Chem 29(10):2135–2142
Zanardi-Lamardo E, Mitra S, Vieira-Campos AA et al (2019) Distribution and sources of organic contaminants in surface sediments of Hooghly river estuary and Sundarban mangrove, eastern coast of India. Mar Pollut Bull 146:39–49
Zuloaga O, Prieto A, Usobiaga A et al (2009) Polycyclic aromatic hydrocarbons in intertidal marine bivalves of Sunderban mangrove wetland, India: an approach to bioindicator species. Water Air Soil Pollut 201:305–318

Chapter 8
Characterizing the Human Health Risk Along with the Bioaccumulation of Heavy Metals in the Aquatic Biota in the East Coastal Waters of the Indian Peninsula

Shresthashree Swain, Deepak Kumar Das, Anushka Seal, Abhra Chanda, and Sourav Das

Abstract Several trace metal contaminants accumulate in estuarine water column and sediments, sometimes reaching concentrations beyond the threshold, and impart severe ill effects to the surrounding biota. In this regard, we have discussed the state of the art of some of the most commonly studied heavy metals like As (arsenic), Cu (copper), Co (cobalt), Ni (nickel), Cd (cadmium), Cr (chromium), Zn (zinc), Pb (lead), and Fe (iron) in the transitional water mass of the estuaries along the east coast of India, in this chapter. Besides the pollution scenario, we have assessed the plausible adverse effects these pollutants are posing at present or likely to pose in the future to the coastal communities and the local biodiversity. The aquatic food chain in these estuarine and coastal waters is badly affected due to the potential of heavy metals to get bioabsorbed, bioaccumulated, and biomagnified to higher life-forms. We collated the observations on these metal levels in several marine floras like mangroves and seagrasses and faunas like gastropods, bivalves, mollusks, and fishes in this review. Overall, this chapter has assimilated all the facts published so far regarding heavy metal contamination on the east coast of India.

Keywords Heavy metals · Trace metals · Estuarine water · Estuarine sediments · Marine biota · Mangroves · Fishes · Human health risk · Toxicity · Anthropogenic activity

---

S. Swain
Department of Chemistry, University of Calcutta, Kolkata, West Bengal, India

D. K. Das
Calcutta National Medical College and Hospital, Kolkata, West Bengal, India

A. Seal
Indian Institute of Technology (Indian School of Mines), Dhanbad, Jharkhand, India

A. Chanda (✉) · S. Das
School of Oceanographic Studies, Jadavpur University, Kolkata, West Bengal, India

S. Das, T. Ghosh (eds.), *Estuarine Biogeochemical Dynamics of the East Coast of India*, https://doi.org/10.1007/978-3-030-68980-3_8

8.1 Introduction

Water pollution is a burning environmental issue in India. Natural phenomena and a suite of anthropogenic disturbances both deteriorate the surface water quality (Aktar et al. 2010; Lokhande et al. 2011; Pandey et al. 2009; Singh et al. 2020). Improper management of aquatic resources, absence of any government intervention, and human-generated waste products are some of the crucial factors that deteriorate the water quality (Mehta 2012). Almost one-fifth of the global population belongs to India; however, they have access to only 4% of the global functional water bodies. Water is a universal polar solvent and it can dissolve several substances. Thus, contamination of water takes place very easily (Tiwari and Ali 1988; Aktar et al. 2010; Singh et al. 2020). Untreated sewage is one of the chief concerns that leads to most of the aquatic pollution in India (CPCBE&F 2008). Agricultural runoff and indiscriminate discharges from small-scale industries also pose severe problems. Overall, the unregulated discharge from industrial sector, domestic sewage without any treatment, and solid wastes pollute the lentic as well as lotic water bodies of India (Water Quality Database of Indian Rivers, MoEF 2016).

Land degradation is the fundamental global process regulated by variations in the climate, water, soil, and several anthropogenic practices, which in turn pollutes the rhizosphere and the adjacent water columns (Riva et al. 2017; Salih et al. 2017; Wang et al. 2017; Marchi et al. 2018; Kumar et al. 2020). Inhabitants living near the estuaries transform the riparian zones, which translates into a greater threat to estuarine regions (Li et al. 2014). The degree of deterioration has always been higher in the estuarine reaches, as pollutants that find its way to the upstream riverine reaches eventually end up in the estuarine waterways (Bhattacharya et al. 2008; Hejabi et al. 2011; Singh et al. 2020). Once the pollutants enter the estuarine environment, most of the heavy metals tend to attach with the fine-grained particles and accrue in the sediment, which leads to unwanted effects on a suite of living forms (Bibi et al. 2007; Kumar et al. 2020).

The major cities and industrial hubs are polluting estuaries by releasing untreated waste into it from several sources such as leather manufacturing factories, tanneries, textile mills, distilleries, chemical plants, and slaughterhouses (Bhatnagar et al. 2013; Madhulekha 2016). The industries unquestionably contribute the most in polluting the estuaries due to widespread discharge of toxic metals and substances that originate as by-products from several manufacturing units (Madhulekha 2016; Hussain and Rao 2018). Heavy metals not only accumulate, but these tend to biomagnify to higher trophic levels, and this phenomenon poses severe threat to human beings (Paul 2017).

Heavy metals are natural components of the Earth's crust. These substances do not undergo degradation or destruction easily. Heavy metals are toxic, are bioaccumulative, and do not undergo biodegradation within the estuarine ecosystems (Li et al. 2014), which leads to undesirable consequences on several life-forms including human beings (Brady et al. 2014). Heavy metal pollution has long become a global issue. The distribution of these metals, their impact on living organisms, and their source of origin, at both local and global levels, have become worthy of investigation (Brady et al. 2014; Duodu et al. 2017; Kumar et al. 2019a; Kumar et al. 2020). Some of these metals, which usually exist in trace quantities in the environment such as arsenic (As), cadmium (Cd), chromium (Cr), copper (Cu), cobalt (Co), lead (Pb), mercury (Hg), and nickel (Ni), pose significant threats to several living organisms owing to their potential in bioaccumulation and biomagnification (Fang et al. 2016; Kumar et al. 2019b; Kumar et al. 2020).

High heavy metal concentration in the estuarine ecosystem is an indicator of anthropogenic activities, and it causes colossal damage to the ecosystems (Kumar et al. 2020). Consequently, it has become crucial to characterize the heavy metals in several compartments of an estuarine ecosystem. These heavy metals exist in the open environment in diverse forms like organic, exchangeable, bound, and residual species (Kumar et al. 2020). The continuous discharge of heavy metals into estuarine ecosystems alters the biogeochemical properties of the water (Banerjee et al. 2016). The top layer of the sediment profile, the water column, and the biota that thrives within an estuary are the prime reservoirs of heavy metals (Maiti and Chowdhury 2013). These heavy metals upon consumption through various foods may lead to the formation of gastrointestinal, reproductive, neurological, tumors, muscular, and hereditary disorders in human beings (Matos et al. 2017; Dwivedi et al. 2018; Genthe et al. 2018). Thus, in this chapter, we have collated and discussed the present state of the art of heavy metal contamination in the water column, sediments, and marine organisms adjoining the estuaries of the east coast of India.

8.2 Heavy Metal Contamination in Estuarine Water

A vast majority of the studies trying to characterize the heavy metal contamination focused on the sediments and marine biota. However, reports of quantifying these metal levels in the water column of the estuaries are scarce. Samanta and Dalai (2018) recently quantified the Co, Ni, and Cu levels in the estuarine water column of Hooghly (Table 8.1). They emphasized that the concentrations of all these three heavy metals are substantially higher in the lower and middle reaches of the Hooghly estuary. They reported that in the low-saline regions of the estuary, the contributions of groundwater and anthropogenic activities are much higher. However, closer to the river mouth, interaction with suspended solute particles plays a crucial role in the riverine heavy metal fluxes to the adjacent Bay of Bengal. In almost the same period, Mitra et al. (2018) reported the concentrations of a wide array of heavy metals in the surface water of the Hooghly River estuary, covering the freshwater, the brackish water, and the saline water of the Hooghly. Mitra et al. (2018) observed that Al, Fe, Ni, Pb, and Mn concentrations were beyond the limit prescribed by the World Health Organization (WHO) and, hence, can inflict substantial non-carcinogenic risk to human beings, whereas the toxic levels of other heavy metals like Cr, Cd, and As pose a cancer risk, especially to children compared to adults. In the adjoining estuaries within the Sundarban mangroves, Bhattacharya et al. (2015)

Table 8.1 A comparison of mean dissolved metal concentrations (in μg l^{-1}) in the estuarine water column along the east coast of India

Estuaries	Fe	Al	As	Mn	Zn	Cr	Cu	Co	Ni	Pb	Cd	Authors
Hooghly estuary	–	–	–	–	–	–	60	6.2	50	–	–	Samanta and Dalai (2018)
Hooghly freshwater	53,726	78,109	3.8	817	66	62	47	17	40	30	1.9	Mitra et al. (2018)
Hooghly brackish water	55,863	78,138	3.4	812	64	62	46	17	41	31	2.2	Mitra et al. (2018)
Hooghly saline water	55,901	74,764	2.1	811	59	58	45	17	42	29	1.8	Mitra et al. (2018)
Sundarban estuaries	–	–	–	223	57	39	34	21	36	15	5.5	Bhattacharya et al. (2015)
Devi estuary	98.2	–	–	7.2	15.4	2.5	5.6	3.8	9.4	5.6	0.8	Sundaray et al. (2012)
Mahanadi estuary	100.6	–	–	20.7	61.0	5.2	7.7	6.6	18.9	9.2	2.2	Sundaray et al. (2012)
Mahanadi and tributaries	746		2.8	–	37	29	17	–	19.8	4	0.2	Hussain et al. (2020)
Godavari and tributaries	1–240	–	0.04–9.31	–	0.2–94.23	0.13–13.67	0.04–114.84	–	20–75.25	0.01–7.41	0.001–1.59	Hussain et al. (2017)
Cauvery estuary	140	–	–	5.98	7.6	0.58	0.21	0.03	0.76	0.36	0.07	Raju et al. (2013)

observed substantially high heavy metal concentrations, especially in sites situated in the fringe or periphery of the mangrove forest, which received substantial quantities of industrial and metropolitan sewage load from the city of Kolkata. They further reported that dermal absorption of heavy metals like Pb, Cr, and Cd leads to a plausible risk of cancer, especially for those people who continually come into close contact with these waters, like fishermen and agricultural laborers. Sundaray et al. (2012) characterized the concentrations of a wide variety of heavy metals in the Mahanadi estuary and Devi estuary. Compared to Hooghly and Sundarban, these metal levels were much less in these two estuaries of Odisha. Sundaray et al. (2012) mainly attributed the effluents from the fertilizer plants; agricultural runoff, especially during the monsoon months; and the municipal sewage disposal from Sambalpur, Cuttack, and Paradeep port townships behind these pollutant repositories in the Mahanadi river estuary. Cd, Ni, and Pb exceeded the maximum permissible limits in some of the polluted stations in the Mahanadi estuary. Hussain et al. (2020) carried out a recent investigation in the Mahanadi estuary and the associated tributaries. They observed that the Fe, Cr, and Cu concentrations have substantially increased than those reported by the earlier studies in this region. Hussain et al. (2020) also attributed the wastewater discharge from the adjoining industries and municipal townships to the increased concentration of heavy metals in the Mahanadi estuary. Hussain et al. (2017) researched the Godavari estuary, where they characterized the As, Cd, Cr, Cu, Fe, Pb, Ni, and Zn in the Godavari along with the tributaries of Godavari River estuary. The degree of heavy metal contamination in this region was much less than that observed in the northern estuaries like Hooghly and Mahanadi. As, Cd, Cr, Fe, and Zn were within the acceptable limits proposed by the Bureau of Indian Standards (BIS). Cu and Ni were the principal polluters in the Godavari River system, with values beyond the threshold observed in many of the sampled stations. Among all the estuaries, the water column of the Krishna estuary, in this regard, was found to be the least polluted from heavy metals (Table 8.1). Raju et al. (2013) reported that Fe, Mn, Zn, Cr, Cu, Co, Ni, Pb, and Cd concentrations in the Krishna estuary were within the prescribed limits of both BIS and WHO. Thus, we can infer that the northern estuaries along the east coast of India are comparatively much more polluted than the southern estuaries.

8.3 Heavy Metal Contamination Estuarine Sediments

Sediments act as one of the most preferred repository sites for heavy metals. The east coast of India has many mangrove forests, and these sites act as sinks for a wide variety of heavy metals. Chowdhury et al. (2015) observed substantially high concentrations of As, Cd, Co, Cr, Cu, Fe, Hg, Mn, Ni, Pb, and Zn in the mangrove sediments of Sundarban (Table 8.2). The periphery of Sundarban mangroves, though encompass a large area of pristine mangroves, is densely populated with human beings and imposes substantial anthropogenic stress on the mangrove environment. Chowdhury and Maiti (2016) reported high Cd, Pb, and Cr contamination in the

Table 8.2 A comparison of mean metal concentrations (in μg g^{-1}) in the estuarine sediments along the east coast of India

Estuaries	Fe	Al	As	Mn	Zn	Cr	Cu	Co	Ni	Pb	Cd	Authors
Sundarban	2942	–	3.82	647.2	34.4	28.3	38.29	7.67	34.5	15.8	0.21	Chowdhury et al. (2015)
Sundarban periphery	–	–	–	170	90.8	33.8	–	–	41.5	37.3	1.7	Chowdhury and Maiti (2016)
Sundarban	41,412	–	–	1191	88.3	–	60.6	–	47.5	52.9	0.48	Chowdhury et al. (2017)
Mahanadi estuary	60,033			951	117	69	44	41	47	123	2.9	Sundaray et al. (2011)
Dhamra estuary	55,294	–	–	546	77.9	96.9	23.5	12.7	48.3	12.5	1.03	Panda et al. (2015)
Godavari estuary	–	–	–	–	197	--	25.8	21.8	25.1	14.4	–	Krupadam et al. (2007)
Cauvery estuary	4892	–	–	186	32	15	53	–	18	10	–	Dhanakumar et al. (2013)
Vellar-Coleroon estuary	5426	–	–	156	22.5	34.0	23.7	–	11.0	17.5	1.0	Sundaramanickam et al. (2016)

populous parts of the Sundarban Biosphere Reserve. Spanning over the entire Sundarbans, Chowdhury et al. (2017) also observed significant enrichment of Cu, Ni, Pb, and Hg in the rhizosediments of Sundarban, which are capable of posing a severe threat to the marine biota that thrives within this ecosystem. Sundaray et al. (2011) observed significant enrichment of Cd, Ni, Co, and Pb in the Mahanadi estuary. They inferred that a higher proportion of the sediment-trapped metals are in the bioavailable form, i.e., several biotic flora and fauna can readily uptake these metals and bioaccumulate them, posing a severe health risk to higher life-forms and human beings. Panda et al. (2015), while characterizing the sediment quality of the Dhamra estuary, observed that the sediments adjoining the Dhamra estuary were moderate to heavily polluted by Pb, Cr, Cu, Cd, and Ni. The enrichment factor for Cd was highest among all the other heavy metals in the Dhamra estuary. Sundaramanickam et al. (2016) also observed similar moderate Cd pollution in the sediments of Vellar-Coleroon estuary and the Pichavaram mangrove estuaries. Krupadam et al. (2007) observed significant Ni and Zn accumulation in the surface sediments of the Godavari estuary. However, their depth profile observations indicated significant depletion in metal content with depth. This observation showed that most heavy metal accumulation occurred in the recent past in the Godavari estuary. Further south, in the Cauvery estuarine sediments, Dhanakumar et al. (2013) observed low to high risk of Cu, Zn, and Pb accumulation, with an overall much less heavy metal concentration than that observed in the northern estuaries. Figure 8.1 shows the mean concentration of the most commonly studied heavy metals in the water column and sediments of the major estuaries along the east coast of India.

8.4 Heavy Metal Contamination Through Food Chain

Several marine flora and fauna along the east coast of India bioaccumulate substantial quantities of heavy metals. Mangroves, in this regard, are considered by many as an efficient living tool for phytoremediation of heavy metals (Chowdhury et al. 2015) (Table 8.3). While working in the Indian Sundarban mangroves, Chowdhury et al. (2015) observed that out of the many mangrove plant species, *Excoecaria agallocha* showed the highest magnitude of the bio-concentration factor for Cd. This observation led the authors to infer that this plant species can effectively remove Cd from sediment and water. However, analyzing nine dominant mangrove plant species from the Sundarbans, Chowdhury et al. (2017) inferred that essential micronutrients such as Mn, Fe, Zn, Cu, Co, and Ni were present in substantial quantities in all the plant organs compared to the non-essential ones, such as Cr, As, Pb, Cd, and Hg. Apart from mangroves, another crucial flora that grows in patches along the east coast of India, i.e., seagrasses (viz., the species *Cymodocea serrulata* and *Syringodium isoetifolium*), accumulates a substantial quantity of heavy metals like Fe, Mn, Zn, and Cu near the Palk Strait in the southern tip of the east coastline (Govindasamy et al. 2011). Besides the floral species, many marine faunas from this region exhibited heavy metal pollution. De et al. (2010) reported varying heavy

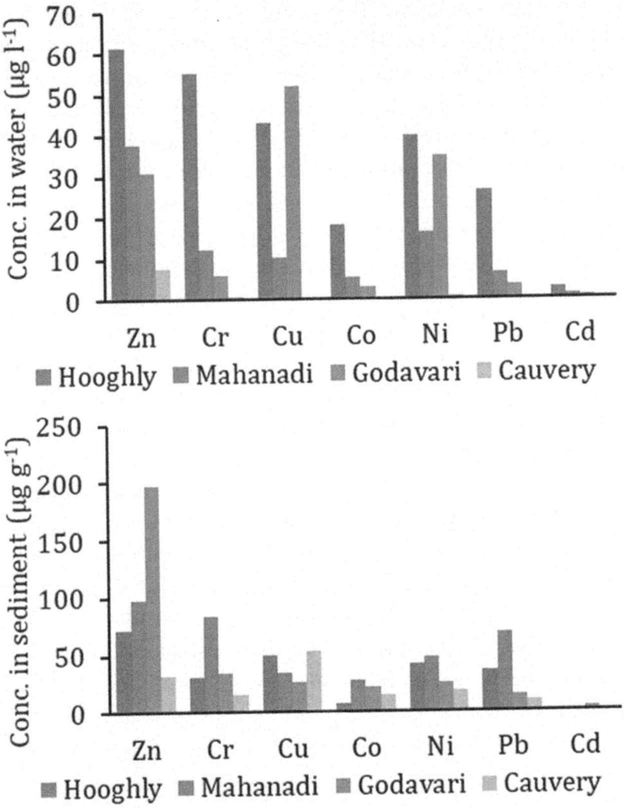

Fig. 8.1 Mean concentrations of Zn, Cr, Cu, Co, Ni, Pb, and Cd observed in the sediments and water column of Hooghly, Mahanadi, Godavari, and Cauvery estuaries

metal content in the muscle tissue of five important marine fishes, namely, *Pampus argenteus*, *Liza parsia*, *Glossogobius* sp., *Setipinna phasa*, and *Cynoglossus* sp. Among the studied heavy metals, the concentrations of Cu, Pb, and Zn were substantially higher than the rest of the heavy metals. Saha et al. (2006) reported significant heavy metal accumulation from the bivalves (*M. pinguis*) and Polychaeta (*M. indicus*, *G. sootai*, *P. cultrifera*, and *L. notocirrata*) that thrive in the Sundarban, which in general followed a hierarchy of Zn > Mn > Cu > Cr > Se > Hg. Giri and Singh (2014), while working in the Subarnarekha estuary, observed that heavy metals like As, Cr, and Cu were substantially high in shrimps, which are edible. These reports showed that shrimps from a few locations of this estuary are severely detrimental to human health upon regular consumption. Mohapatra et al. (2009) observed substantially high Cu and Zn levels in the muscle tissues of edible mud crab *Scylla serrata*, found along the Mahanadi estuary. However, the concentrations of toxic metals like Pb and Se were much less. On the contrary, Karar et al. (2019) observed significant bioaccumulation of Pb and Cd in the hepatopancreas of the Blue

Table 8.3 A list of mean metal concentrations (in μg g^{-1} dw) in few marine flora and fauna along the east coast of India

Estuaries	Type	Fe	As	Mn	Zn	Cr	Cu	Co	Ni	Pb	Cd	Authors
Mangrove young leaf	Plant	106.8	0.05	206.2	21.9	1.19	13.39	1.19	1.55	1.32	0.4	Chowdhury et al. (2015)
Mangrove mature leaf	Plant	91.7	0.1	357.7	21.2	1.15	12.44	0.65	1.38	1.65	0.37	
Mangrove tree bark	Plant	332.9	0.04	138.2	17.5	1.83	12.98	1.17	1.28	2.32	0.62	
Mangrove pneumatophores	Plant	406.5	0.06	195.0	15.5	2.17	16.3	1.02	1.59	2.78	0.18	
Mangrove plants (as a whole)	Plant	1376	2.8	3858	56.3	2.8	34.4	12.5	11.2	0.63	0.51	Chowdhury et al. (2017)
Pampus argenteus	Fish	–	–	–	23.4	BDL	16.2	–	3.41	12.4	0.99	De et al. (2010)
Liza parsia	Fish	–	–	–	30.0	1.0	32.0	–	2.25	18.6	1.2	
Glossogobius sp.	Fish	–	–	–	12.1	BDL	21.4	–	2.2	14.3	0.62	
Setipinna phasa	Fish	–	–	–	44.1	1.02	24.9	–	3.17	18.3	1.03	
Cynoglossus sp.	Fish	–	–	–	44.7	3.89	47.9	–	3.69	19.9	1.0	
P. indicus	Fish	143.33	0.48	–	52.66	3.75	44.09	0.57	13.72	0.20	0.32	Giri and Singh (2014)
M. gulio	Fish	92.02	0.08	–	57.39	0.59	3.66	0.27	4.31	0.22	0.04	
P. conchonius	Fish	129.83	0.16	–	67.95	1.08	4.36	0.74	6.45	0.14	0.03	
L. calbasu	Fish	66.88	0.19	–	71.46	0.53	3.95	0.04	8.20	0.11	0.01	
L. rohita	Fish	107.05	0.37	–	51.80	0.55	8.74	0.23	2.42	0.16	0.04	
L. bata	Fish	104.35	0.36	–	51.43	0.98	17.37	0.42	2.88	0.18	0.22	
M. indicus	Polychaete	–	–	49.12	51.09	31.72	15.66	–	–	–	–	Saha et al. (2006)
G. sootai	Polychaete	–	–	35.54	40.18	9.12	12.94	–	–	–	–	
P. cultrifera	Polychaete	–	–	41.17	37.60	10.53	9.89	–	–	–	–	
L. notocirrata	Polychaete	–	–	55.64	45.79	14.09	10.56	–	–	–	–	
M. pinguis	Bivalve	–	–	44.26	117.3	5.54	13.17	–	–	–	–	
Portunus pelagicus	Crab	–	–	–	–	0.83	21.06	–	–	1.67	1.01	Karar et al. (2019)
Scylla serrata	Crab	171.4	–	11.2	312.6	–	132.3	–	–	0.23	–	Mohapatra et al. (2009)

swimmer crab *Portunus pelagicus* in the northern Bay of Bengal off the Hooghly-Matla estuary to the offshore transition zone. Viswanathan et al. (2013) also observed similar high levels of Pb and Cd in the hepatopancreas of the Blue swimmer crab in the Ennore estuary down south along the east coast of India. Karar et al. (2019) further inferred that salinity played a crucial role in regulating the metal deposition rate in these crabs. Higher salinity disfavored heavy metal accumulation, whereas the same crabs under freshwater conditions (i.e., during monsoon season) accumulated substantial quantities of several trace metals. In the Pulicat Lake, near the Ennore estuary situated in the Andhra Pradesh, Batvari et al. (2008) observed significant levels of Pb and Fe accumulation in the gills of two fish species, namely, *Carangoides malabaricus* and *Belone stronglurus*. However, the metal levels in the edible muscle tissues were below the threshold. In the very same Pulicat Lake, Batvari et al. (2016) studied and reported crab and shrimp species like *Scylla serrata* and *Penaeus* sp. had substantially elevated levels of Pb. Similarly, in the Machilipatnam coast of Andhra Pradesh, Krishna et al. (2014) observed substantial accumulation of Pb, Ni, and Cd in the marine fish *Liza macrolepis*.

8.5 Human Health Hazards Due to Heavy Metal Exposure

Heavy metal contamination through the ecological food chain can significantly affect the human health via several pathways. Ingestion of heavy metals via daily dietary intake can manifest as both non-cancerous effects and cancerous effects in a human being. Trace metals and metalloids, such as arsenic (As), chromium (Cr), manganese (Mn), zinc (Zn), and copper (Cu) of geogenic and anthropogenic origin, are incessantly entering the aquatic environs and have become an issue of severe concern to human health and other biotic life-forms (Rahman et al. 2013, 2019) (Fig. 8.2). For instance, Cr is recognized for causing several adverse pulmonary health hazards, like fibrosis, lung inflammation, carcinoma, and emphysema (Forti et al. 2011), whereas high Cu intake can lead to serious health complications, for example, kidney and liver failure (WHO 1995; Rahman et al. 2019). Kumar and Riyazuddin (2010), in this regard, observed that tannery effluents from Chennai significantly polluted the groundwaters with Cr. Such Cr-enriched groundwater can meet the nearby rivers and pose a significant threat to other marine biotas as well.

Rajamohan et al. (2010) observed significant Cu concentrations from the effluent of a nuclear power plant on the Kalpakkam coast. Lead (Pb) causes pathological and neurotic changes in various body parts and the central nervous system, which lead to the delay in the development of mental acumen in children. Chakraborty et al. (2015) reported that the overall Indian estuarine sediments have substantial quantities of lead. Inorganic arsenic is a well-known class I carcinogen, which leads to several carcinomas in humans (Biswas et al. 2019). Excessive consumption of inorganic arsenic can manifest severe health hazards, i.e., cancer and non-cancer. Hyperpigmentation, skin cancer, keratosis, and vascular complications are some of the common diseases that arise due to the ingestion of As (USEPA 2017). However,

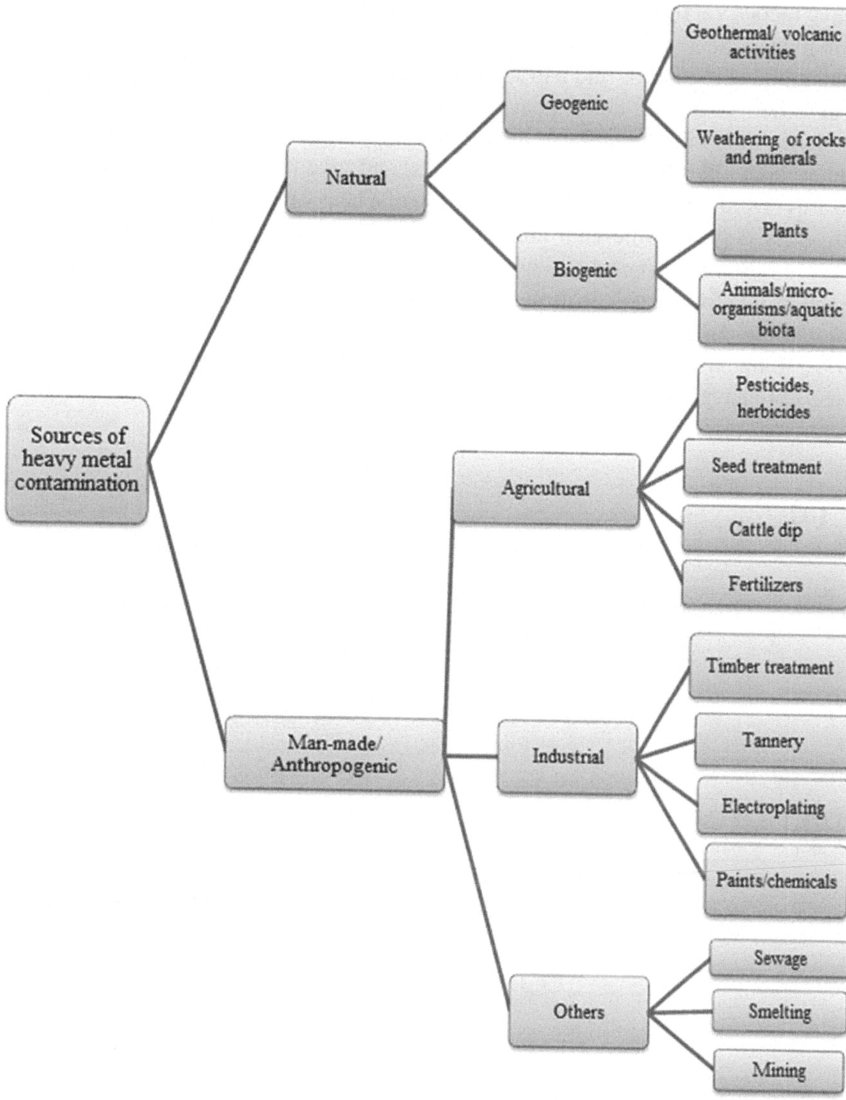

Fig. 8.2 Schematic diagram showing the channels and pathways through which heavy metals are introduced to the open environment (Sources: Authors)

arsenic pollution is mostly found to be due to geogenic reasons and is available in higher concentrations in the groundwater of West Bengal. Zn and its composites are responsible for the reduction in erythrocyte life-span. Again, Zn is less poisonous to fishes than As, Cr, and Cu (Dara 1998). Mn concentrations beyond a certain threshold can prove to be toxic, especially under acidic conditions (Rahman et al. 2019). A high intake of Mn can cause central nervous system disorders. Daily intake of heavy metals via food consumption is the principal path of metal exposure for most individuals. Several studies envisaged that the consumption of heavy metal-contaminated foodstuffs and fish species significantly affect human health (Rahman et al. 2019). As we all know, heavy metals are non-biodegradable and have a long residence time in nature. These metals find their way to several vital organs in the human body, like liver, kidneys, and bones, and give rise to several severe health hazards (Duruibe et al. 2007). The ingestion of a heavy metal-contaminated diet can extremely deplete some vital nutrients in the human body, which leads to severe damage in immunological defenses, impaired psychosocial behavior, and disabilities associated with malnutrition. Hence, the risk assessment of these heavy metals through the daily dietary intake is a significant issue in recent times (Islam et al. 2017; Rahman et al. 2019). Figure 8.3 illustrates the most common causes and effects of heavy metal toxicity.

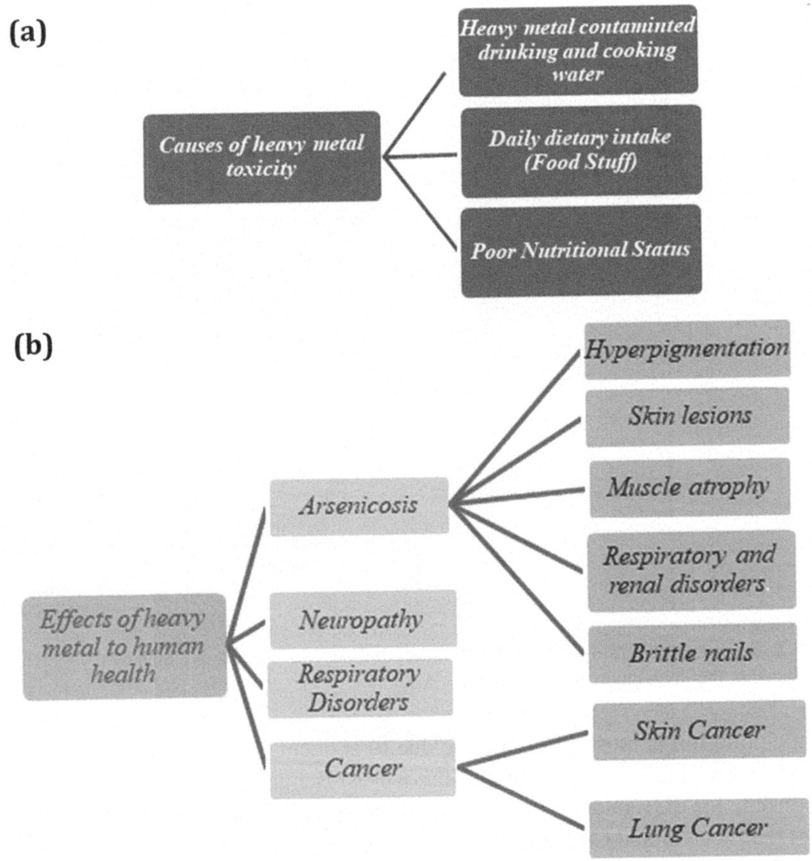

Fig. 8.3 Flowchart diagram showing the (**a**) most common causes and (**b**) effects of heavy metals to human health. (Sources: Authors)

8.5.1 Human Health Risk Assessment

We can estimate the degree of human health risk of heavy metals due to ingestion, inhalation, and dermal contact by several established protocols. We can compute the non-carcinogenic as well as the carcinogenic human health risk of all the heavy metals by estimating the average daily intake rate of heavy metals. USEPA (2015) has furnished the reference dose values for each heavy metal and the carcinogenic slope factor for all those heavy metals, which are proven carcinogens. Target hazard quotient (THQ) >1 for any metal implies considerable non-carcinogenic risk. Similarly, cancer risk (CR) $>10^{-4}$ for a given metal denotes a high probability of developing any form of cancer within the lifetime of a person. Kumar et al. (2020), while working on the sediment heavy metal concentration in many of the Indian estuaries, observed that Fe, Cu, Co, Mn, Ni, Pb, As, and Cr pose a significant non-carcinogenic threat. They also reported that both adults and children are most susceptible to developing cancer due to cadmium toxicity, followed by Cr and As. They also inferred that ingestion (mainly by contaminated food consumption) is the main pathway by which heavy metals enter the human body.

8.6 Management and Mitigation Options

According to Bhowmick et al. (2018), the population in heavy metal-affected regions of India is severely malnourished or undernourished, so they are highly vulnerable to heavy metal toxicity. Healthy and nutritious foods that are rich in antioxidants, vitamins, and proteins should be encouraged as an adjunct to diet. Malnourished villagers residing in heavy metal endemic regions should receive priority. Social awareness should be spread regarding the better supplementation of their daily dietary intake with locally grown and procured foodstuff. Every year, a large number of people perish due to cancers triggered by chronic heavy metal intoxication. The situation is perilous; therefore, our governments and the people have to come together and need to find appropriate solutions to combat this situation.

The government should promote research that can potentially discover affordable solutions on a massive scale. We have shown through Fig. 8.4 an example of initiatives taken by the West Bengal state government and the local population that is a surface water body free of arsenic and widely used for drinking. Tap water supplied by the municipality should meet the standards of drinking water for carcinogens and other noxious compounds. It should even make the reporting of arsenical diseases essential like polio and hepatitis so that we can keep track of the growing number of arsenic patients in the country. These steps would help us to control such incidences and reduce the severity of pollution scenarios with time.

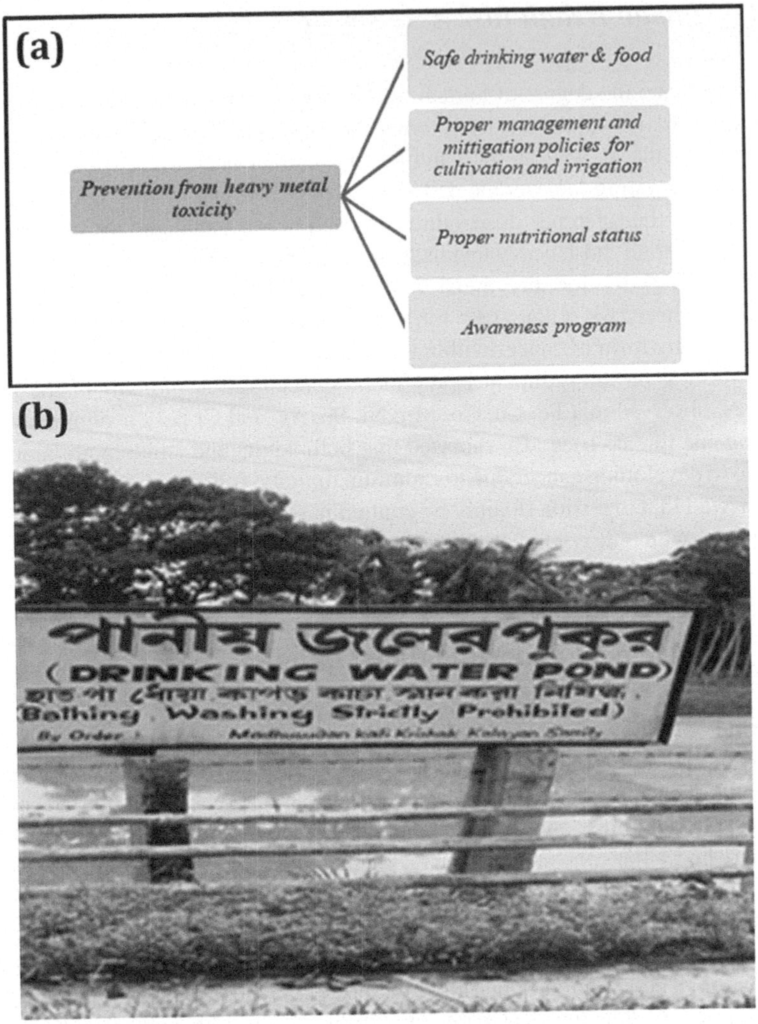

Fig. 8.4 (**a**) The key pathways to prevent heavy metal toxicity and (**b**) an initiative by the West Bengal Govt. to provide drinking water free of arsenic to rural people living near the coastal margin. (Sources: Authors)

8.7 Conclusion

Overall, we can infer that the estuaries along the east coast of India are substantially polluted by heavy metals. The northern estuaries with high freshwater discharge like Hooghly and Mahanadi are more polluted than the southern estuaries, as these estuaries receive substantial quantities of industrial and domestic effluents. Cd has the lowest concentrations, and still, it seems to be one of the most potent heavy metals that have exhibited substantial enrichment in the estuarine sediments and can pose a carcinogenic risk to human health. Besides this, other heavy metals like Pb, Cu, Cr, and Ni also cause potential hazards to several marine floras and faunas thriving in and around the estuaries. Arsenic, though not found in substantial quantity in many of the estuarine sediments, is found beyond the thresholds in many of the groundwater aquifers, especially in West Bengal, making this area one of the potent arsenic belts in the world. Since these pollutants can pose serious health hazards to human beings, mitigation of this problem deserves special attention at present.

References

Aktar MW, Paramasivam M, Ganguly M et al (2010) Assessment and occurrence of various heavy metals in surface water of Ganga River around Kolkata: a study for toxicity and ecological impact. Environ Monit Assess 160:207–213

Banerjee S, Kumar A, Maiti SK et al (2016) Seasonal variation in heavy metal contaminations in water and sediments of Jamshedpur stretch of Subarnarekha River, India. Environ Earth Sci 75(3):265

Batvari BPD, Kamala-Kannan S, Shanthi K et al (2008) Heavy metals in two fish species (*Carangoides malabaricus* and *Belone stronglurus*) from Pulicat Lake, north of Chennai, southeast coast of India. Environ Monit Assess 145:167–175

Batvari BPD, Sivakumar S, Shanthi K et al (2016) Heavy metals accumulation in crab and shrimps from Pulicat lake, North Chennai coastal region, southeast coast of India. Toxicol Ind Health 32:1–6

Bhatnagar MK, Raviraj S, Sanjay G, Prachi B (2013) Study of tannery effluents and its effects on sediments of river ganga in special reference to heavy metals at Jajmau, Kanpur, India. J Environ Res Development 8:56–59

Bhattacharya AK, Mandal SN, Das SK (2008) Heavy metals accumulation in water, sediment and tissues of different edible fishes in upper stretch of Gangetic West Bengal. Trends Appl Sci Res 3:61–68

Bhattacharya BD, Nayak DC, Sarkar SK et al (2015) Distribution of dissolved trace metals in coastal regions of Indian Sundarban mangrove wetland: a multivariate approach. J Clean Prod 96:233–243

Bhowmick S, Pramanik S, Singh P et al (2018) Arsenic in groundwater of West Bengal, India: a review of human health risks and assessment of possible intervention options. Sci Total Environ 612:148–169

Bibi MH, Ahmed F, Ishiga H (2007) Assessment of metal concentrations in lake sediments of Southwest Japan based on sediment quality guidelines. Environ Geol 52:625–639

Biswas A, Swain S, Chowdhury NR et al (2019) Arsenic contamination in Kolkata metropolitan city: perspective of transportation of agricultural products from arsenic-endemic areas. Environ Sci Pollut Res 26:22929–22944

Brady JP, Ayoko GA, Martens WN, Goonetilleke A (2014) Enrichment, distribution and sources of heavy metals in the sediments of Deception Bay, Queensland, Australia. Marine Pollution Bull 81(1):248–255

Chakraborty S, Chakraborty P, Nath BN (2015) Lead distribution in coastal and estuarine sediments around India. Mar Pollut Bull 97:36–46

Chowdhury A, Maiti SK (2016) Assessing the ecological health risk in a conserved mangrove ecosystem due to heavy metal pollution: a case study from Sundarbans Biosphere Reserve, India. Human Ecol Risk Assess Int J 22(7):1519–1541

Chowdhury R, Favas PJ, Jonathan MP et al (2017) Bioremoval of trace metals from rhizosediment by mangrove plants in Indian Sundarban Wetland. Mar Pollut Bull 124(2):1078–1088

Chowdhury R, Favas PJ, Pratas J et al (2015) Accumulation of trace metals by mangrove plants in Indian Sundarban wetland: prospects for phytoremediation. Int J Phytoremediation 17(9):885–894

CPCBE&F (2008) Central Pollution Control Board, Ministry of Environment & Forests Report. Government of India

Dara SS (1998) Environmental chemistry and pollution control. S Chand Publishing, India

De TK, De M, Das S et al (2010) Level of heavy metals in some edible marine fishes of mangrove dominated tropical estuarine areas of Hooghly River, North East Coast of Bay of Bengal, India. Bull Environ Contam Toxicol 85(4):385–390

Dhanakumar S, Murthy KR, Solaraj G, Mohanraj R (2013) Heavy-metal fractionation in surface sediments of the Cauvery River Estuarine Region, Southeastern Coast of India. Arch Environ Contam Toxicol 65:14–23

Duodu GO, Goonetilleke A, Ayoko GA (2017) Potential bioavailability assessment, source apportionment and ecological risk of heavy metals in the sediment of Brisbane River estuary, Australia. Mar Pollut Bull 117:523–531

Duruibe JO, Ogwuegbu MOC, Egwurugwu JN (2007) Heavy metal pollution and human biotoxic effects. Int J Phys Sci 2:112–118

Dwivedi S, Mishra S, Tripathi RD (2018) Ganga water pollution: a potential health threat to inhabitants of Ganga basin. Environ Int 117:327–338

Fang H, Huang L, Wang J et al (2016) Environmental assessment of heavy metal transport and transformation in the Hangzhou Bay, China. J Haz Mat 302:447–457

Forti E, Salovaara S, Cetin Y (2011) In vitro evaluation of the toxicity induced by nickel soluble and particulate forms in human airway epithelial cells. Toxicol Vitr 25:454–461

Genthe B, Kapwata T, Roux WL et al (2018) The reach of human health risks associated with metals/metalloids in water and vegetables along a contaminated river catchment: South Africa and Mozambique. Chemosphere 199:1–9

Giri S, Singh AK (2014) Assessment of human health risk for heavy metals in fish and shrimp collected from Subarnarekha River, India. Int J Environ Health Res 24(5):429–449

Govindasamy C, Arulpriya M, Ruban P et al (2011) Concentration of heavy metals in seagrasses tissue of the Palk Strait, Bay of Bengal. Int J Environ Sci 2:145–153

Hejabi AT, Basavarajappa HT, Karbassi AR, Monavari SM (2011) Heavy metal pollution in water and sediments in the Kabini river, Karnataka, India. Environ Monit Assess 182:1–13

Hussain J, Dubey A, Hussain I et al (2020) Surface water quality assessment with reference to trace metals in River Mahanadi and its tributaries, India. Appl Water Sci 10(8):1–12

Hussain J, Husain I, Arif M, Gupta N (2017) Studies on heavy metal contamination in Godavari river basin. Appl Water Sci 7(8):4539–4548

Hussain J, Rao NP (2018) Status of trace and toxic metals in Indian Rivers. Ministry of Water Resources, Government of India, River Data Compilation-2. Directorate Planning and Development Organization, New Delhi 110066

Islam GMR, Habib MR, Waid JL (2017) Heavy metal contamination of freshwater prawn (*Macrobrachium rosenbergii*) and prawn feed in Bangladesh: a market-based study to highlight probable health risks. Chemosphere 170:282–289

Karar S, Hazra S, Das S (2019) Assessment of the heavy metal accumulation in the Blue Swimmer Crab (Portunus pelagicus), northern Bay of Bengal: role of salinity. Mar Pollut Bull 143:101–108

Krishna PV, Jyothirmayi V, Rao KM (2014) Human health risk assessment of heavy metal accumulation through fish consumption, from Machilipatnam coast, Andhra Pradesh, India. Int Res J Public Environ Health 1:121–125

Krupadam RJ, Ahuja R, Wate SR (2007) Heavy metal binding fractions in the sediments of the Godavari estuary, East Coast of India. Environ Model Assess 12:145–155

Kumar AR, Riyazuddin P (2010) Chromium speciation in groundwater of a tannery polluted area of Chennai City, India. Environ Monit Assess 160:579. https://doi.org/10.1007/s10661-008-0720-9

Kumar V, Parihar RD, Sharma A et al (2019b) Global evaluation of heavy metal content in surface water bodies: a meta-analysis using heavy metal pollution indices and multivariate statistical analyses. Chemosphere 236. https://doi.org/10.1016/j.chemosphere.2019.124364
Kumar V, Sharma A, Kaur P et al (2019a) Pollution assessment of heavy metals in soils of India and ecological risk assessment: a state-of-the-art. Chemosphere 216:449–462
Kumar V, Sharma A, Pandita S et al (2020) A review of ecological risk assessment and associated health risks with heavy metals in sediment from India. Int J Sediment Res 35:516–526
Li Z, Ma Z, van der Kuijp TJ et al (2014) A review of soil heavy metal pollution from mines in China: pollution and health risk assessment. Sci Total Environ 468:843–853
Lokhande RS, Singare PU, Pimple DS (2011) Pollution in water of Kasardi river flowing along Taloja industrial area of Mumbai, India. World Environ 1:6–13
Madhulekha SA (2016) Evaluation of water quality in river Ganga due to contaminant of heavy metals, Kanpur, India. Int J Innov Trend Eng 20:97–100
Maiti SK, Chowdhury A (2013) Effects of anthropogenic pollution on mangrove biodiversity: a review. J Environ Prot 4:1428. https://doi.org/10.4236/jep.2013.412163
Marchi M, Ferrara C, Biasi R et al (2018) Agro-forest management and soil degradation in Mediterranean environments: towards a strategy for sustainable land use in vineyard and olive cropland. Sustainability 10(7):2565. https://doi.org/10.3390/su10072565
Matos LA, Cunha ACS, Sousa AA et al (2017) The influence of heavy metals on toxicogenetic damage in a Brazilian tropical river. Chemosphere 185:852–859
Mehta P (2012) Impending water crisis in India and comparing clean water standards among developing and developed nations. Arch Appl Sci Res 4(1):497–507
Mitra S, Sarkar SK, Raja P, Biswas JK, Murugan K (2018) Dissolved trace elements in Hooghly (Ganges) river estuary, India: risk assessment and implications for management. Mar Pollut Bull 133:402–414
Mohapatra A, Rautray TR, Patra AK et al (2009) Elemental composition in mud crab Scylla serrata from Mahanadi estuary, India: in situ irradiation analysis by external PIXE. Food Chem Toxicol 47:119–123
Panda SK, Sangita S, Kar RN, Panda CR (2015) Temporal and spatial variation of sediment associated biotic communities with heavy metal contamination gradient at Dhamra estuary, Odisha. Indian J Geol Mar Sci 44:1884–1893
Pandey J, Shubhashish K, Pandey R (2009) Metal contamination of Ganga River (India) as influenced by atmospheric deposition. Bull Environ Contam Toxicol 83:204–209
Paul D (2017) Research on heavy metal pollution of river Ganga: a review. Ann Agrar Sci 15(2):278–286
Rahman MM, Asaduzzaman M, Naidu R (2013) Consumption of arsenic and other elements from vegetables and drinking water from an arsenic-contaminated area of Bangladesh. J Hazard Mater 262:1056–1063
Rahman MS, Hossain MS, Ahmed MK et al (2019) Assessment of heavy metals contamination in selected tropical marine fish species in Bangladesh and their impact on human health. Environ Nanotechnol Monit Manag 11:100210. https://doi.org/10.1016/j.enmm.2019.100210
Rajamohan R, Rao TS, Anupkumar B et al (2010) Distribution of heavy metals in the vicinity of a nuclear power plant, east coast of India: with emphasis on copper concentration and primary productivity. Indian J Mar Sci 39:182–191
Raju KV, Somashekar RK, Prakash KL (2013) Spatio-temporal variation of heavy metals in Cauvery River basin. Proc Int Acad Ecol Environ Sci 3(1):59–75
Riva MJ, Daliakopoulos IN, Eckert S et al (2017) Assessment of land degradation in Mediterranean forests and grazing lands using a landscape unit approach and the normalized difference vegetation index. Appl Geogr 86:8–21
Saha M, Sarkar SK, Bhattacharya B (2006) Interspecific variation in heavy metal body concentrations in biota of Sunderban mangrove wetland, Northeast India. Environ Int 32(2):203–207

Salih AM, Ganawa ET, Elmahl AA (2017) Spectral mixture analysis (SMA) and change vector analysis (CVA) methods for monitoring and mapping land degradation/desertification in arid and semiarid areas (Sudan), using Landsat imagery. Egypt J Remote Sens Space Sci 20:S21–S29

Samanta S, Dalai TK (2018) Massive production of heavy metals in the Ganga (Hooghly) River estuary, India: global importance of solute-particle interaction and enhanced metal fluxes to the oceans. Geochim Cosmochim Acta 228:243–258

Singh V, Ngpoore NK, Chand J, Lehri A (2020) Monitoring and assessment of pollution load in surface water of river ganga around Kanpur, India: a study for suitability of this water for different uses. Environ Technol Innov 100676. https://doi.org/10.1016/j.eti.2020.100676

Sundaramanickam A, Shanmugam N, Cholan S et al (2016) Spatial variability of heavy metals in estuarine, mangrove and coastal ecosystems along Parangipettai, southeast coast of India. Environ Pollut 218:186–195

Sundaray SK, Nayak BB, Kanungo TK, Bhatta D (2012) Dynamics and quantification of dissolved heavy metals in the Mahanadi river estuarine system, India. Environ Monit Assess 184(2):1157–1179

Sundaray SK, Nayak BB, Lin S, Bhatta D (2011) Geochemical speciation and risk assessment of heavy metals in the river estuarine sediments—a case study: Mahanadi basin, India. J Haz Mat 186:1837–1846

Tiwari TN, Ali M (1988) Water quality index for Indian rivers. In: Trivedy RK (ed) Ecology and pollution of Indian rivers, 1st edn. Ashish, New Delhi, pp 271–286

USEPA (2015) U. S. Environmental Protection Agency. Risk based screening table generic, summary table. United States Environmental Protection Agency. Retrieved from http://www.epa.gov/risk/riskbased-screening-table-generictables. Accessed 10 Nov 2020

USEPA (2017). https://cfpub.epa.gov/ncea/iris/search. Accessed on 18 Jan 2017

Viswanathan C, Azhaguraj R, Selvanayagam M, Raffi SM (2013) Heavy metal levels in different tissues of the Blue Swimming crab (*Portunus pelagicus*, Portunidae) collected from Ennore Estuary. Int J Res Fish Aquac 3:1–6

Wang X, Cheng H, Li H et al (2017) Key driving forces of desertification in the Mu Us Desert, China. Sci Report 7(1):3933. https://doi.org/10.1038/s41598-017-04363-8

Water Quality Database of Indian rivers, MoEF. Retrieved 15 September 2016

WHO (1995) Lead. Geneva: Environmental Health Criteria. WHO, Geneva, Switzerland

Chapter 9
Geostatistical Analysis of Suspended Particulate Matter Along the North-Western Coastal Waters of Bay of Bengal

Atreya Basu, Sayan Mukhopadhaya, Kaushik Gupta, Debasish Mitra, Shovan Lal Chattoraj, and Anirban Mukhopadhyay

Abstract A geostatistical approach was used to envisage the spatial variability of the suspended particulate matter (SPM) concentration in the coastal waters of the Bay of Bengal. The study was conducted in-between Subarnarekha and Rasulpur estuary. Semivariogram analysis followed by ordinary kriging unfolded the environmental factors influencing spatial variation of the SPM concentration. An alongshore increasing trend of SPM concentration was observed toward the eastern sector of the study area (67–168 mg/l), with maximum concentration observed in the Junput–Rasulpur belt. While the western sector, from the Subarnarekha estuary mouth to Tajpur, showed relatively low SPM concentration (≤16–46 mg/l). The high-energy regime of the Hugli estuarine flow was found to be the dominant factor behind the high SPM concentration in the eastern sector. The fluvial activity of the tidal inlets and subsurface topography resulted in the intermittent occurrence of high SPM concentration in the alongshore direction. This was explained by the observed cyclicity and nugget effect in the semivariogram. The in situ analysis revealed a dominancy and uniform distribution of muddy suspended sediments in the eastern sector compared to the sandy nature of suspended sediments in the western sector. The eastern zone was marked by deposition and pollution-prone regime, whereas the western zone was found to be in a predominant erosional regime. The geostatistical approach provided an advantage over the classical statistical method, as it included the aspect of spatial distribution, with the production of near accurate cartography of SPM concentration and its associated uncertainty.

A. Basu
Centre for Earth Observation Science (CEOS), University of Manitoba, Winnipeg, Canada

Indian Institute of Remote Sensing, Dehradun, Uttarakhand, India

S. Mukhopadhaya
BASF Digital Farming GmbH, Köln, Germany

K. Gupta · A. Mukhopadhyay (✉)
Centre for Earth Observation Science (CEOS), University of Manitoba, Winnipeg, Canada

D. Mitra · S. L. Chattoraj
Indian Institute of Remote Sensing, Dehradun, Uttarakhand, India

S. Das, T. Ghosh (eds.), *Estuarine Biogeochemical Dynamics of the East Coast of India*, https://doi.org/10.1007/978-3-030-68980-3_9

Keywords Suspended Particulate Matter · Geostatistics · Univariate · Subarnarekha · Bay of Bengal · Semivariogram · Kriging · Coastal Water

9.1 Introduction

On a global scale, coastal waters are characterized by the presence of suspended particulate matter (SPM), which serves as a carrier of resuspended pollutant and nutrient in the coastal environment (Healy et al. 2002; Turner and Millward 2002; Webster and Lemckert 2002; Winterwerp and Van Kesteren 2004). Any type of coastal ecological study, therefore, requires stringent monitoring of SPM variability in terms of both space and time. For not only the ecological study but also coastal engineering and management, SPM variability monitoring is of utmost importance (De Kok 1992).

The inference of spatial distribution maps, of SPM, is the most common and convenient way of coastal environmental monitoring. The regional physiography inadvertently influences the cartography. Variation in regional geography and sea-bed geology of a dynamic coastal zone add complexities to methodologies involved in the cartographical procedure. These cartographical complexities are well addressed when parameterization of the dynamic variables is considered. The use of geographic information system (GIS) and its associated databases have proved to serve better in the case of comparative cartographical study. Nevertheless, the problem persists in correlating these GIS-processed cartographical products (maps). This might be due to several drawbacks associated with methodologies involved to create the maps, or due to data procurement error in the complex coastal waters (Méar et al. 2006).

Geostatistical analysis deals with spatially autocorrelated data, thereby characterizing the regular component of variation of natural objects (regionalized variable) (Kuzyakova et al. 2001; Bohling 2005). Quantification of the spatial autocorrelation at a required scale of interest also helps in estimation and subsequent minimization of the error. This is the basic goal of a geostatistical study: enabling an unbiased and minimal error estimation. Regionalized variables blend in the effect of surrounding environmental phenomena. Therefore, modelling the regionalized variables for a proper spatial representation would lead to the extraction of essential environmental information. Geostatistics helps to geo-visualize the environmental scenario based on the spatial variation of the regionalized variables. The effect of distances in between and among the regionalized variable plays a major role in geostatistics. The parameterization is solely user-dependent and is not limited by classical statistical laws and physical considerations (Desbarats 1996; Bellehumeur et al. 2000), thus adding dynamic character to the process (Méar et al. 2006). It includes tools that produce maps by analyzing spatial data structure through the characterization of spatial autocorrelation (Davis and Sampson 1986; Bohling 2005). Variogram analysis has been the choice of geo-statisticians for quantifying spatial variability of environmental phenomenon for many years (Gringarten and Deutsch 2001). Regionalized variables can be quantitatively mapped using a variogram as a geostatistical tool (Matheron 1963). The nature of spatial dispersion normally is revealed through a variogram analysis (Anckar et al. 1998; Western et al. 1998; Aubry 2000; Ouyang et al. 2002; Burak et al. 2010; Legendre and Legendre 2012; Wang et al. 2014). By implementation of the variogram model, optimal interpolation (e.g., kriging) along with stochastic simulation (generation of multi-equiprobable images of the regionalized variables) can be performed to generate the best linear unbiased estimate at each location (Van Groenigen et al. 1999; Bastante et al. 2005; Bohling 2005; Oliver and Webster 2014).

Continuous progress in the field of automatic calculation has caused a surge in the use of geostatistics for spatial analysis of environmental data (Chang et al. 1998; Goovaerts 1999; Atkinson and Lewis 2000; Batista et al. 2001; Saito and Goovaerts 2001; Caeiro et al. 2003; Leecaster 2003; Li and Lu 2010; Vannametee et al. 2014; Malone et al. 2016). Likewise, in the field of marine sedimentology and aquatic ecology, geostatistics is a powerful tool to represent the regionalized phenomenon (e.g., coastal hydrodynamics, erosion, accretion, etc.) in the form of a spatial variability map. In this study, the suspended sediments in the nearshore waters of the Bay of Bengal was given the focal attention. Suspended and resuspended sediments can integrate contaminants, which could provide valuable information on the source, throughput, and sink of pollutants, whose monitoring and assessment are essential from an ecological point of view. Quantification of the fine fraction of sediment is a necessity as it has the property to get itself associated with the contaminants (Forstner et al. 1982; Birch and Taylor 2000). In this study, the SPM concentration was interpolated and the error arising from this interpolation was estimated for a coastal environmental assessment. This present study, centered on, SPM concentration variability monitoring to explain the regionalized coastal phenomenon affecting the environmental state of the Bay of Bengal coastal waters.

9.2 Materials and Methods

9.2.1 Study Area

The Bay of Bengal (BOB), a semi-enclosed tropical basin, lies toward the northeastern part of the Indian Ocean. "The initial Paleocene–Eocene collision of India with the subduction zone of the northern side of Tethys Ocean created the BOB basin" (Curray and Moore 1974; Curray et al. 2002; Lee and Lawyer 1995; Alam et al. 2003). Its seabed topography is mainly dominated by fine-grained to coarse-grained sediments of the Bengal Fan. The continental slope of eastern India borders the western margin of the Bengal Fan, while its eastern margin lies in the northern end of the Sunda trench and the accretionary prism of the Sunda subduction zone. This accretionary prism holds much of the Bengal Fan sediments. The unique geographical location of the bay, mainly its proximity to the equator and voluminous freshwater influx into the bay system makes the circulation pattern complicated. BOB experiences a seasonal reversal of the monsoon (Gopalan et al. 2000).

Over the Pleistocene period, the quantitative amount of sandy bed load, carried by the River Ganges, had filled the Bengal Delta. Subsequently, those sediments were distributed in the entire BOB to form the largest submarine canyon of the world (Coleman 1969; Thorne et al. 1993). The nearshore shelf waters located west of the Bengal Delta were chosen as our study area (Fig. 9.1). This region is under the influence of two major estuarine systems: the Hugli (a distributary of the River Ganges) to the east and the Subarnarekha to the west. River Hugli carries a huge amount of sediment load through its course down the hills of the Himalayas into the plains of West Bengal. Similarly, the River Subarnarekha, having its origin in the Indian state of Jharkhand passes through the mineral-rich belt of the Singbhum shear zone, also carries a substantial amount of sediment load. Both the rivers converge its flow in the BOB shelf waters, thereby adding to the sediment concentration of the shelf bed. From several studies, it has been observed that there has been a recent trend of increasing heavy metal pollution in the waters of the estuaries and open coast of the Indian state of West Bengal (Guhathakurta and Kaviraj 2000; Guzzella et al. 2005; Sarkar et al. 2007; Mitra et al. 2010). This increasing trend might be attributed to the heavy sediment load carried by the two rivers. River Hugli passes through the most urbanized and industrialized region of West Bengal, thereby adding heavy metal pollutants to the shelf waters. Rasulpur estuarine flow in BOB is influenced by the Hugli estuarine flow as it lies adjacent to the Hugli estuary mouth on the open coast. The Subarnarekha River has its course through the copper and uranium mining region, which also might contribute to the pollution of the coastal waters. Local hydrodynamic processes and wave action enhances the distribution of these pollutants aided by the suspended/resuspended sediments of the coastal waters. So, monitoring of suspended/resuspended sediment concentration will indirectly reflect the potential pollutant dispersal pattern enabling proper coastal ecological management.

9.2.2 Data

One hundred (100) surface water samples were collected from the coastal (nearshore) waters following a random stratified sampling technique. Alongshore and toward offshore sampling was performed to represent the best possible SPM spatial variability within the sample set (Fig. 9.2). Ebb tidal period was selected as the time domain for sample collection as it is convenient to easily deduce the estuarine outflow effects into coastal waters, which might become quite complicated during floods tidal period, where an inflow of seawater into the estuaries take place. Individual sample collection locations were geocoded using Garmin GPS eTrex 10 having a spatial accuracy of 3 m. The collected samples, in each of 500 ml, seawater rinsed, open screw cap bottle was subjected to SPM concentration retrieval by the weighing method (Zhang et al. 2010). Vacuum filtration system with Whatman™

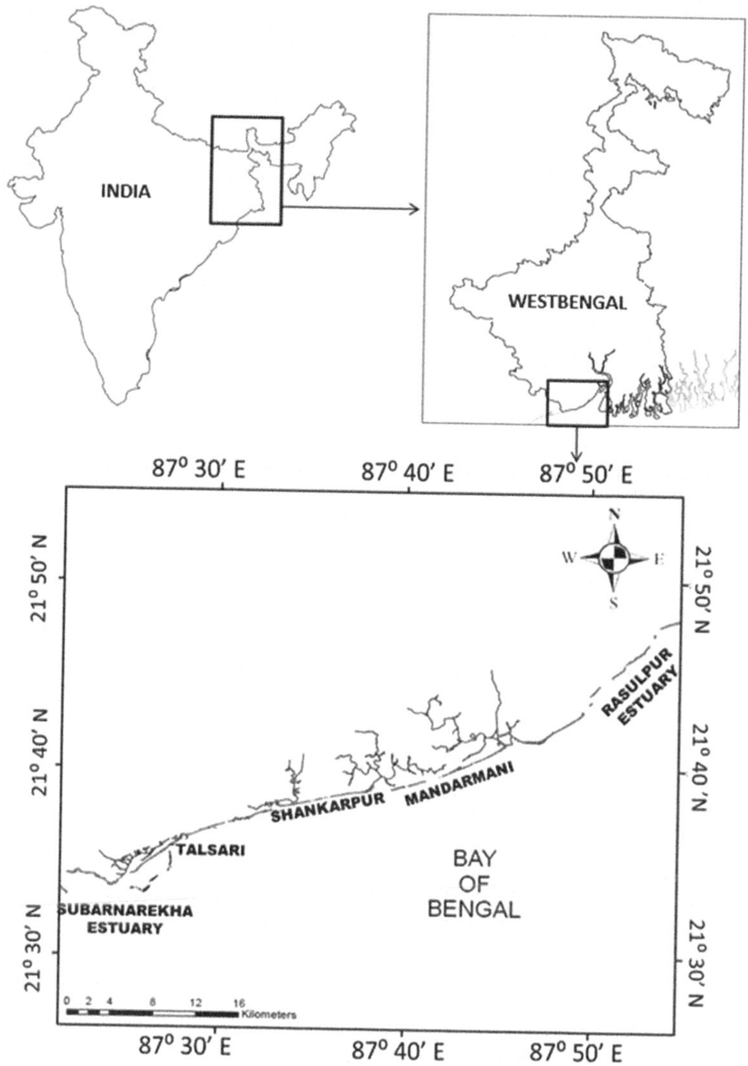

Fig. 9.1 Map showing the study area along the coast of West Bengal, India

934-AH™ Glass Microfiber Filter (1.5 μm pore size) was used to retain the fine fraction of silt and clay as the residue after filtration. This residue was then washed with 100 ml of distilled water over three times to remove any fraction of salt present within it, which could have induced error in the result of the geostatistical analysis, as dissolved salts do not account for major pollutant dispersal in the coastal waters. An analytical accuracy of ±0.04 mg/500 ml was ascertained while the accuracy of ±0.03 mg was observed due to the error arising from the weighing process. The retrieved and geocoded SPM concentration data were treated as the regionalized variable for this study. Environmental parameters influencing the spatial

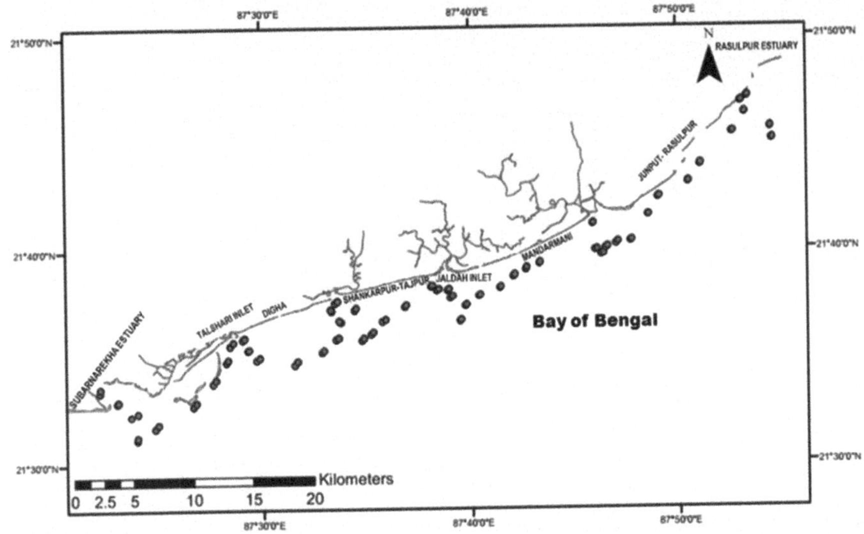

Fig. 9.2 The figure shows 100 nearshore water sampling points in the study area. A near to "z" pattern of water sample collection was followed to represent the variability in SPM concentration

distribution of the SPM are encompassed within the regionalized phenomenon (Aubry 2000; Ouyang et al. 2002; Méar et al. 2006).

9.2.3 *Geostatistical Approach*

Geostatistical analysis performs well, producing a good cartographic estimation if the frequency distribution data of the regionalized variable assumes a Gaussian nature (Webster and Oliver 1992; Ouyang et al. 2002; Schnabel et al. 2004). But in nature non-normality of the data sets are common. Various factors might be responsible for the induction of skewness into the SPM data set. Reasons like analytical error, sampling scale error (duration of sampling is longer than the operating time scale of environmental mechanism, e.g., accretion, erosion, etc.), erroneous sample collection, a human error might be responsible for the induction of skewness in the dataset. The original data was found to be skewed (Figs. 9.3 and 9.4). So, for a proper geostatistical analysis, the Gaussian transformation of the data set was performed to remove the non-normality based on the Shapiro–Wilcoxon test result, using the IBM SPSS statistics trial version software. The non-normality also induces a trend in the data set if it has an organized structure, which makes the regionalized variable non-stationary (Méar et al. 2006). So a trend analysis was carried out to remove the trend, if present, to obtain an unbiased data set. The unbiased data set was then subjected to an anisotropic analysis, which reflected the variance of SPM concerning distance and direction. This analysis was performed to deduce a

Fig. 9.3 Histogram of the positively skewed SPM concentration of the original field dataset

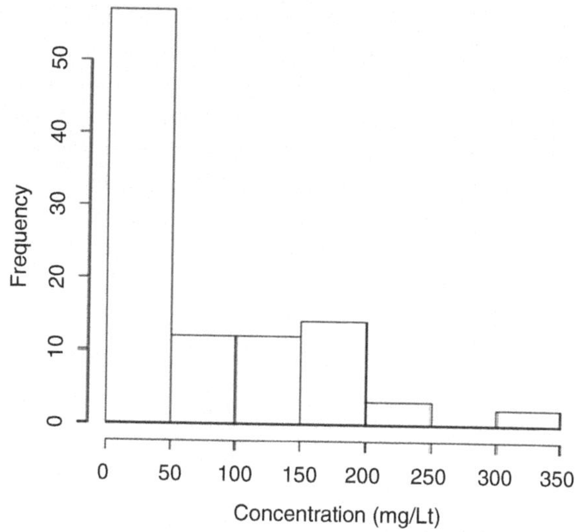

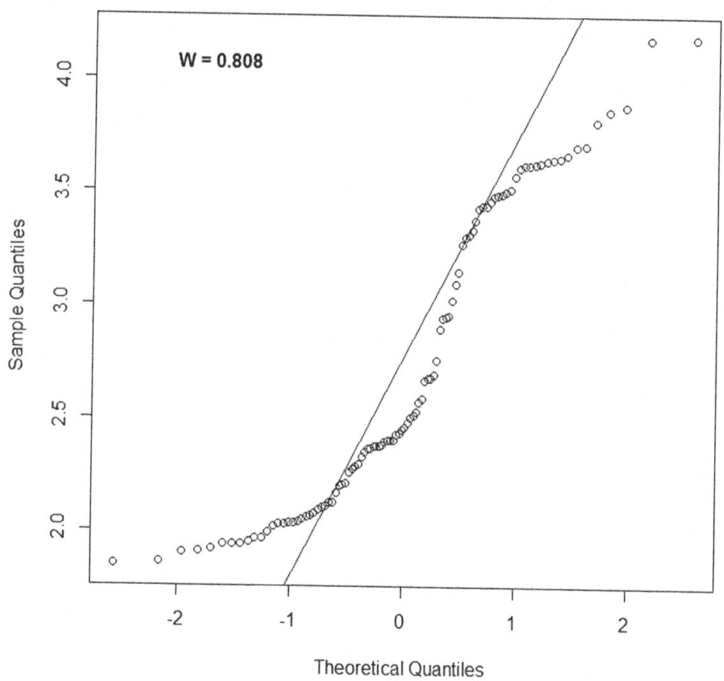

Fig. 9.4 Q–Q plot of the SPM concentration of the original field dataset. The plot depicts non-normality with a Shapiro–Wilk test statistics value (W) of 0.808

comprehensive understanding of the complex mechanism (sediment resuspension, settling of SPM, erosion, local hydrodynamics, anthropogenic influence, etc.) and their mutual interaction influencing the spatial variability of SPM concentration.

A semivariographic study was conducted to reveal the random and structured aspect of spatial distribution (Méar et al. 2006) for each of the dataset. This included the original transformed data of 100 sampled locations, as well as for sub datasets. Experimental semivariogram was first established followed by the choice of the method of interpolation. Isaaks and Srivastava (2001) defined the experimental semivariogram as follows:

$$\gamma(h) = \frac{1}{2} E\left\{\left[Z(x+h) - Z(x)\right]^2\right\} \tag{9.1}$$

$$\gamma(h) = \frac{1}{2N(h)} \sum_{Nh}^{h=1} \left[Z(x+h) - Z(x)\right]^2 \tag{9.2}$$

The major consideration adopted for this study was the conservation of the spatial variation of the fields of measurement. This was achieved by using ordinary kriging as the method of interpolation (Desbarats 1996; Goovaerts 1997, 1999; Vanderborght et al. 1997; Bellehumeur et al. 2000). This ensured the non-underestimation of the error variances, a key aspect of this study. Ordinary kriging utilized the values of the normalized data to reproduce the statistics such as the histograms and semivariograms for geo-visualization purposes (Asselman 1999). A final map was produced by averaging the 100 simulations generated, covering the entire study area, with each simulation producing an individual map. R-Software (http://www.r--project.org) was used for the computation of the classical statistics (mean, median, mode, etc.), normality, and geostatistical analysis.

9.3 Results and Discussion

9.3.1 Field Data Analysis

The sampled SPM concentration was in the range of 11.80 mg/l to 306.40 mg/l, thereby representing a variability of the SPM concentration within our original dataset. Thus, enabling a proper geostatistical analysis. A lower concentration of the SPM was observed toward the western sector of the study area, with a gradually increasing concentration toward the eastern sector (Fig. 9.5). The maximum concentration was observed in the Junput and Rasulpur belt in the east, while the minimum concentration was observed near the Subarnarekha estuary mouth in the west. This eastwardly increasing trend might be attributed to the intensive sediment load carried by the largest estuarine system of India, the Hugli, which lies toward the eastern sector of the study area, near to the Rasulpur estuary mouth. A considerable high amount of SPM concentration was observed near the estuary mouths, viz.

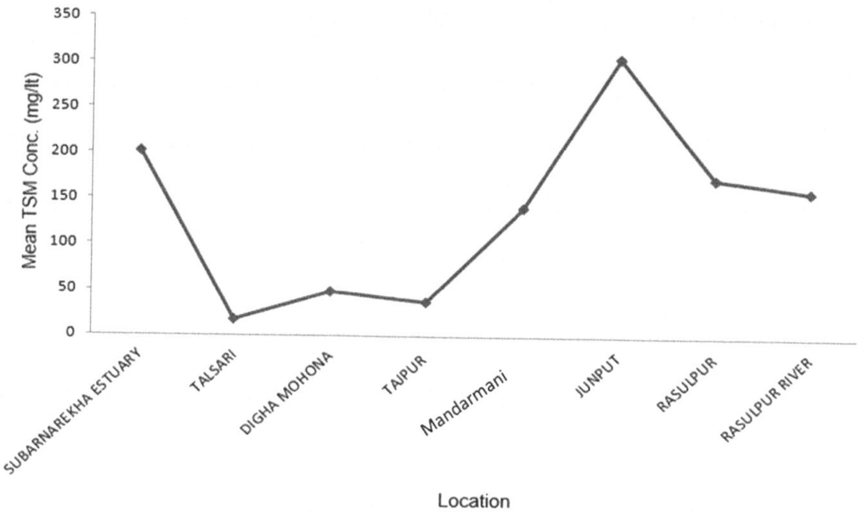

Fig. 9.5 Alongshore variation in the field SPM concentration

Subarnarekha estuary, Digha inlet, Pichhabani inlet, and Rasulpur estuary, indicating a high amount of fluvial activity. Hugli estuarine flow dynamics dominated till the Mandarmani belt, whereas the Subarnarekha estuarine flow dynamics had a negligible impact towards its east. This aspect can be validated by the low SPM concentration in the Talshari–Tajpur nearshore waters, compared to the very high SPM concentration in the rest of the coastal belt of the study area. A rise of SPM concentration was observed from the Junput to Rasulpur belt, while a fall in the SPM concentration was observed from the Subarnarekha estuary mouth to the Talshari inlet (Fig. 9.5). The reason might be attributed to the subsurface topography and the current flow-wave energy regimes operating in these regions. In a general overview, the fraction of clay, silt, and fine sand constituted the SPM in the coastal waters of the study area. Fine sand–sized sediments in suspension were found to be dominant in most of the coastal waters of the study area. This confirms the single provenance of these sediments. Sandy sediment load carried by the Hugli estuarine system, which was deposited on the continental shelf over the Pleistocene period, is the main source of the suspended sediments (Coleman 1969; Thorne et al. 1993). As per the Hjulström curve theory, it was evident that higher flow energy was required for the suspension of larger grain–sized sediments. The rise in SPM concentration toward the Rasulpur belt indicated a high flow energy regime in this part of the study area. SPM concentration remained considerably low in the nearshore waters of the Subarnarekha estuary mouth to Tajpur. This region is said to be impacted upon by the dual effects of tidal currents and wave energy. But the tidal current speed in the western sector is expected to be extremely low as compared to the eastern sector, due to the voluminous and high flow rate of the Hugli River. The Hugli estuary has a larger cross-sectional area than the Subarnarekha estuary, thereby, allowing greater volumes of water to flow. So, the Subarnarekha estuarine flow had a negligible impact towards its east, which resulted in a decreased suspension of sediments in the Talshari–Tajpur belt. Field observations revealed a significant absence of waves beyond the Mandarmani belt toward Rasulpur. This sector of coastal waters is dominated by the strong tidal flow of the Hugli estuary, thereby comprising a high-energy environment, which led to the increased suspension of the fine sediments. Mostly clay- and silt-dominated suspended sediments were observed in the Rasulpur region, indicating deposition of the sandy sediments. Therefore, the eastern sector of the study area signified a depositional environment, whereas the western sector was supposed to be in an erosional regime, due to the dual impact of waves and tidal currents. SPM concentration decreased toward offshore, which goes with the general scientific agreement that toward offshore, as depth increases, lower suspensions of sediments are found in a normal weather condition. From the analysis of the field data, it was obvious that the regionalized variable assumed for this study, that is, SPM concentration encompassed the regional environmental phenomenon. So this regionalized variable was considered to be suitable for its application in the geostatistical analysis for estimation and uncertainty assessment, finally drawing inference on the impact of potential environmental promoters aiding in polluting the nearshore waters of the study area.

9.3.2 *Data Treatment*

The original dataset of the retrieved SPM concentration was found to be positively skewed (skewness = 1.187) with a mean and median value of 75.41 mg/l and 36.30 mg/l, respectively. The normality test (Shapiro–Wilk test) produced a test statistics value (W), which was less than one ($W = 0.808$), within a 95% confidence interval (Fig. 9.4). This value led to the rejection of the null hypothesis, that is, the original dataset was not a Gaussian distribution and implied a necessary transformation of the dataset. Double logarithmic and double square root transformation was applied on the original dataset. It was observed that though the double logarithmic transformation led to near normal distribution (skewness = 0.05; $W = 0.93$), trimodal peaks were observed. There might be an occurrence of improper geostatistical analysis if the transformed dataset was divided into three parts due to a smaller number of data points. A minimum of 30 data points is an essential criterion to perform any geostatistical analysis. So, this transformation approach was rejected (Journel and Huijbregts 1978; Pannatier 2012). Normalization involving double square root transformation produced a result close to normality (skewness = 0.50; $W = 0.90$), with a bimodal frequency distribution curve (Fig. 9.6). Therefore, this transformed variable (SPM concentration) was used for geostatistical analysis. For further validation and cross-validation purpose, the transformed SPM concentration data was divided into two parts, viz. first 68 data points, which represented the western sector of the study area and the rest 32 data points, which represented the eastern sector of the study area. This division of the data points was based on the range of

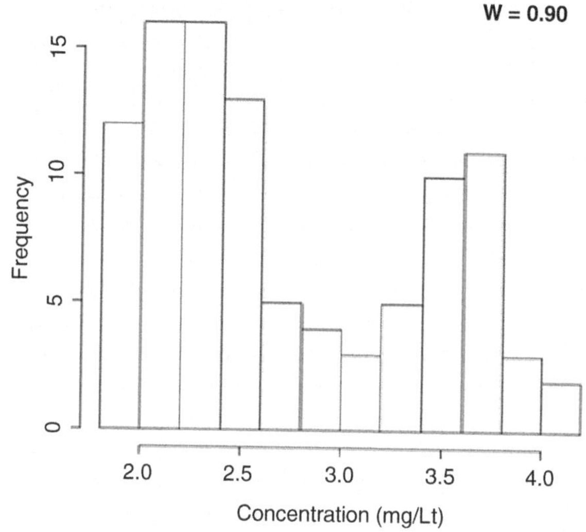

Fig. 9.6 Histogram of the square root–square root data transformation of the original SPM concentration dataset. The histogram depicts a bimodal distribution with a normality value (Shapiro–Wilk test statistics value) of 0.90

the bimodal frequency distribution histogram (Fig. 9.6). For each of these divided datasets normality test was carried out to assess the skewness of the dataset. The dataset has 68 data points that showed a skewness of 1.03 and a test statistics value (*W*) of 0.85, while the dataset of 32 data points showed a negative skewness of −0.371 and a *W* value of 0.901. The *W* value for both the divided datasets indicated a near-normal distribution, with the dataset having 32 data points showed a nearer normal distribution than the other. The more uniform SPM concentration distribution in the eastern sector of the study area revealed the significant role of the tidal flow of the Hugli estuarine system, dominating the suspended sediment distribution. Also, the eastern sector had a dominancy of the clay-silt sized sediments with a very little sand-sized suspended sediment content. The uniformity in the SPM content was also responsible for the strong near-normal distribution. The western sector had energy sources both from the tidal current and waves, for driving bed sediments into suspension and resuspension, which might be responsible for the less uniform nature of SPM concentration. Moreover, the presence of both sand- and mud-sized suspended sediments could have made the SPM concentration distribution less Gaussian in nature. The frequency distribution and normality study of the divided datasets reflected on the potential erosional and depositional regimes of the study area. A uniform distribution might indicate a less erosive regime, where deposition might be dominant. While a non-uniform distribution might indicate an erosional regime. Accordingly, the eastern sector of the study area, that is, beyond Mandarmani to Rasulpur is less prone to erosion and vice versa for the western sector.

9.3.3 Trend Analysis and Anisotropy

The semivariogram of the transformed 100 data points (Fig. 9.7) does not clearly show a continuously increasing trend, and hence no discernable trend was observed. This allowed the use of geostatistical tools without further transformations (Saito and Goovaerts 2001; Méar et al. 2006). Generally, a variogram is represented graphically showing the quantitative statistics which is characterized by the spatial continuity of the data taken into consideration. Often, different length scales according to different directions are shown by a variogram, which is known as geometric anisotropy. Considering a linear variogram, geometric anisotropy will be appearing at different slopes in different directions. Further, to investigate the anisotropy of an omnidirectional variogram, the tolerance was set at 45 degrees. The directional variogram plots the results with the set tolerance of 45 degrees, with directions 0°, 45°, 90°, and 135° (Fig. 9.8). It appears that for directions at 45° and 90° the variogram passes through a maximum of the data points, whereas for directions at 0° and 135° the variogram passes through very few data points. This clearly shows evidence of geometric anisotropy as the range changes considering the same sill parameter. At a scale of the study area, the anisotropy is oriented toward the ENE direction (Fig. 9.8). This orientation corresponds to the strongest gradient of high SPM concentration in the ENE direction (Méar et al. 2006). This can be validated from the field data analysis, where the highest SPM concentration was observed in the eastern sector of the study area. In addition, field observations reveal the presence of muddy depositional features in the Junput–Rasulpur belt, for example, wide and long mudflats are the most common geomorphological features in this coastal belt. Thus, the anisotropic analysis confirms the existence of the depositional regime in the eastern sector of the study area.

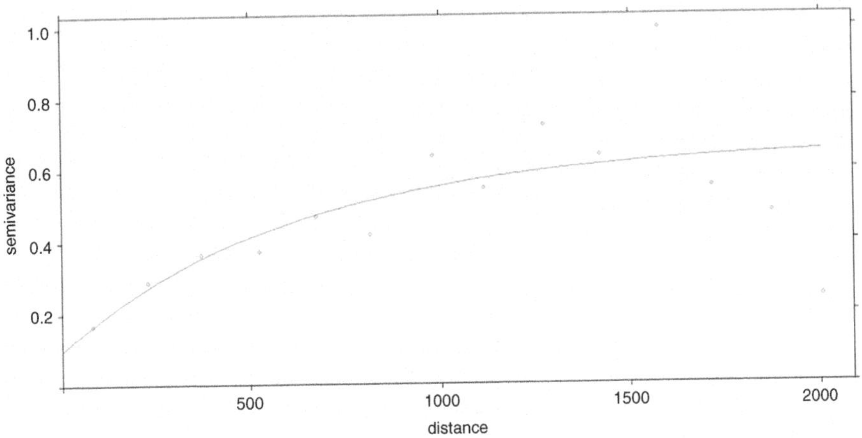

Fig. 9.7 Semi-variogram of the transformed 100 data points (exponential model, cutoff = 6000, width = 150, nugget = ~0.01, sill = ~0.065, range = 2000)

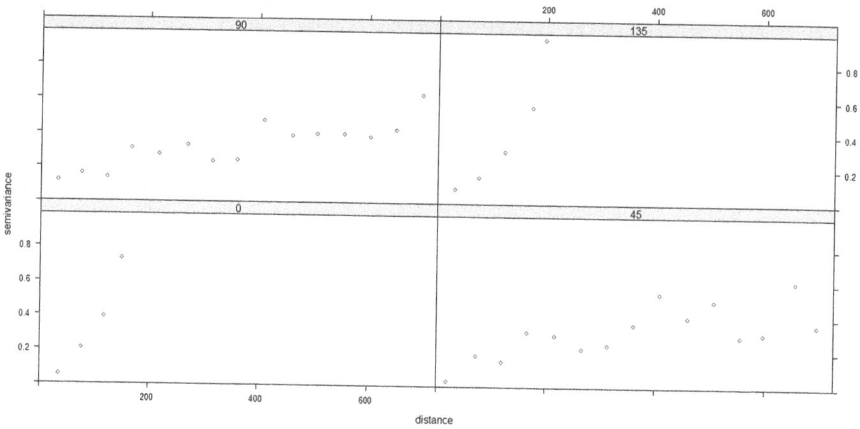

Fig. 9.8 Directional semivariogram of the transformed 100 data points at 0°, 45°, 90° and 135°

9.3.4 *Semivariogram Analysis*

The semivariogram of the 100 (Fig. 9.7) and 68 data points (Fig. 9.9) showed a nugget effect at 0.10 and 0.035, respectively. No nugget effect was observed for the semivariogram of the 32 data points (Fig. 9.10). Though the nugget effects are not strong in nature, this effect characterizes variations in the amount of the SPM concentration over a very short distance (Méar et al. 2006). The sources of these variations might be of natural or artificial origin, which might result from analytical errors or the subsurface morphology, or the local energy regimes prevailing in this region (Gringarten and Deutsch 2001). The absence of the nugget effect in the semivariogram of the 32 data points indicated a uniform distribution of SPM concentration, thereby confirming the active role of a high-energy environment (Hugli estuarine flow) in the uniform distribution of suspended sediments. In an overall scenario, low nugget effect in the entire study area depicted a nearly uniform distribution of suspended sediments. The western sector of the study area (68 data points) showed a slightly more nugget effect than the eastern sector, which might be attributed to the combined effect of the wave and tidal energy influencing the nonuniformity in the concentration of the suspended sediments. Though, it was not possible in this study, to deduce the exact reason behind these effects.

The semivariograms were characterized by the "hole effect," that is, an observed cyclicity in the semivariograms along with the presence of disturbances after the distance where the curve attained a sill. These disturbances weakened gradually (Méar et al. 2006). For the semivariogram of the 100, 68, and 32 data points, the sill was observed at and around 0.065, 0.095, and 0.060, respectively (Figs. 9.7 and 9.9a, b). This cyclicity confirms the increased SPM concentration at frequent intervals in the ENE direction. This suggested a high fluvial activity near the mouth of the tidal inlets. This can also be validated from the field data analysis where high SPM concentration was observed near the mouth of the estuaries and tidal inlets.

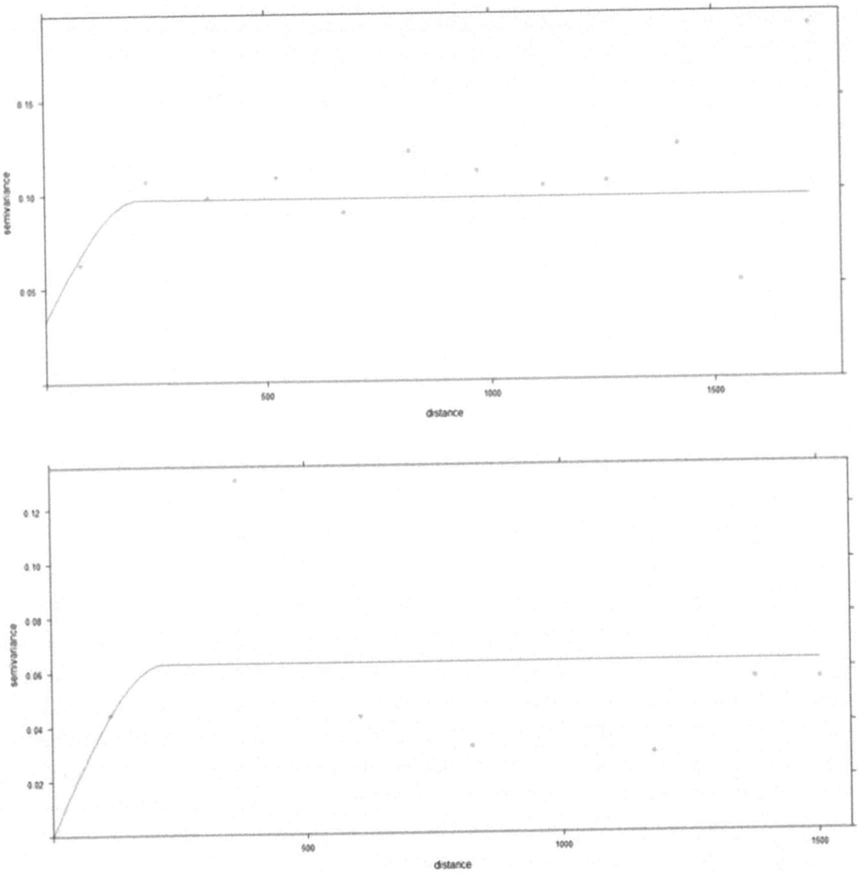

Fig. 9.9 Semivariogram of (**a**) 68 data points (exponential model, cutoff = 3500, width = 150, nugget = ~0.035, sill = ~0.095, range = 200); (**b**) 32 data points (exponential model, cutoff = 1600, width = 250, nugget = 0.0, sill = ~0.06, range = 250)

The physiographic characteristics of the study area also showed the presence of numerous, closely located, tidal inlets, thereby asserting to cyclicity observed in the semivariograms (Fig. 9.1).

The mean squared error of the three datasets of 100, 68, and 32 data points were found close to zero, that is, 0.07, 0.35, and 0.02, respectively. This suggested an exponential model fitting, as the best fit model (Mukhopadhaya 2016). Low RMSE values resulted from this model fitting, with values of 0.27, 0.59, and 0.16 for the semivariogram of 100, 68, and 32 data points, respectively. Cross-validation was performed among the three datasets to assess the accuracy of the model fitting. The 68 data point model, fitted on the 32 data points, produced a RMSE value of 0.24, while the vice versa produced a RMSE value of 0.42. Similarly, RMSE values of 0.78 and 0.29 were observed for the 68 data point model and 32 data point model

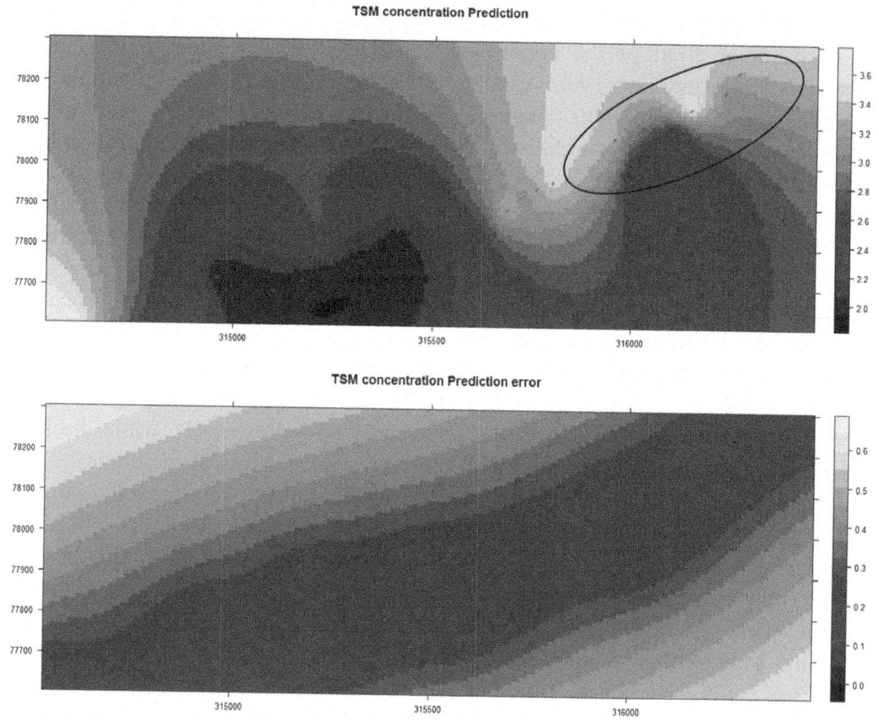

Fig. 9.10 (**a**) Estimation map of the SPM concentration of the study area. The circle represents the eastern sector of the study area with high SPM concentration; (**b**) Uncertainty Map of the SPM concentration of the study area. (The coordinates are in EPSG system)

fitting on the 100 data points, respectively. The low RMSE values among the datasets suggested the stability of the exponential model of the 100 data point, thus enabling proper interpolation and estimation of the SPM concentration in the study area.

9.3.5 Interpolation and Uncertainty Assessment

Significant information present within the dataset governs the choice of the interpolation method. From the semivariogram analysis, high SPM concentration separated by variable distances in the alongshore direction and its eastwardly increasing trend was found to be the most critical and important result. For a proper geostatistical interpolation, the interpolation operator using their principal characteristics (distance from the shore, concentration, geographical localization, hydrodynamic and wave energy regime, etc.) must visualize these results (Méar et al. 2006). In the case of this study, only one field variable was available (SPM concentration) and therefore ordinary kriging was chosen as the interpolation method (Matheron 1963). A stochastic method of the conditional simulation was applied in this study (Pebesma and Wesseling 1998; Vann et al. 2002). Normally a single simulation produces a poor-quality result (Vann et al. 2002). In kriging, the total variance of the data is reproduced both quantitatively through the histogram and spatially through the semivariogram (Méar et al. 2006). Asselman (1999) underlined that for proper geostatistical analysis, a series of simulations need to be produced, all different, to take into consideration, the spreading of the group of variographic points. Ten simulations of the transformed variable (SPM concentration) were calculated based on the majority of data variance values, which resulted in 10 different equiprobable distributions. All these 10 distributions were averaged to produce a single, robust SPM concentration estimation (Fig. 9.10a) and uncertainty map (Fig. 9.10b).

In the SPM concentration map (Fig. 9.10a), the highest concentration of sediments estimated are denoted with the pink, red, and yellow color, while the low sediment concentration is denoted with the blue shades. Most of the highest concentrations of the sediments are observed toward the north of the sampled points. This is the landward region, which will normally have a high concentration of sediments, which proves that the exponential model considered for interpolation was well correlated with the physiography of the study area. So, the geographical localization aspect was considered in kriging. The circled location in Fig. 9.10a represents the Mandarmani–Junput–Rasulpur belt (eastern sector of the study area), which has the highest concentration of SPM near the shoreline. The SPM concentration amounted up to 2.8–3.6 mg/l, which with reverse transformation would amount to ~67–168 mg/l. This high concentration can be attributed to the large amount of sediment load carried by the Hugli estuarine system and the depositional regime operating within this coastal belt. This is the major reason behind the increased concentration of sediment suspension and resuspension in this region. Normally as the distance from the shore increases toward offshore, the SPM concentration decreases. But in the Rasulpur estuarine region, high sediment concentration continued offshore. This is because this offshore region lies near to the mouth of the Hugli estuary where there is always a high instantaneous concentration of suspended sediments. A similar scenario of high offshore suspended sediment concentration is observed in the Mandarmani region, though the actual reason for it was not established in this study. This might be attributed to the influence of subsurface topography and the hydrodynamic-wave energy regimes prevailing in this zone. Very low concentration of SPM was observed in the western sector of the study area ranging from ≤2 to 2.6 mg/l (~ ≤ 16–46 mg/l, with reverse transformation), which gradually decreases further offshore. This low amount suspension of sediments might be attributed to the low-energy regimes prevailing in this region, compared to the high-energy regime of the Hugli estuarine flow. Due to the fluvial activity, the estuarine and the tidal inlet mouths have a relatively high SPM concentration than the coastal waters adjacent to it. The uncertainty assessment map (Fig. 9.10b) revealed an increased uncertainty in the landward side of the sample points, represented by the pink, red, and yellow color. This observation was obvious as terrigenous sedimentology is absolutely different from the nearshore bed sedimentology, thus further confirming the accuracy of the exponential model. In the nearshore waters' uncertainty was found to be very low (0.0–0.2 mg/l), but toward offshore the uncertainty increased up to >0.6 mg/l. Near to the sample points, the geostatistical interpolator can conveniently visualize the principal characteristics thereby performing a good interpolation. This resulted in a lower uncertainty value in the nearshore waters. But the visualization power of the interpolator exponentially decreased with increasing distance from the shore both landward and seaward, as per the exponential model considered for this study. This resulted in an increased uncertainty toward offshore. A total of 25 random points were chosen based on proximity to the shoreline to estimate SPM concentration value and corresponding uncertainty (Table 9.1). The results clearly estimated a high SPM concentration near to the shoreline with low uncertainty and vice versa for the offshore points (Points 21–25).

Table 9.1 SPM prediction and uncertainty

Location	Latitude	Longitude	Estimation (mg/l)	Uncertainty
Point 1	21°33′32.21"N	87°22′50.46″E	3.643675	0.157875
Point 2	21°37′54.10"N	87°33′1.55″E	2.669945	0.1707792
Point 3	21°38′33.09"N	87°38′2.28″E	2.3665	0.1462062
Point 4	21°41′12.31"N	87°46′26.17″E	3.05934	0.1543263
Point 5	21°47′11.55"N	87°53′54.80″E	3.57687	0.1679237
Point 6	21°35′13.38"N	87°26′34.27″E	2.688678	0.2209525
Point 7	21°37′8.02"N	87°30′41.44″E	2.300609	0.2270635
Point 8	21°38′28.08"N	87°35′56.78″E	2.343078	0.2053525
Point 9	21°39′52.87"N	87°42′44.16″E	3.386699	0.1713768
Point 10	21°43′55.25"N	87°50′57.17″E	3.29412	0.1379519
Point 11	21°32′42.57"N	87°23′34.22″E	3.484629	0.1442196
Point 12	21°37′11.89"N	87°33′36.37″E	2.531489	0.1290741
Point 13	21°38′22.71"N	87°38′59.15″E	2.5474	0.1373568
Point 14	21°40′52.84"N	87°46′13.13″E	3.081888	0.1433142
Point 15	21°46′37.18"N	87°54′36.05″E	3.534607	0.1840622
Point 16	21°30′38.74"N	87°24′8.83″E	2.475146	0.1815113
Point 17	21°35′31.49"N	87°34′20.13″E	2.00279	0.149901
Point 18	21°36′52.54"N	87°39′58.67″E	2.438923	0.1466496
Point 19	21°39′57.73"N	87°47′59.80″E	2.293205	0.1613187
Point 20	21°44′44.49"N	87°49′14.53″E	3.412383	0.1759059
Point 21	21°46′22.49"N	87°51′45.53″E	3.082138	0.6774902
Point 22	21°48′01.49"N	87°54′54.53″E	2.976366	0.5817625
Point 23	21°51′14.49"N	87°56′19.53″E	3.22282	0.4597023
Point 24	21°53′57.49"N	87°57′39.53″E	2.638559	0.5473473
Point 25	21°56′21.49"N	87°59′22.53″E	2.76949	0.6803108

9.4 Conclusions

Univariate geostatistical analysis, using SPM concentration as the regionalized variable, gave an insight into the environmental promoters, influencing regional sedimentology of the nearshore waters of the NE coast of Bay of Bengal. Semivariogram analysis, followed by ordinary kriging, showed a high dependency of SPM concentration on geographical localization, energy regimes. The coastal waters in between the Subarnarekha and Rasulpur estuary can be delineated into two zones, based on the SPM concentration and energy regimes. The western sector was characterized by a considerably low SPM concentration with a low-energy regime, whereas the eastern sector had just the opposite scenario. The Junput–Rasulpur belt was found to be in a depositional environment. While the western sector of Subarnarekha mouth to Tajpur was found to be in the erosional regime. Due to the huge sediment load carrying capacity of the Hugli estuary and high-energy environment of the region of Junput–Rasulpur, increased SPM concentration was observed. In addition, the Hugli estuary carries a substantial amount of anthropogenic pollutants. High SPM concentration increases the pollutant-carrying capacity in the coastal waters. Therefore, the Junput–Rasulpur belt is the most anthropogenic pollution prone zone in the study area. The coastal waters of the western sector of Subarnarekha to Tajpur inlet is less prone to anthropogenic pollution, due to low SPM concentration and constant wave action. Wave action might be held responsible for redistribution of the suspended sediments, rather than concentrating it in one direction.

This study was conducted to assess the efficacy of the geostatistical tool in estimating SPM concentration in the complex nearshore waters. Sound cartography was produced, estimating the SPM concentration, with the uncertainty assessment. Toward offshore and landward, uncertainty increased due to poor performance of the model, which increased exponentially with distance from the shore. Therefore, distance from the shoreline is a major factor controlling the geostatistical analysis of the SPM concentration. Though this study was conducted with 100 data points, a more precise and accurate geostatistical output can be obtained if a greater number of data points are available (Webster and Oliver 1992, 1993; Gascuel-Odoux and Boivin 1994; Goovaerts 1997). This study assumed a constant mean and variance across the study area, for which ordinary kriging was the choice of interpolation, but in nature, a dynamic system exists in the coastal waters. The dependency of one variable with another is impossible to establish with the ordinary kriging approach. It is recommended that more field variables should be obtained to properly understand the dependency of one variable on another, for inferring a comprehensive environmental scenario. Co-kriging can be used as an interpolation method as it allows consideration of other environmental variables. This study further recommends a temporal study based on the variance maps produced by the geostatistical analysis, to assess the deposition and erosion rates (Méar et al. 2006).

Acknowledgments Our sincerest gratitude goes toward the Director, Indian Institute of Remote Sensing, Indian Space Research Organization, for providing the moral support for this project. We would also like to thank Dr. Sandip Kumar Mukhopadhyay, Associate Professor of Marine Science Department, University of Calcutta, and Dr. Punarbasu Chaudhuri, Associate Professor of Environmental Science Department, University of Calcutta, for providing the laboratory support. Lastly, we acknowledge the relentless support of Dr. Sugata Hazra, School of Oceanographic Studies, for being supportive enough for the entire period of this research work.

References

Alam M, Alam MM, Curray JR, Chowdhury MLR, Gani MR (2003) An overview of the sedimentary geology of the Bengal Basin in relation to the regional tectonic framework and basin-fill history. Sediment Geol 155(3):179–208

Anckar E, Dennegård B, Nyborg MR, Kuijpers A (1998) Marine pollution pattern of Skagerrak and Kattegat—a geostatistical evaluation. GFF 120(3):285–291

Asselman NE (1999) Grain-size trends used to assess the effective discharge for floodplain sedimentation, River Waal, the Netherlands. J Sediment Res 69(1)

Atkinson PM, Lewis P (2000) Geostatistical classification for remote sensing: an introduction. Comput Geosci 26(4):361–371

Aubry P (2000) Le traitement des variables régionalisées en écologie: apports de la géomatique et de la géostatistique (Doctoral dissertation, Université Claude Bernard-Lyon I)

Bastante FG, Taboada J, Alejano LR, Ordóñez C (2005) Evaluation of the resources of a slate deposit using indicator kriging. Eng Geol 81(4):407–418

Batista AC, Sousa AJ, Batista MJ, Viegas L (2001) Factorial kriging with external drift: a case study on the Penedono region, Portugal. Appl Geochem 16(7):921–929

Bellehumeur C, Marcotte D, Legendre P (2000) Estimation of regionalized phenomena by geostatistical methods: lake acidity on the Canadian Shield. Environ Geol 39(3–4):211–220

Birch GF, Taylor SE (2000) The use of size-normalised procedures in the analysis of organic contaminants in estuarine sediments. Hydrobiologia 431(2–3):129–133

Bohling G (2005) Introduction to geostatistics and variogram analysis. Kansas Geol Sur 2

Burak DL, Fontes MP, Santos NT, Monteiro LVS, de Sousa Martins E, Becquer T (2010) Geochemistry and spatial distribution of heavy metals in Oxisols in a mineralized region of the Brazilian Central Plateau. Geoderma 160(2):131–142

Caeiro S, Painho M, Goovaerts P, Costa H, Sousa S (2003) Spatial sampling design for sediment quality assessment in estuaries. Environ Model Softw 18(10):853–859

Chang YH, Scrimshaw MD, Emmerson RHC, Lester JN (1998) Geostatistical analysis of sampling uncertainty at the Tollesbury Managed Retreat site in Blackwater Estuary, Essex, UK: kriging and cokriging approach to minimise sampling density. Sci Total Environ 221(1):43–57

Coleman JM (1969) Brahmaputra River: channel processes and sedimentation. Sediment Geol 3(2–3):129–239

Curray JR, Moore DG (1974) Sedimentary and tectonic processes in the Bengal Deep-sea Fan and Geosyncline. In: Burk CA, Drake CL (eds) Continental margins. Springer, New York, pp 617–627

Curray JR, Emmel FJ, Moore DG (2002) The Bengal Fan: morphology, geometry, stratigraphy, history and processes. Mar Pet Geol 19(10):1191–1223

Davis JC, Sampson RJ (1986) Statistics and data analysis in geology, vol 646. Wiley, New York

De Kok JM (1992) A three-dimensional finite difference model for computation of near-and far-field transport of suspended matter near a river mouth. Cont Shelf Res 12(5–6):625–642

Desbarats AJ (1996) Modeling spatial variability using geostatistical simulation. In: Geostatistics for environmental and geotechnical applications. ASTM International

Forstner U, Calmano W, Schoer J (1982) Metals in sediments from the Elbe, Weser and Ems estuaries and from the German Bight: grain size effects and chemical forms. Thalassia Jugosl 12:30–36
Gascuel-Odoux C, Boivin P (1994) Variability of variograms and spatial estimates due to soil sampling: a case study. Geoderma 62(1–3):165–182
Goovaerts P (1997) Kriging vs stochastic simulation for risk analysis in soil contamination. In: geoENV I—Geostatistics for environmental applications. Springer, Netherlands, pp 247–258
Goovaerts P (1999) Geostatistics in soil science: state-of-the-art and perspectives. Geoderma 89(1):1–45
Gopalan AKS, Krishna VV, Ali MM, Sharma R (2000) Detection of Bay of Bengal eddies from TOPEX and in situ observations. J Mar Res 58(5):721–734
Guhathakurta H, Kaviraj A (2000) Heavy metal concentration in water, sediment, shrimp (Penaeus monodon) and mullet (Liza parsia) in some brackish water ponds of Sunderban, India. Mar Pollut Bull 40(11):914–920
Gringarten E, Deutsch CV (2001) Teacher's aide variogram interpretation and modeling. Math Geol 33(4):507–534
Guzzella L, Roscioli C, Vigano L, Saha M, Sarkar SK, Bhattacharya A (2005) Evaluation of the concentration of HCH, DDT, HCB, PCB and PAH in the sediments along the lower stretch of Hugli estuary, West Bengal, Northeast India. Environ Int 31(4):523–534
Healy T, Wang Y, Healy JA (eds) (2002) Muddy coasts of the world: processes, deposits and function, vol 4. Gulf Professional Publishing
Isaaks EH, Srivastava RM (2001) An introduction to applied geostatistics. 1989. *New York, USA: Oxford University Press. Jones DR, A taxonomy of global optimization methods based on response surfaces*. J Glob Optim 23:345–383
Journel AG, Huijbregts CJ (1978) Mining geostatistics. Academic press
Kuzyakova IF, Romanenkov VA, Kuzyakov YV (2001) Application of geostatistics in processing the results of soil and agrochemical studies. Eurasian Soil Sci 34(11):1219–1228
Leecaster M (2003) Spatial analysis of grain size in Santa Monica bay. Mar Environ Res 56:67–78
Legendre P, Legendre LF (2012) Numerical ecology, vol 24. Elsevier
Li S, Lu W (2010) Automatic fit of the variogram. In: Information and Computing (ICIC), 2010 Third International Conference on, vol 4. IEEE, pp 129–132
Malone BP, Jha SK, Minasny B, McBratney AB (2016) Comparing regression-based digital soil mapping and multiple-point geostatistics for the spatial extrapolation of soil data. Geoderma 262:243–253
Matheron G (1963) Principles of geostatistics. Econ Geol 58(8):1246–1266
Méar Y, Poizot E, Murat A, Lesueur P, Thomas M (2006) Fine-grained sediment spatial distribution on the basis of a geostatistical analysis: example of the eastern Bay of the Seine (France). Cont Shelf Res 26(19):2335–2351
Mitra A, Mondal K, Banerjee K (2010) Concentration of heavy metals in fish juveniles of Gangetic Delta of West Bengal, India. Res J Fish Hydrobiol 5(1):21–26
Mukhopadhaya S (2016) Rainfall mapping using ordinary kriging technique: case study: Tunisia. J Basic Appl Eng Sci 3(1):1–5
Oliver MA, Webster R (2014) A tutorial guide to geostatistics: computing and modelling variograms and kriging. Catena 113:56–69
Ouyang Y, Higman J, Thompson J, O'Toole T, Campbell D (2002) Characterization and spatial distribution of heavy metals in sediment from Cedar and Ortega rivers subbasin. J Contam Hydrol 54(1–2):19–35
Pannatier Y (2012) VARIOWIN: software for spatial data analysis in 2D. Springer Science & Business Media
Pebesma EJ, Wesseling CG (1998) Gstat: a program for geostatistical modelling, prediction and simulation. Comput Geosci 24(1):17–31
Saito H, Goovaerts P (2001) Accounting for source location and transport direction into geostatistical prediction of contaminants. Environ Sci Technol 35(24):4823–4829

Sarkar SK, Saha M, Takada H, Bhattacharya A, Mishra P, Bhattacharya B (2007) Water quality management in the lower stretch of the river Ganges, east coast of India: an approach through environmental education. J Clean Prod 15(16):1559–1567

Schnabel U, Tietje O, Scholz RW (2004) Uncertainty assessment for management of soil contaminants with sparse data. Environ Manag 33(6):911–925

Thorne CR, Russell AP, Alam MK (1993) Planform pattern and channel evolution of the Brahmaputra River, Bangladesh. Geol Soc Lond, Spec Publ 75(1):257–276

Turner A, Millward GE (2002) Suspended particles: their role in estuarine biogeochemical cycles. Estuar Coast Shelf Sci 55:857–883

Van Groenigen JW, Siderius W, Stein A (1999) Constrained optimisation of soil sampling for minimisation of the kriging variance. Geoderma 87(3):239–259

Vanderborght J, Jacques D, Mallants D, Tseng PH, Feyen J (1997) Analysis of solute redistribution in heterogeneous soil. In: geoENV I—Geostatistics for environmental applications. Springer, Netherlands, pp 283–295

Vann J, Bertoli O, Jackson S (2002, March) An overview of geostatistical simulation for quantifying risk. In: Proceedings of Geostatistical Association of Australasia Symposium' Quantifying Risk and Error, vol 1, p 1

Vannametee E, Babel LV, Hendriks MR, Schuur J, de Jong SM, Bierkens MFP, Karssenberg D (2014) Semi-automated mapping of landforms using multiple point geostatistics. Geomorphology 221:298–319

Wang L, Coles NA, Wu C, Wu J (2014) Spatial variability of heavy metals in the coastal soils under long-term reclamation. Estuar Coast Shelf Sci 151:310–317

Webster R, Oliver MA (1992) Sample adequately to estimate variograms of soil properties. J Soil Sci 43(1):177–192

Webster R, Oliver MA (1993) How large a sample is needed to estimate the regional variogram adequately? In: Geostatistics Tróia'92. Springer, Netherlands, pp 155–166

Webster T, Lemckert C (2002) Sediment resuspension within a microtidal estuary/embayment and the implication to channel management. J Coast Res 36:753–759

Western AW, Blöschl G, Grayson RB (1998) Geostatistical characterisation of soil moisture patterns in the Tarrawarra catchment. J Hydrol 205(1):20–37

Winterwerp JC, Van Kesteren WG (2004) Introduction to the physics of cohesive sediment dynamics in the marine environment, vol 56. Elsevier

Zhang M, Tang J, Dong Q, Song Q, Ding J (2010) Retrieval of total suspended matter concentration in the Yellow and East China Seas from MODIS imagery. Remote Sens Environ 114(2):392–394

Chapter 10
Multiple Facets of Aquatic Pollution in the Estuarine and Continental Shelf Waters Along the East Coast of India

Anirban Akhand, Abhra Chanda, and Sourav Das

Abstract Aquatic pollution often takes place in estuaries, especially the ones flowing through large cities and industrial setups due to the discharge of various chemical substances. The estuaries situated along the east coast of India are no exception in this regard. This chapter has discussed five broad types of pollution that occur in the east coast estuaries of India, namely eutrophication, algal bloom, fecal coliform, high biochemical oxygen demand, and petroleum hydrocarbon pollution. Land-derived runoff along with the pore water from mangrove vegetation often enhances the nutrient levels in these estuaries, which in turn has favored harmful algal blooms. Indiscriminate discharge of domestic sewage makes some of the estuaries active hotspots of coliform pollution, deteriorating the entire water quality. Besides this, the municipal wastewater coupled with some industry effluents enhances the organic matter load along with petroleum hydrocarbon load beyond thresholds. All these forms of pollution have become essential aspects of estuarine biogeochemistry, which need comprehensive study and monitoring. A proper characterization of the dynamics of these pollutants has become an utmost necessity. The present chapter has collated and discussed all the principal findings reported, in this regard, in the estuaries along the east coast of India.

Keywords Eutrophication · nutrient pollution · algal bloom · fecal coliform · organic pollution · BOD · COD · petroleum hydrocarbon

---

A. Akhand (✉)
Coastal and Estuarine Environment Research Group, Port and Airport Research Institute, Yokosuka, Kanagawa, Japan

A. Chanda · S. Das
School of Oceanographic Studies, Jadavpur University, Kolkata, West Bengal, India

S. Das, T. Ghosh (eds.), *Estuarine Biogeochemical Dynamics of the East Coast of India*, https://doi.org/10.1007/978-3-030-68980-3_10

10.1 Introduction

A substantial share of the continental weathering products coupled with the anthropogenic input ends up in the rivers and drains to their respective estuaries and eventually to the coastal seas in the form of a multitude of dissolved and particulate matter (Kanuri et al. 2020). Over the past decades, the rivers and estuarine banks throughout the world have witnessed a significant rise in the population, which in turn has led to increased discharge of domestic and industrial effluents in these regions (Lotze et al. 2006). With the increasing population pressure vis-à-vis the enhanced load of such effluents, the overall water quality of the estuaries and nearshore coastal waters all around the world are getting jeopardized due to the enhancement in concentrations of an array of pollutants (Jha et al. 2015; Soo et al. 2016). Apart from abiotic pollutants, several biotic communities are also known to pollute the estuarine environments (Batabyal et al. 2014).

Besides heavy metals and persistent organic pollutants, several other types of aquatic pollution have become a concern for environmentalists all through the world. Essential nutrients, especially nitrogen and phosphorus, which are abundant in sufficient quantity in the rivers and estuaries, often lead to aquatic pollution when their concentrations increase abnormally beyond a certain threshold, mostly because of agricultural runoffs and domestic effluents full of inorganic nitrogen and phosphorus (Wang et al. 2019). Such excessive presence of nutrients leads to eutrophication, which usually disrupts the balance of the ecological food chain by favoring excessive growth of several noxious algae, reducing the transparency of the water column, and creating hypoxic or even sometimes anoxic environments leading to the loss of several species (Hobbie et al. 2017). As a consequence of eutrophication, "algal blooms" have emerged as another potential threat to the well-being of the aquatic ecosystems all over the world, which usually denotes the abnormal growth of several toxic and harmful algae comprising many dinoflagellates, diatoms, and cyanobacteria (Anderson et al. 2012; Raven et al. 2020). In addition to these, discharge of untreated domestic sewage also exposes a substantial quantity of fecal matters to the rivers and estuaries, which acts as a growth culture medium for several bacterial and viral pathogens (Frena et al. 2019). We often quantify the degree of such sewage contamination by measuring the fecal coliform bacterial count in the aquatic column (Shah et al. 2007).

The estuaries often receive a substantial quantity of organic matter load, derived from the remnants of plants, animals, and wastes that finds its way in the estuaries. These organic substances continuously change speciation as these are acted upon by a spectrum of microbes to draw nutrition from these matters (Supriyantini et al. 2019). To quantify the load of such organic matter, biochemical oxygen demand (BOD) and chemical oxygen demand (COD) are two of the most crucial parameters that serve as a proxy (Gaspar et al. 2018). Another aspect of organic matter pollution, which deserves a separate category, from the perspective of marine pollution, is that of oil pollution. Oil pollution has become a significant concern for marine scientists, as the past decades have witnessed several events of oil spills, increased shipping, and water-based transport activities. The tourism sector, as well as several industrial activities like pyrolysis and combustion of fossil fuels, injects an assembly of petroleum hydrocarbons to the estuarine water column (Venkatachalapathy et al. 2011). The recalcitrant part of these petroleum hydrocarbons often ends up accumulating in several marine organisms and causes toxic effects, which has made this pollution an ecological and human health issue (Venkatachalapathy et al. 2011).

The east coast of India, in this regard, has several large and small estuaries, which drain into the Bay of Bengal, and this coastline has one of the largest cities and metropolis along with a large number of industrial belts, thus making these estuaries prone to all these types of aquatic pollution. The present chapter has discussed and collated all the principal findings observed concerning these pollutions in the estuaries and nearshore waters along the east coast of India.

10.2 Data Mining

The authors have extracted the data and principal findings mostly from various journal publications, and few conference proceedings, and edited book volumes available on the Internet. Search engines like Google Scholar, Science Direct, and Scopus were used to retrieve the existing literature. We used keywords and phrases like "nutrient pollution," "eutrophication," "cultural eutrophication," "algal blooms," "harmful algae," "biodegradable organic matter," "BOD," "COD," "fecal coliform," "Hooghly estuary," "Mahanadi estuary," "Godavari estuary," "east coast of India," "coastal waters," "Bay of Bengal," "biogeochemistry," and "physicochemical parameters" to refine the searches.

10.3 Eutrophication and Drivers of Nutrient Pollution

The term "eutrophication" originated from the Greek word *eutrophos*, which means well nourished. Eutrophication is a condition of the water body where the water bodies become enriched with high minerals and nutrients, which in turn induces excessive growth of algae. Mukhopadhyay et al. (2006) studied the lateral transport and dynamics of nutrients in the Hooghly estuary. They reported that mangrove-dominated estuarine systems in Sundarbans act as a source of nutrients, and the input of litter and sediment-associated nutrients released during estuarine transport controls the source strength of nutrients. They also revealed that the occurrence of light-limited turbid condition controls the autotrophic production and the net ecosystem metabolism is heterotrophic in nature, whereas regeneration of nutrients from organic matter originated from an external source. Manna et al. (2010) reported eutrophication in the tidal creek of Sundarban estuary, which indicates the deteriorated status of the water quality. They further suggested that the enhanced instability with coastal eutrophication in Sundarbans has far-reaching consequences for many aspects of mangrove ecosystem function under present and future climate conditions. The eutrophication that takes place because of human-induced or anthropogenic activities is popularly known as cultural eutrophication. De et al. (2011) reported cultural eutrophication from the Sundarbans and northeast coast of India. They revealed that the increased anthropogenic input of excess nutrients to the estuarine waters of Sundarbans resulted in the increased mean numerical abundance and three times an increase in the number of phytoplankton species. They also asserted that in the Sundarbans ecosystem, increasing the trend of nitrogen to phosphate ratio attributed to increased diatom abundance and subsequent deposition of biogenic silicate to the sediment. Das et al. (2017), however, reported that the estuarine to offshore transition zone remained phosphate limited during the post-monsoon season and light-limited for the rest of the year, which prevented this region from experiencing eutrophication. Das et al. (2015) reported that the N:P ratio in this region significantly deviated from the ideal Redfield ratio, which could be another reason behind the low chlorophyll content in this region. Prasad (2012) reported that the ecological status of Indian mangroves, including Sundarbans, Bhitarkanika, Coringa, and Pichavaram situated on the east coast of India, is highly disturbed by anthropogenic impacts through additional inputs of nutrients (Table 10.1). Prasad (2012) also asserted that the long-term nutrient analysis in Pichavaram mangrove water explains that the significant increase in dissolved nutrients since the 1980s originated from nonpoint sources like agriculture and aquaculture, which altered biogeochemical processes in this ecosystem. Dissolved inorganic nitrogen to silicate ratios indicates that the terrestrial weathering supplies substantially high silica to mangrove waters, high phosphorus to nitrogen ratios indicates high phosphate loadings along with nitrogen from both point, and nonpoint sources to the coastal mangrove waters of India.

Table 10.1 Mean nutrient concentrations in the major mangrove estuaries along the east coast of India

Study site	Sampling time	Nitrate (μM)	Phosphate (μM)	Silicate (μM)	Authors
Hooghly estuary	1999–2001	13.0–35.25	1.51–2.64	65.48–150.90	Mukhopadhyay et al. (2006)
Coast of Sundarbans	2008–2009	11.39–22.82	1.29–2.25	65.55–92.39	De et al. 2011
Sundarban estuaries	Data collated from several studies carried out in past three decades	14.06	1.23	65.2	Prasad (2012)
Bhitarkanika estuary		20.48	3.16	466	
Coringa estuary		15.93	2.92	75.5	
Pichavaram estuary		600	82.1	126	
Mangalavanam estuary		25.4	5.4	14.9	

10.4 Algal Bloom

An algal bloom is a rapid increase or accumulation in the algal population, which may occur in the marine environment and freshwater environment. This phenomenon is generally associated with nutrient pollution and eutrophication. Red tides, blue-green algae, and cyanobacteria are quite common examples of harmful algal blooms and may lead to several health hazards like fish die-offs. Several researchers have reported algal blooms from the east coast of India. Biswas et al. (2004) reported that ten species of phytoplankton (belonging to diatoms, dinoflagellates, and blue-green algae) reached the bloom proportion through an increased supply of nutrients from deep water by mixing in the water column in Sundarban. After a decade, Biswas et al. (2014) reported a negative effect on the tintinnid microzooplankton community caused by bloom-forming centric diatoms from the western part of the Indian Sundarbans (coastal area of Sagar Island). They suggested that the aquaculture activities and river effluents act as prime drivers of algal bloom formation. Akhand et al. (2012) indicated a reduced number of the indicator dinoflagellate species, *Ceratium symmetricum* in the Hooghly–Matla estuary due to increased warming. Apart from reporting eutrophication in the Sundarbans, Manna et al. (2010) reported the presence of toxic Dinoflagellates and Cyanophyceae in the creeks of Sundarbans as well. Srichandan et al. (2019) worked on the dynamics of phytoplankton with environmental variables in the Mahanadi estuary, the Chilika lagoon, and coastal waters off Gopalpur covering ten sampling stations and all seasons. They reported dinophyta bloom by causative species *Noctiluca scintillans* during pre-monsoon and bacillariophyta bloom by causative species *Asterionellopsis glacialis* during monsoon. They also depicted the lagoon and estuary as hypereutrophic during all seasons, whereas the coastal water was highly eutrophic during monsoon and pre-monsoon. Sarma et al. (2010) reported that the river discharge has a significant influence on nutrients loading to the Godavari estuary, situated in the southeastern coast of India; they also suggested that the increase in phytoplankton biomass and bloom development has been limited by suspended matter during peak discharge period and by phosphate during no discharge period. Instead of identifying different hydrological factors as drivers of algal bloom in previous studies, Acharyya et al. (2012) revealed a meteorological factor responsible for phytoplankton bloom formation in the Godavari estuary. They stated that the reduced precipitation from 2007 to 2009 over the Indian subcontinent caused reducing in river water discharge during the peak discharge period, which in turn slowed the flushing time of the estuary from 1.2 days to 6.3 days, respectively. Therefore, intense phytoplankton blooms take place by an increase in instability of water column, reduced suspended material load, which suggests resilience toward the unwanted phytoplankton bloom and overall health of the Indian estuaries, can be altered by the variability in monsoon rainfall, and dam regulated discharge.

10.5 Fecal Coliform Pollution

Fecal coliforms are a facultatively anaerobic gram-negative bacterium that generally originates in the intestines of warm-blooded animals and a significant microbial indicator to monitor water quality (Brenner et al. 1993; Grant 1997). The number of fecal coliform study is much less on the east coast of India than that of other water-quality parameters. Hooghly and northeast coast of India received comparatively more attention in this regard. In an exhaustive study in the endemic cholera belt of West Bengal, India, Batabyal et al. (2014) studied cultivable Vibrio and total bacteria in the water–sediment interface. They reported a benthopelagic coupling of Vibrio dynamics controlled by resuspension of sediment and turbidity variability. They further inferred that in Diamond Harbour station of Hooghly River, a salinity increase from 0.6 to 7.9 led to a 1000-fold amplification of initial cultivable coliforms. However, a higher predominance of coliforms in the Kolkata metropolis occurred due to greater disposal of untreated sewage into the river. Basu et al. (2013) also reported the presence of a higher number of coliform bacteria, *Streptococcus* sp., and *Escherichia coli* than the standard limit while worked in the Hooghly estuary. In another study on tintinnids (microzooplankton) composition and production conducted in the Hooghly estuary, Rakshit et al. (2014) reported seasonal variability of fecal coliform and found that it was abruptly high during monsoon season and low during pre-monsoon season in different stations studied. In a recent study, Mitra et al. (2018) reported that the surface water of the Hooghly estuary exceeded the permissible BIS drinking water–level limit for fecal coliform, and the elevated levels of coliform might get prone to a public health risk to a large section of people who regularly get exposed to the contaminated water by bathing. Sinha et al. (2020) observed that the total coliform count in the waters adjoin the Sagar Island after the event of Ganga Sagar Mela (a Hindu ceremony of mass bathing, where thousands of Hindu pilgrims take a holy dip) increases 10–15 times compared to the concentration observed in other times of the year. Samantray et al. (2009) studied the water quality of the three major rivers and canals of the Orissa coast: Mahanadi River and its distributaries, Taldanda Canal, and Atharbanki River. They reported higher fecal coliform pollution in the Atharbanki River than the other two rivers and canal studied; and indicated that the deterioration of the water quality in the rivers due to industrialization and human activities.

10.6 Organic Matter Pollution: High BOD and COD

Biochemical oxygen demand (BOD) and chemical oxygen demand (COD) are two fundamental parameters, which give us an idea about the organic matter content of any water body. BOD values enable us to quantify the organic matter content that can be degraded or oxidized by microbial activity, whereas the COD values indicate the total organic matter that can be oxidized by chemical treatment. Most of the

BOD and COD observations along the estuaries of the east coast have been made in the Hooghly estuary, followed by Mahanadi and Godavari. Reports of BOD and COD from estuaries down further south are very scarce. The BOD and COD levels recorded in the aquatic column along the east coast of India are tabulated in Table 10.2. As early as in the year 1970, Basu et al. (1970) carried out oxygen consumption measurements in the Hooghly and the adjoining Matla estuary in the Sundarban and reported the ranges from 0.7 to 6.0 mg l^{-1} in Hooghly and 0.2 to 4.3 mg l^{-1} in Matla estuaries, which according to the authors were much low in both the estuaries.

Table 10.2 Observed BOD and COD levels as either mean ± standard deviation or minimum–maximum in the major estuaries along the east coast of India

Study site	Sampling time	BOD (mg l^{-1})	COD (mg l^{-1})	Authors
Haldia channel of Hooghly estuary	1997, 1998	0.67–4.17	–	Sadhuram et al. (2005)
Babughat, Hooghly River estuary	2002, 2003	2.20–5.95	7.6–16.4	Sarkar et al. (2007)
Diamond Harbor, Hooghly estuary	2002, 2003	1.18–3.40	9.36–24.8	Sarkar et al. (2007)
Ganga Sagar, Hooghly estuary	2002, 2003	0.75–2.82	24.5–80.4	Sarkar et al. (2007)
Naihati, Hooghly River estuary	2010	8.47 ± 1.78 (4.5–15.2)	74.3 ± 18.5 (26.2–120.3)	Basu et al. (2013)
Hooghly River estuary limnetic regions to river mouth	2014, 2015	0.62–21.4	–	Mitra et al. (2018)
Hooghly River b/w Kolkata and Howrah	2017, 2018	3.08 ± 0.58 to 4.58 ± 0.94	–	Chanda et al. (2020)
Mahanadi River estuary (freshwater zone)	2002	0.29–5.06	–	Panda et al. (2006)
Mahanadi River estuary (saline zone)	2002	0.38–6.87	–	Panda et al. (2006)
Mahanadi estuary (lower stretch)	2006	2.6–6	–	Samantray et al. (2009)
Atharbanki River	2006	9.8–18.0	–	Samantray et al. (2009)
Paradeep, Mahanadi estuary	2009	1.3–2.8	–	Naik et al. (2013)
Gautami-Godavari estuary	September 2001	1.52–2.8	–	Tripathy et al. (2005)
Kakinada Bay	September 2001	2.88–5.85	–	Tripathy et al. (2005)
Coringa River	September 2001	3.68–6.12	–	Tripathy et al. (2005)
Vasishta Godavari estuary	2016, 2017	1.33 ± 1.66 to 1.80 ± 1.30	–	Rao et al. (2019)

However, they did not report their results in terms of BOD or COD. In the following years, with the increase in industrial activities in and around Haldia and Diamond Harbor region, the BOD values increased rapidly in the Hooghly estuary. Sinha et al. (1998) reported that in the year 1980, the sewage outfall zones near Diamond Harbor and Jigerkhali obelisk had BOD values as high as 40 g l^{-1}. However, observations made in the mid of the Haldia channel in the year 1997–98 in all the three seasons showed that BOD varied between 0.67 and 4.17 mg l^{-1}, which was well within the threshold limit of >5.0 mg l^{-1}. Sarkar et al. (2007), in the year 2002–03, characterized the monthly and spatial variability of both BOD and COD along the tract of Hooghly estuary. Their observed levels of BOD and COD showed increased values of BOD near the Babughat region. Sarkar et al. (2007) attributed the effect of domestic sewage from the adjoining twin cities of Kolkata and Howrah behind such high BOD levels near Babughat, which decreased steadily toward the river mouth near Ganga Sagar station. Recently, Chanda et al. (2020) also observed very high BOD levels of 3.08 ± 0.58 mg l^{-1}, 3.86 ± 1.15 mg l^{-1}, and 4.58 ± 0.94 mg l^{-1} in the pre-monsoon, the monsoon, and the post-monsoon seasons, respectively, in the Hooghly River flowing between the polluted cities of Kolkata and Howrah. However, near the river mouth, Sarkar et al. (2007) observed the highest range of COD values. However, the authors argued that such high COD ranges could not be due to pollution, but the effect of mangroves in disseminating organic matter–rich waters in the adjacent water column. Mitra et al. (2018) reported high values of BOD as well from the same location of Ganga Sagar, and they attributed the abundance of organic matter from the upstream region coupled with microbial degradation behind such high BOD values. Sinha et al. (2020) observed an increase in BOD of 1.3 to 2.5 times after the event of Ganga Sagar Mela, which takes place every year. Compared to Hooghly estuary, the ranges of BOD observed in the stretch of Mahanadi estuary were almost similar, and the maximum values recorded were, to some extent, higher than that observed in Hooghly. Panda et al. (2006) exhaustively sampled the entire stretch of Mahanadi River estuary and reported BOD in the ranges of 0.29 to 5.06 mg l^{-1} and 0.38 to 6.87 mg l^{-1} in the freshwater and saline water stretch of the river estuary, respectively. Panda et al. (2006) attributed a long list of industrial and domestic activities from the Paradeep, Cuttack, Sambalpur, and Hirakund regions behind such high BOD concentrations in the Mahanadi. Samantray et al. (2009) observed a similar high range of BOD in the Mahanadi estuary (2.6–6 mg l^{-1}) and anomalously high BOD values in the adjoining Atharbanki River ranging from 9.8 to 18.0 mg l-1. Samantray et al. (2009) attributed the effluents of Paradeep Port Township and several fertilizer industries behind such abnormally high concentrations of BOD. However, in the Mahanadi River mouth, Naik et al. (2013) reported much-reduced levels of BOD ranging from 1.3 to 2.8 mg l^{-1}, mostly due to dilution with marine water. This observation shows that though the inner estuarine regions of Mahanadi often experience high BOD load, the coastal waters of Bay of Bengal adjacent to this estuary are not to that extent polluted, like that of Hooghly. Compared to the northern estuaries along the east coast of India, much fewer studies took place on the southern ones. However, Tripathy et al. (2005) recorded substantially high BOD concentrations from the Gautami–Godavari estuary (1.52–2.8 mg l^{-1}), Kakinada Bay (2.88–5.85 mg l^{-1}), and the adjoining

Coringa River (3.68–6.12 mg l^{-1}). Agricultural and domestic wastes from the Kakinada township are mainly responsible for such magnitudes. However, in the counterpart of Vasistha Godavari estuary, Rao et al. (2019) observed much less BOD levels, sampling all-round the year (1.33 ± 1.66 mg l^{-1} to 1.80 ± 1.30 mg l^{-1}).

10.7 Petroleum Hydrocarbon Pollution

Table 10.3 shows the PHC levels in the water, sediment, and biological organisms recorded along the east coast of India. The sediments of Hooghly estuary sampled in the year 1984–85 recorded the presence of petroleum-derived hydrocarbons (PHC) (Ghosh and Choudhury 2001), with magnitudes ranging between 100 ppm and 310 ppm. Higher concentrations of PHC were observed by Ghosh and Choudhury (2001) during the monsoon seasons, signifying that land-derived input, which enhanced the water-discharged PHC in the sediments of this region. Sadhuram et al. (2005), on the contrary, while sampling the waters in the Haldia channel of Hooghly estuary, observed PHC ranging from below detection level (BDL) to 89 mg m^{-3}, with higher concentrations in the winter compared to summer. This observation showed that high temperature and evaporation played a significant role in removing the PHCs from the water column, which was evident from its lower concentration during the summer. Panigrahy et al. (2014) sampled in several coastal transects adjoining the states of West Bengal and Odisha during the post-monsoon seasons of the years 2002 to 2009. They observed PHC concentrations of 1.10 to 2.15, 0.07 to 2.57, 0.04 to 1.58, 0.69 to 2.30, 0.45 to 3.42, 0.27 to 3.75, 0.14 to 2.72, 0.07 to 1.66, and 0.48 to 2.96 mg m^{-3} in Digha, Gopalpur, Chilika, Rushikulya, Konark, Puri, Paradeep, Dhamra, and Chandipur, respectively. They also observed PHC concentrations of 1.17 to 18.5, 0.45 to 5.48, and 0.26 to 0.90 in the Hooghly, Mahanadi, and Dhamra estuaries. Panigrahy et al. (2014) inferred from their yearlong study that due to intense port activities, the PHC levels in the Hooghly estuary were significantly high than all the other transects. Kiran et al. (2015), while analyzing three sediment cores off the Godavari estuary, observed low PHC concentrations of 0.014 to 0.704 ppm.

However, in the further south along the Chennai coast, Venkatachalapathy et al. (2010) observed substantially high PHC concentrations in the sediments collected from Adyar (7.26–16.83 ppm) and Kuvam (5.5–39.72 ppm) estuaries, and they attributed the shipping activities of Chennai port and wastewater discharge behind such high magnitudes. Venkatachalapathy et al. (2012) measured the PHC levels in and around the Pichavaram mangroves in the Vellar–Coleroon estuarine complex and observed that the non-mangrove region (1.05–3.26 ppm) had lower values than the mangrove region (3.15–6.26 ppm). However, the coastal region (3.21–7.71 ppm) had the highest concentration in this estuarine complex. Very few studies examined the PHC content in the biotas from the east coast of India. However, Veerasingam et al. (2011a, b) reported the PHC content from 10 fish species (0.52–2.05 ppm wet weight) and eight mollusk species (2.44–6.04 ppm wet weight) along the Tamil Nadu coast. They found that the PHC levels in both fishes and mollusks did not indicate significant bioaccumulation.

Table 10.3 Petroleum hydrocarbon levels observed in the coastal sediments (in mg kg^{-1}), waters (in mg m^{-3}) and marine organisms (in mg kg^{-1} wet weight) along the east coast of India

Study site	Sampling time	Medium of sampling	PHC	Authors
Hooghly estuary	1984, 1985	Sediments	100–310	Ghosh and Choudhury (2001)
Haldia channel of Hooghly estuary	1997, 1998	Water	BDL–89	Sadhuram et al. (2005)
Digha Gopalpur Chilika Rushikulya Konark Puri Paradeep Dhamra Chandipur Hooghly estuary Mahanadi estuary Dhamra estuary	2002–2009 (only post-monsoon season)	Water	1.10 to 2.15 0.07 to 2.57 0.04 to 1.58 0.69 to 2.30 0.45 to 3.42 0.27 to 3.75 0.14 to 2.72 0.07 to 1.66 0.48 to 2.96 1.17 to 18.5 0.45 to 5.48 0.26 to 0.90	Panigrahy et al. (2014)
Godavari estuary	Not specified	Sediment	0.014 to 0.704	Kiran et al. (2015)
Adyar estuary	July 2008	Sediment	7.26–16.83	Venkatachalapathy et al. (2010)
Kuvam estuary	July 2008	Sediment	5.5–39.72	Venkatachalapathy et al. (2010)
Pichavaram mangroves, Vellar–Coleroon estuary	July 2009	Sediment	1.05–7.71	Venkatachalapathy et al. (2012)
Tamil Nadu coast	Not specified	10 fish species	0.52–2.05	Veerasingam et al. (2011a)
Tamil Nadu coast	March 2010	8 mollusk species	2.44–6.04	Veerasingam et al. (2011b)

10.8 Summary and Conclusion

Analyzing all the collated data and observations made by earlier researchers, we summarized that many of the estuaries along the east coast of India undergo transient eutrophication due to excessive nutrient load from land runoff and sewage. In this regard, the mangrove-dominated estuaries are worst affected, owing to the nutrient addition to the adjacent water column from mangrove-derived pore water. As an immediate consequence of such enhanced nutrient levels, several workers observed harmful algal blooms in many of these estuaries, though these were transient and did not exist all through the year. Factors like freshwater discharge, nitrogen to phosphorus ratio, and construction of dams affected the occurrence of algal blooms. Industrial activities, along with the dissemination of domestic sewage, jeopardized the water quality of many of these estuaries. Many estuarine reaches recorded the presence of fecal coliform in such ranges that even bathing in those waters is expected to cause a human health hazard. Owing to increasing anthropogenic activities, the organic matter load along with petroleum hydrocarbons has increased substantially in these estuaries. Overall, the abundance of these pollutants was higher in the northern estuaries along the east coastline compared to the southern estuaries.

References

Acharyya T, Sarma VVSS, Sridevi B et al (2012) Reduced river discharge intensifies phytoplankton bloom in Godavari estuary. India Mar Chem 132:15–22

Akhand A, Maity S, Mukhopadhyay A et al (2012) Dinoflagellate *Ceratium symmetricum* pavillard (Gonyaulacales: ceratiaceae): its occurrence in the Hooghly-Matla estuary and offshore of Indian Sundarban and its significance. J Threat Taxa 4(7):2693–2698

Anderson DM, Cembella AD, Hallegraeff GM (2012) Progress in understanding harmful algal blooms: paradigm shifts and new technologies for research, monitoring, and management. Annu Rev Marine Sci 4:143–176

Basu AK, Ghosh BB, Pal RN (1970) Comparison of the polluted Hooghly estuary with the unpolluted Matlah estuary, India. J Water Pollut Control Fed 42:1771–1781

Basu S, Banerjee T, Manna P et al (2013) Influence of physicochemical parameters on the abundance of Coliform bacteria in an Industrial Site of the Hooghly River, India. Proc Zool Soc 66(1):20–26

Batabyal P, Einsporn MH, Mookerjee S et al (2014) Influence of hydrologic and anthropogenic factors on the abundance variability of enteropathogens in the Ganges estuary, a cholera endemic region. Sci Total Environ 472:154–161

Biswas H, Mukhopadhyay SK, De TK et al (2004) Biogenic controls on the air—water carbon dioxide exchange in the Sundarban mangrove environment, northeast coast of Bay of Bengal, India. Limnol Oceanogr 49(1):95–101

Biswas SN, Rakshit D, Sarkar SK et al (2014) Impact of multispecies diatom bloom on plankton community structure in Sundarban mangrove wetland, India. Marine Pollut Bull 85(1):306–311

Brenner KP, Rankin CC, Roybal YR et al (1993) New medium for the simultaneous detection of total coliforms and *Escherichia coli* in water. Appl Environ Microbiol 59:3534–3544

Chanda A, Das S, Bhattacharyya S et al (2020) CO_2 effluxes from an urban tidal river flowing through two of the most populated and polluted cities of India. Environ Sci Pollut Res 27:30093–30107

Das S, Chanda A, Giri S et al (2015) Characterizing the influence of tide on the physico-chemical parameters and nutrient variability in the coastal surface water of Northern Bay of Bengal during winter season. Acta Oceanol Sin 34:102–111

Das S, Giri S, Das I et al (2017) Nutrient dynamics of northern Bay of Bengal (nBoB)—emphasizing the role of tides. Reg Stud Mar Sci 10:116–134

De TK, De M, Das S et al (2011) Phytoplankton abundance in relation to cultural eutrophication at the land-ocean boundary of Sunderbans, NE Coast of Bay of Bengal, India. J Environ Stud Sci 1(3):169–180

Frena M, Santos APS, Souza MR et al (2019) Sterol biomarkers and fecal coliforms in a tropical estuary: seasonal distribution and sources. Mar Pollut Bull 139:111–116

Gaspar FL, Pinheiro BR, Noriega CED et al (2018) Alkalinity, inorganic carbon and CO_2 flux variability during extreme rainfall years (2010–2011) in two polluted tropical estuaries NE Brazil. Braz J Oceanogr 66:115–130

Ghosh PB, Choudhury A (2001) Seasonal variation of aliphatic hydrocarbons in sediments of Hooghly estuary, north-east coast of India. Trop Ecol 42(1):133–136

Grant MA (1997) A new membrane filtration medium for simultaneous detection and enumeration of Escherichia coli and total coliforms. Appl Environ Microbiol 63:3526–3530

Hobbie SE, Finlay JC, Janke BD et al (2017) Contrasting nitrogen and phosphorus budgets in urban watersheds and implications for managing urban water pollution. Proc Natl Acad Sci 114(16):4177–4182

Jha DK, Devi MP, Vidyalakshmi R et al (2015) Water quality assessment using water quality index and geographical information system methods in the coastal waters of Andaman Sea, India. Mar Pollut Bull 100(1):555–561

Kanuri VV, Saha R, Raghuvanshi SP et al (2020) Sewage fluxes and seasonal dynamics of physicochemical characteristics of the Bhagirathi-Hooghly River from the lower stretch of River Ganges, India. Chem Ecol 36(1):30–47

Kiran R, Krishna VG, Naik BG et al (2015) Can hydrocarbons in coastal sediments be related to terrestrial flux? A case study of Godavari river discharge (Bay of Bengal). Curr Sci 108:96–100

Lotze HK, Lenihan HS, Bourque BJ et al (2006) Depletion degradation, and recovery potential of estuaries and coastal seas. Science 312(5781):1806–1809

Manna S, Chaudhuri K, Bhattacharyya S, Bhattacharyya M (2010) Dynamics of Sundarban estuarine ecosystem: eutrophication induced threat to mangroves. Saline Syst 6(1):8. http://www.salinesystems.org/content/6/1/8

Mitra S, Ghosh S, Satpathy KK et al (2018) Water quality assessment of the ecologically stressed Hooghly River Estuary, India: a multivariate approach. Mar Pollut Bullet 126:592–599

Mukhopadhyay SK, Biswas H, De TK, Jana TK (2006) Fluxes of nutrients from the tropical river Hooghly at the land–ocean boundary of Sundarbans, NE Coast of Bay of Bengal, India. J Mar Syst 62(1–2):9–21

Naik S, Mahapatro D, Behera DP et al (2013) Spatio-temporal study of zooplankton Community in Mahanadi Estuary, Bay of Bengal. Int J Ecosyst 3(6):185–195

Panda UC, Sundaray SK, Rath P et al (2006) Application of factor and cluster analysis for characterization of river and estuarine water systems–a case study: Mahanadi River (India). J Hydrol 331(3–4):434–445

Panigrahy PK, Satapathy DR, Panda CR, Kar RN (2014) Dispersion pattern of petroleum hydrocarbon in coastal water of Bay of Bengal along Odisha and West Bengal, India using geospatial approach. Environ Monit Assess 186(12):8303–8315

Prasad MBK (2012) Nutrient stoichiometry and eutrophication in Indian mangroves. Environ Earth Sci 67(1):293–299

Rakshit D, Biswas SN, Sarkar SK et al (2014) Seasonal variations in species composition, abundance, biomass and production rate of tintinnids (Ciliata: Protozoa) along the Hooghly (Ganges) River Estuary, India: a multivariate approach. Environ Monit Assess 186(5):3063–3078

Rao CV, Chari N, Muralikrishna R (2019) The impact of shrimp pond effluent on water quality of Vasishta Godavari estuary with respect to brackish water aquaculture, East Coast of India. Egypt J Aquat Biol Fish 23(3):245–255

Raven JA, Gobler CJ, Hansen PJ (2020) Dynamic CO_2 and pH levels in coastal, estuarine, and inland waters: theoretical and observed effects on harmful algal blooms. Harmful Algae 91:101594. https://doi.org/10.1016/j.hal.2019.03.012

Sadhuram Y, Sarma VV, Murthy TR, Rao BP (2005) Seasonal variability of physico-chemical characteristics of the Haldia channel of Hooghly estuary, India. J Earth Syst Sci 114(1):37–49

Samantray P, Mishra BK, Panda CR, Rout SP (2009) Assessment of water quality index in Mahanadi and Atharabanki Rivers and Taldanda Canal in Paradip area, India. J Human Ecol 26(3):153–161

Sarkar SK, Saha M, Takada H et al (2007) Water quality management in the lower stretch of the river Ganges, east coast of India: an approach through environmental education. J Clean Prod 15(16):1559–1567

Sarma VVSS, Prasad VR, Kumar BSK et al (2010) Intra-annual variability in nutrients in the Godavari estuary, India. Cont Shelf Res 30(19):2005–2014

Shah VG, Dunstan RH, Geary PM et al (2007) Evaluating potential applications of faecal sterols in distinguishing sources of faecal contamination from mixed faecal samples. Water Res 41(16):3691–3700

Sinha PC, Rao YR, Dube SK, Murthy CR (1998) A numerical model for residual circulation and pollutant transport in a tidal estuary (Hooghly) of NE Coast of India. Indian J Mar Sci 27:129–137

Sinha R, Das S, Ghosh T (2020) Pollution and its consequences at Ganga Sagar mass bathing in India. Environ Dev Sust 22(2):1413–1430

Soo CL, Ling TY, Lee N, Apun K (2016) Assessment of the characteristic of nutrients, total metals, and fecal coliform in Sibu Laut River, Sarawak, Malaysia. Appl Water Sci 6(1):77–96

Srichandan S, Baliarsingh SK, Prakash S et al (2019) Seasonal dynamics of phytoplankton in response to environmental variables in contrasting coastal ecosystems. Environ Sci Pollut Res 26(12):12025–12041

Supriyantini E, Yulianto B, Santoso A et al (2019) Organic matter concentrations in Morosari River Estuary, Sayung, Demak, Central Java. In IOP Conference Series: Earth and Environmental Science (Vol. 246, No. 1, p. 012039). IOP Publishing

Tripathy SC, Ray AK, Patra S, Sarma VV (2005) Water quality assessment of Gautami—Godavari mangrove estuarine ecosystem of Andhra Pradesh, India during September 2001. J Earth Syst Sci 114(2):185–190

Veerasingam S, Venkatachalapathy R, Raja P et al (2011a) Petroleum hydrocarbon concentrations in ten commercial fish species along Tamilnadu coast, Bay of Bengal, India. Environ Sci Pollut Res 18(4):687–693

Veerasingam S, Venkatachalapathy R, Sudhakar S, Raja P, Rajeswari V (2011b) Petroleum hydrocarbon concentrations in eight mollusc species along Tamilnadu coast, Bay of Bengal, India. J Environ Sci 23(7):1129–1134

Venkatachalapathy R, Veerasingam S, Ramkumar T (2010) Petroleum hydrocarbon concentrations in marine sediments along Chennai coast, Bay of Bengal, India. Bull Environ Contam Toxicol 85(4):397–401

Venkatachalapathy R, Veerasingam S, Basavaiah N et al (2011) Environmental magnetic and petroleum hydrocarbons records in sediment cores from the north east coast of Tamilnadu, Bay of Bengal. India Mar Pollut Bull 62(4):681–690

Venkatachalapathy R, Veerasingam S, Rajeswari V (2012) Distribution and origin of petroleum hydrocarbons in Pichavaram mangrove swamp along Tamilnadu coast, Bay of Bengal, India. Geochem Int 50(5):476–480

Wang J, Fu Z, Qiao H, Liu F (2019) Assessment of eutrophication and water quality in the estuarine area of Lake Wuli, Lake Taihu, China. Sci Total Environ 650:1392–1402

Chapter 11
Nutrient Cycling and Seasonal Dynamics of Primary Production in Nearshore Waters of East Coast of India

Rajdeep Roy, Ravidas Krishna Naik, Priya M. D'Costa, P. V. Nagamani, and S. B. Choudhury

Abstract The five major rivers (Ganges, Mahanadi, Godavari, Krishna, and Cauvery) flowing in the north eastern Indian Ocean greatly influence the biogeochemical cycling nearshore waters of the east coast of India. For example, estuaries which are located at the interface between land and ocean are an area of intense recycling of organic matter with a large salinity gradient. The estuarine environment is generally complex, as local circulations and mixing affect the chemistry and deposition of organic matter, thereby controlling the estuarine–coastal nutrient budget. The fluvial inputs are major sources of nutrients to the Bay of Bengal (BoB) which also regulates the phytoplankton dynamics in both estuaries and coastal waters. The annual supply of nutrients by the Ganges and Brahmaputra rivers to the BoB is estimated to be 133×10^9 mol year^{-1} which is ~2% of the riverine input to the world ocean. Both river runoff and precipitation are more intense during the southwest monsoon (SWM), resulting in lowering of surface salinity to 3 to 7 units in the Bay of Bengal. This also creates a strong seasonality in phytoplankton primary production in both estuaries and coastal waters. In general, primary production in the nearshore waters of eastern India has been found to covary with reduced suspended material and stability of water column with three seasonal peaks at few locations. Although it was hypothesized that the biological productivity of BoB is low in comparison to its adjacent basin due to various reasons such as a narrow shelf, cloud cover during the summer monsoon, turbidity resulting from sediment influx, etc. However, recent studies suggest the presence of cyclonic eddies can enhance primary production in BoB due to the entrainment of nutrient-rich waters

R. Roy (✉) · P. V. Nagamani · S. B. Choudhury
Indian Space Research Organization, National Remote Sensing Centre, Hyderabad, India
e-mail: rajdeep_roy@nrsc.gov.in

R. K. Naik
National Centre for Polar and Ocean Research (NCPOR), Ministry of Earth Sciences, Vasco-da-Gama, Goa, India

P. M. D'Costa
School of Earth, Ocean and Atmospheric Sciences, Goa University, Taleigao Plateau, Goa, India

S. Das, T. Ghosh (eds.), *Estuarine Biogeochemical Dynamics of the East Coast of India*, https://doi.org/10.1007/978-3-030-68980-3_11

within the oligotrophic mixed layer. All these suggest that factors controlling primary production in BoB remain poorly understood. As growing industrial activities and regulated discharge through dams can influence the riverine nutrient load, it is expected plankton's primary productivity and estuarine carbon budget may need periodic reassessment in near future.

Keywords Estuaries · Bay of Bengal · Ganges · Nutrients · Primary Production, carbon

11.1 Introduction

The five major rivers (Ganges, Mahanadi, Godavari, Krishna, and Cauvery) flowing in the north eastern Indian Ocean greatly influence the biogeochemical cycling of the estuarine environment. Estuaries are commonly described as semi-enclosed bodies of water, situated at the interface between land and ocean, where seawater is gradually diluted by the inflow of freshwater. The estuarine environment is generally complex, as local circulations and mixing effects the chemistry and deposition organic matter thereby controlling the estuarine–coastal nutrient budget. Along the east coast of India, the Hugli estuary forms largest delta in the world due to confluence of the distributaries of the Ganges and Brahmaputra (Fig. 11.1). The lower section of the estuary stretches through the rich mangrove forest, known as the Sunderban, covering the Matla River and therefore sometimes referred to as Hugli–Matla estuarine system although hydrologically they are completely different. The inland water connection provides a shipping channel to the major port (Kolkata port trust) through Kolkata and Haldia docks in the upper and lower sections, respectively. While Hugli estuary is shallow, funnel shaped, and considered as a positive estuary that traces its path through the southern region of the West Bengal state of India, several geomorphological changes has been recorded in recent times (Jayappa et al. 2006). The Matla River, on the other hand, no longer receives freshwater influx either from the River Hugli or Bidyadhari and thus becomes an enclosed tidal inlet of the sea with limited wave action and water movements. The region generally becomes filled with seawater during high tide, and most of the water gets drained away toward the sea at low tide, leaving a narrow stream of 0.9–1.2 m water in some places (Sarkar et al. 2004). The Hugli estuary receives a perennial freshwater discharge from the Ganges, and its lower stretches act as an open estuary throughout the year connecting to the coastal Bay of Bengal (BoB). The depth of the Hugli Estuary varies along the channel from ~21 m at Diamond Harbour to ~8 m at the mouth of the estuary (Central Inland Fisheries Research Institute 2012). The seasonality in this region can be described as spring inter-monsoon (February–May), summer monsoon (June–September), fall inter-monsoon (October–November), and northeast monsoon (December–January), respectively. The highest river discharge (3000 ± 1000 m^3 s^{-1}) in the Hooghly estuary is observed during the summer

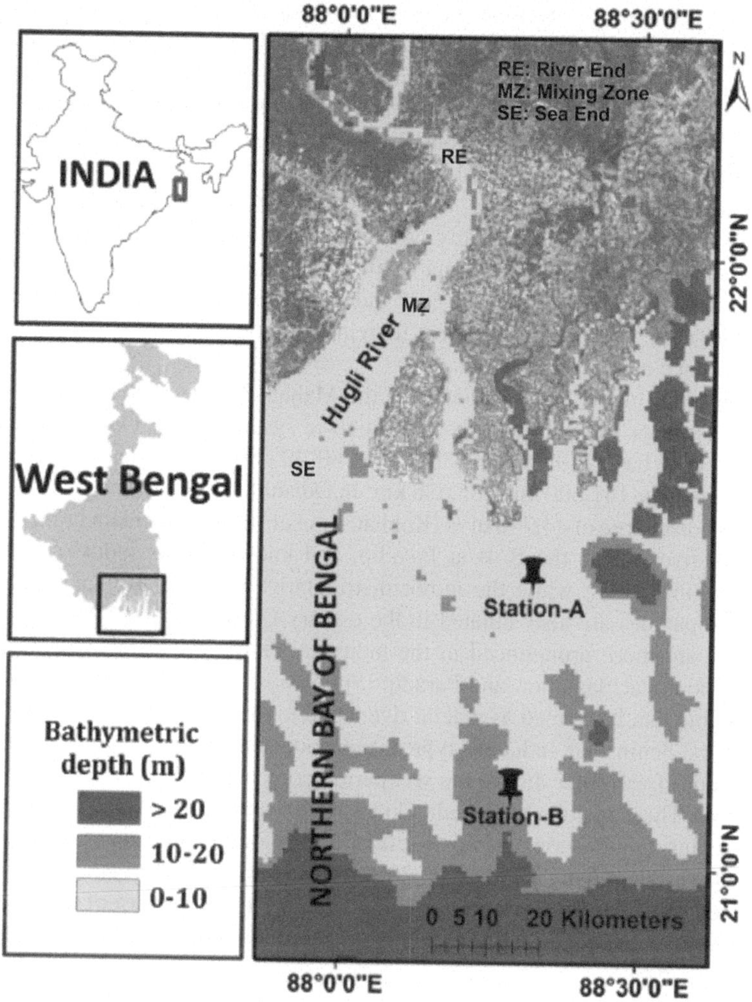

Fig. 11.1 Shows the stations at the river end, mixing zone, and sea end. The seasonal distribution of nutrients and their fluxes at these locations are described along with influence of tidal fluctuations at Station A and Station B offshore. (Reproduced from Das et al. 2017 with permission from Elsevier)

monsoon season and reduces to the minimum (1000 ± 80 m^3 s^{-1}) during the pre-monsoon season (Mukhopadhyay et al. 2006; Ray et al. 2015). The tidal regime of the estuary is strong and semi-diurnal in nature, with the tide height ranging from 5.2 m during spring to 1.8 m during the neap tide period and a velocity as high as 6 knots with significant influence the currents and the resultant circulation pattern (Sadhuram et al. 2005; Mukhopadhyay et al. 2006). The tidal flow and the opposing freshwater flow bring in a huge volume of organic and inorganic suspended sediments that result in high turbidity of the water column. The fluvial inputs are major sources of nutrients to the BoB; however, most of the nutrients are found to be removed within the estuary itself (Sarma et al. 2009). The annual supply of nutrients by the Ganges and Brahmaputra rivers to the BoB is 133 × 10^9 mol $year^{-1}$ accounting for 2% riverine input to the world ocean (Sarin et al. 1989). In addition to these, significant agricultural activities along the banks of rivers due to the availability of freshwater have also been reported (Sarma et al. 2014) Isotopic signatures of δ15N Particulate Organic Matter (POM) in the Indian estuaries from the discharge period suggest strong enrichment in highly polluted estuaries (>10‰) compared to relatively clean estuaries (<6‰) (Sarma et al. 2014). Apart from this Hugli estuary receives approximately 13 m^3 sec^{-1} of industrial effluent and urban wastewater (Sadhuram et al. 2005, and references therein) from densely populated and industrial cities such as Kolkata, Haldia.

To the south of the Hugli estuary lies the Mahanadi river system, which is the third-largest river of India and the largest river system in the Odisha state. The Mahanadi river basin (19°20′ to 23°35′N and 80°30′–86°50′E) extends over an area of approximately 141,600 km^2 (65,628 km^2 in Odisha), has a total length of 851 km and a peak discharge of 44,740 m^3/s (Konhauser et al. 1997). The main branch of the Mahanadi river meets the BoB at Paradip, and known for its industrial activity (Radhakrishna 2001), while the minor distributaries meet the BoB near Chilika. Mangrove patches are also situated in the estuary (Sundaray et al. 2006). Human influences are more pronounced in the industrial cities, viz. Sambalpur, Cuttack, Bauda, Choudwar, Jagatpur, and Paradip (Sundaray et al. 2006). While Hugli and Brahmaputra can be termed as glacial rivers due to the influence of snow melt, the rivers in the peninsular India are typical monsoonal rivers, as these rivers receive large amount freshwater during the wet period compare to the dry months (Sarma et al. 2014). These include Mahanadi, Godavari, Krishna and Cauvery. Due to heavy discharge during monsoon period, the estuaries turn into freshwater system with no vertical salinity gradient (Vijith et al. 2009; Sridevi 2013; Sridevi et al. 2015). The discharge of these rivers is in unison with the precipitation pattern of the area, and are therefore called "Monsoonal estuaries" and exhibit a non-steady state behavior (Vijith et al. 2009).

The Godavari estuary in peninsular India is located around 16°15′ N and 82°5′ E, covering an area of 330 km with a basin area of 3.1 × 10^5 km^2. Near the town Rajahmundry the river flow is regulated by a century-old dam at Dowleswaram (Sarma et al. 2014). Strong seasonal variability in nutrient distribution is observed here between monsoon and non-monsoon period (Sarma et al. 2010). While Krishna estuary is about 45 km in length and the estuarine system of Krishna river covers an area of about 320 km^2 with all its four distributaries, the tidal portion extends up to 39 km upstream near Penumudi in the river. It is essentially a shallow one with a mean width of 1.2 km and an average depth of 5–7 m. The estuary has a well-developed sandy coast that experiences longshore drift, build-up spits, and barriers across river mouths, creating a coastal lagoon (Kumari and Rao 2010). It is a meso-tidal estuary with a tide range of 2–3 m and strong tidal current of 1.2 m sec^{-1}. Dissolved and particulate loads of the Krishna river are derived from a variety of igneous, metamorphic, and sedimentary rocks in their catchment areas (Kumari and

Rao 2010). Similarly, the fourth-largest river of southern region, Cauvery, is an easterly flowing river of the peninsular India that runs across three of the southern Indian states, that is, Karnataka, Tamil Nadu, Kerala, and the Union Territory of Puducherry. The river in its 800 km long journey from the Western Ghats traverses through Mysore plateau and finally forms a delta on the eastern coastline of the subcontinent before falling into the Bay of Bengal. The point of origin of Cauvery, Talakaveri, is in the Brahmagiri ranges of the Western Ghats at an elevation of 1341 m (https://www.indiawaterportal.org/). In addition to this, the East India Coastal Currents (EICC) which flows along the coastal BoB also influences the local budget which reverses its direction twice a year. It flows northwards from February to September with a strong peak in March–April and flows southwestward from October to January with strongest flow during January (Schott and McCreary Jr 2001). A study by Bristow et al. (2017) suggests that the amount of nutrients coming from the east coast to India through river discharge can make BoB a dead zone due to increase in primary production and sinking carbon fluxes. All these highlight the complex interaction where anthropogenic perturbation as well as physical process can act together in different time scale thereby influencing the estuarine budget. Thus, Indian estuaries need occasional reassessments of nutrient budget and associated fluxes in order to delineate long-term trends.

11.2 Nutrient Dynamics in and Around East Coast of India

Aquatic dissolved nutrients mainly consist of inorganic nitrogen (nitrate + nitrite), phosphorus, and silica (Si (OH) 4), which generally limit phytoplankton growth in natural waters. Studies on nutrient distribution from the Hugli–Matla estuarine system is limited (Mukhopadhyay et al. 2006; Das et al. 2015, 2017). Further, very few studies highlight the influence of tides on nutrient distribution in this region (Das et al. 2015, 2017). Based on Land-Ocean Interactions in the Coastal Zone (LIOCZ) biogeochemical model, Mukhopadhyay et al. 2006 illustrated that fluxes of inorganic nitrogen were in an order of (65.8 × 10^3 t) followed by phosphorus (12.8×10^3 t). Further silicate discharge by Hugli estuary was found to be in the order of ($Si(OH)_4$ (42.8×10^3 t) where more than 50% of the load was contributed during the monsoon seasons. They also showed that during the estuarine transport, 7.0% of the silicate was removed within the estuary itself; however, there was addition of inorganic nitrogen (59%) and phosphorus (44%), which they attributed to heterotrophic regeneration within the estuary. The seasonal variations of inorganic nitrogen and silicate at three stations (Station 1: sea end; Station 2: mixing zone; Station 3: riverine end) along the Hugli estuarine system is illustrated in Fig. 11.2.

Das et al. 2015 studied biogeochemical characteristics at seven different locations along the Hugli estuarine system during highest high tide (HTT) and lowest low tide (LLT) to understand the influence of the tidal fluxes on nutrient distribution. The data reveal that during the LLT hours, a relative increase of freshwater input in the northern BoB can have elevated the nutrient concentration compared

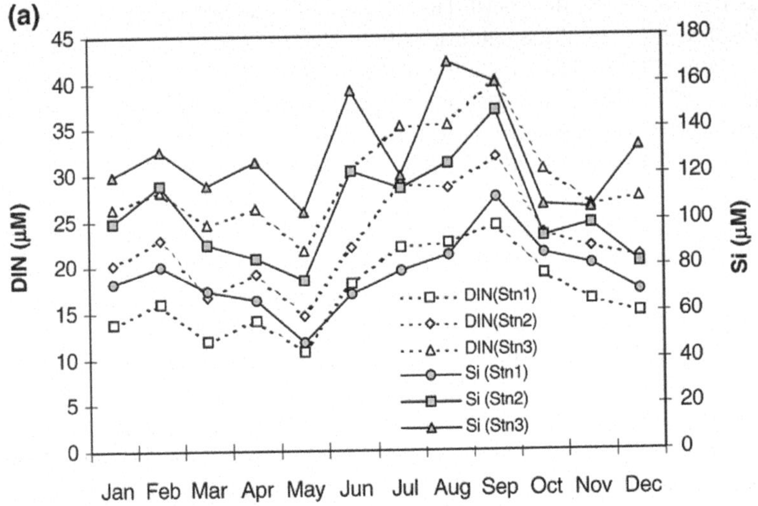

Fig. 11.2 Monthly variation (averaged over 2 years between 1999 and 2001) of (**a**) silicate and dissolved inorganic nitrogen (Reproduced from Mukhopadhyay et al. 2006 with permission from Elsevier)

with that observed during the HHT hours. The ratio of nutrient concentration is found to deviate significantly from the Redfield ratio in the Hugli estuarine system. The abundance of dissolved inorganic phosphorous (DIP) is much higher compared with that of dissolved inorganic nitrogen (DIN) and dissolved inorganic silica (DSi). The anthropogenic sources of DIP from the upstream flow (especially the domestic effluent of several metropolises) can be mainly attributed during those observations. Apart from this, weathering of rocks is found to release soluble alkali phosphate and colloidal calcium phosphate to the rivers which is ultimately drained in estuary can also be a potential source (Ramesh et al. 2015). Further, Das et al. (2017) did a follow-up study during the high tide and low tide conditions (daytime) of the spring and neap phases throughout two annual cycles (February 2013 to January 2015) to understand how tidal processes principally regulate the nutrient variability in short-term scale in the Hugli estuarine region. Influence of tidal fluxes on nutrient distribution along the Hugli estuarine system is illustrated in Fig. 11.3. The results suggest that spring–neap and high tide–low tide contrast was most prominent in pre-monsoon and post-monsoon season. Hugli estuarine system is mainly fed by glacial rivers (Ganges and Brahmaputra); however, few of the other rivers along the east coast can be called as monsoonal rivers as there is significant difference in the discharge rates between monsoon and non-monsoon periods (Sarma et al. 2010). Discharge from the Indian monsoonal rivers is limited to a few months only because these rivers are fed by the SWM-induced precipitation over the catchment. However, glacial rivers also show remarkably high discharge (50% of the annual discharge) during SWM (June–September) due to the influence of monsoonal precipitation over the catchments (Ittekkot et al. 1991). The detailed nutrient budget along with

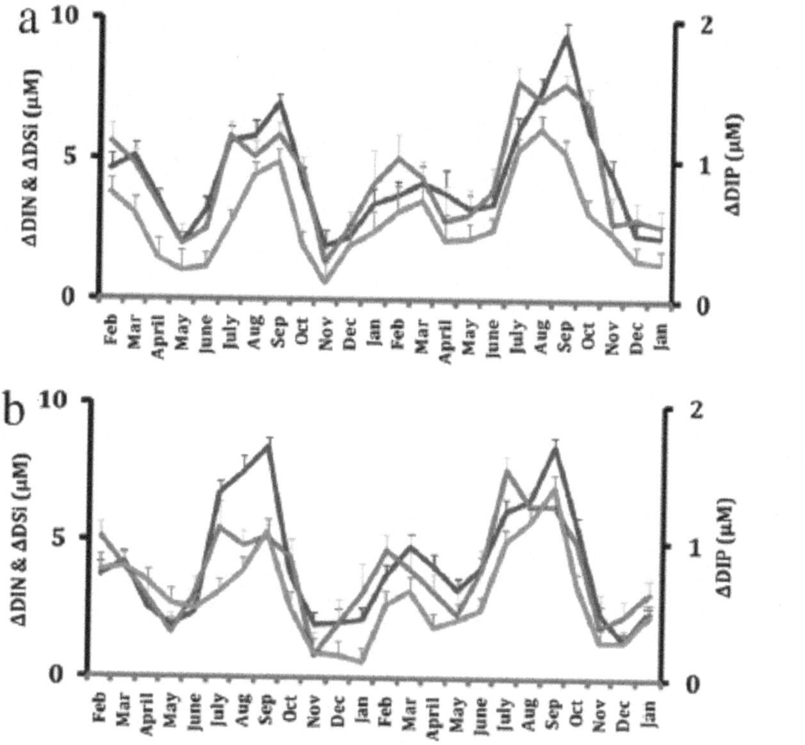

Fig. 11.3 The monthly mean spatial differences (between Station A and Station B) in concentrations of DIP, DIN, and DSi (i.e., ΔDIP, ΔDIN, and ΔDSi) during (**a**) spring tide and (**b**) neap tide. The error bars show the standard deviation from mean during each month. Blue line: ΔDIN; green line: ΔDSi; maroon line: ΔDIP. (Reproduced from Das et al. 2017 with permission from Elsevier)

the catchment area for the east coast of India is illustrated in Table 11.1. Krishna et al. (2016) showed that among the monsoonal rivers in peninsular India, Godavari transports highest amount of nitrate (0.08 Tg yr. $^{-1}$), while the Krishna transports highest amounts of phosphate (0.02 Tg yr.$^{-1}$) and silicate (0.36 Tg yr.$^{-1}$). These rivers export similar amount of dissolved inorganic N (ammonium + nitrite + nitrate) to the BoB (0.12 ± 0.03 Tg yr.$^{-1}$) as compared to that of Arabian Sea; however, the phosphate and silicate load during monsoon is much higher in BoB than Arabian Sea. The monsoonal rivers were found to export 1.03 ± 0.25 Tg yr.$^{-1}$ of silicate to the northern Indian Ocean of which 0.91 ± 0.23 Tg yr.$^{-1}$ enter the BoB (Krishna et al. 2016). Apart from this contribution from atmospheric deposition and submarine groundwater discharge has also been recently quantified. As atmospheric deposition of DIN and DIP from Indo-Gangetic (IGP) plain to BoB remained poorly understood this study is important and timely. Air mass back trajectories suggest conspicuous downwind transport of chemical constituents from the IGP to the BoB during the late northeast monsoon (January–April). Among the water-soluble nitrogen compounds, ammonium was found to dominate 90% of the observed aerosol,

Table 11.1 Catchment area (10^6 km^2), annual mean discharge (km^3), and length (km) of the Indian monsoonal rivers are given. Measured concentrations (mg L^{-1}) in estuaries during southwest monsoon (SWM), and estimated total export (tons year^{-1}) of dissolved inorganic N (nitrite + nitrate + ammonium), phosphate, and silicate to the coastal ocean from the monsoonal rivers were also provided. Export fluxes normalized by catchment area of dissolved inorganic N (kg N km – ^{2} yr.$^{-1}$), phosphate (kg P km^{-2} yr.$^{-1}$), and silicate (kg Si km^{-2} yr.$^{-1}$) from the Indian monsoonal estuaries during SWM are also given from Krishna et al. (2016). Reproduced with permission from Elsevier

Name of the estuary	Catchment area (10^6 km^2)	Mean discharge (km^3)	Length (km)	Dissolved inorganic nitrogen (mg L^{-1})	Phosphate (mg L^{-1})	Silicate (mg L^{-1})	Dissolved inorganic nitrogen (tons yr.$^{-1}$)	Phosphate (tons yr.$^{-1}$)	Silicate (tons yr.$^{-1}$)	Dissolved inorganic nitrogen (kg N km^{-2} yr.$^{-1}$)	Phosphate (kg P km^{-2} yr.$^{-1}$)	Silicate (kg Si km^{-2} yr.$^{-1}$)
Bay of Bengal												
Haldia	0.0102	50.46	–	0.28	0.42	0.74	14,227	20,960	37,157	1395	2055	3643
Subernarekha	0.0193	12.36	446	0.10	0.02	0.79	1250	303	9761	65	16	506
Baitarani	0.0142	28.48	440	0.17	0.08	1.01	4744	2384	28,864	334	168	2033
Mahanadi	0.1416	66.89	858	0.12	1.11	1.62	7800	7071	108,439	55	50	766
Rushikulya	0.009	1.93	175	0.08	0.03	0.45	146	66	865	16	7	96
Vamsadhara	0.011	3.50	254	0.09	0.04	0.67	332	149	2342	30	14	213
Nagavali	0.0094	1.99	256	0.07	0.04	0.39	137	83	786	15	9	84
Godavari	0.313	110.53	1465	0.72	0.12	2.02	79,571	13,192	223,764	254	42	715
Krishna	0.259	69.79	–	0.13	0.30	5.20	9321	21,180	362,875	36	82	1401
Penna	0.055	6.30	597	0.06	0.18	5.74	384	1134	36,186	7	21	658
Ponnayaar	0.016	1.60	396	0.10	0.10	7.56	153	165	12,096	10	10	756
Vellar	0.0086	0.90	210	0.07	0.26	3.92	65	234	3531	8	27	411
Cauvery	0.088	21.35	–	0.08	0.38	3.50	1602	8008	74,725	18	91	849

followed by 10% of other nitrogen compounds (Srinivas et al. 2016), whereas the abundance of soluble DIP varied between 0.4 to 4.8 nmol m^{-3}. Studies such as this suggest complex source of nutrient enrichment within the estuary and should be further investigated for understanding the regional budget.

Further, Rengrajan and Sarma (2015) studied the submarine groundwater discharge (SGD) and its contribution to nutrient fluxes in the BoB. A model based on the decay of ^{224}Ra relative to ^{228}Ra was used to determine apparent water ages of various bays within the estuary. These ages ranged from 2.6 to 4.8 days during November 2011. Knowing the water age, the distribution of radium in the estuary and the radium isotopic composition of groundwater enabled them to calculate SGD fluxes to the estuary and determine the magnitude and seasonal variability in the nutrient fluxes to the estuary associated with SGD. These nutrient fluxes (in units of mmol $m^{-2}d^{-1}$) ranged 1–19 (DIN), 0.6–2.6 (DIP), and 5–40 (DSi) in Gautami–Godavari; 19–40 (DIN), 2.6–5.5 (DIP), and 200 (DSi) in Vasishta Godavari; and 120–140 (DIN), 10 (DIP), and 220 (DSi) in Kakinada bay. It was observed that high SGD fluxes to Kakinada bay contribute significant nutrients in the southern peninsula. It is found that the river-borne nutrients (1.74 ± 0.43 Tg $yr.^{-1}$) support 17% of total primary production (66 Tg $yr.^{-1}$; Fernandes et al. 2008) in the BoB (upper 10 m).

11.3 Primary Productivity Dynamics from Estuarine and Coastal Waters of Bay of Bengal

Ocean biogeochemical processes modulate the ocean ecosystem through gross primary production (GPP) in the ocean surface where inorganic carbon is fixed by phytoplankton in ocean through photosynthetic processes. Determination of GPP is an important function of phytoplankton biomass and its physiological status which can be measured through various techniques (e.g., from carbon isotopes to satellite based) to determine the phytoplankton primary production. Several units have been used to express primary production in literature. However, the most common are the mmol C m^{-2} day^{-1}, mg C m^{-2} day^{-1}, g C m^{-2} y^{-1}, where the use of moles makes the comparison of the stoichiometric ratios easier between nutrients and carbon and most of the chemical oceanographer prefer this. During the measurement of primary production, often the respiration rates are also determined which when subtracted from GPP yields the net primary production (NPP), which is defined as the amount of photosynthetically fixed carbon available to the first heterotrophic level in the aquatic ecosystem. The global oceanic annual net primary production is about an order of 48.5 Gt (1 Gt = 10^{15} grams), which contributes nearly 46.2% of total global annual net primary productivity (marine + terrestrial) (Field et al. 1998). The spatial distribution of NPP is not homogeneous and there are regions of low production such as the central gyres, whereas the high production areas include the estuarine and upwelling regions.

Various methods have been used to estimate primary productivity of oceans. Based on the source of nitrogen, used Primary Production (PP) can be categorized into New Production or Regenerated Production. New Production is defined as the primary production that is based on the nitrate as nitrogen source from diffusion/upwelling from below or from the atmosphere via nitrogen fixation or nitrification whereas Regenerated Production is the production based on ammonium and urea as the source of nitrogen. The ratio of nitrate uptake by the sum of uptake by nitrate and ammonium gives the term "f-ratio," originally coined by Dugdale and Goering (1967). In other words, if we write P = gross production and R = respiration, then we can also approximate f, as $f = (P - R/P)$, which can also be termed as the ration of NPP to GPP. Some of the common methods used are O2 evolution technique, 14C assimilation, ^{15}NO3 assimilation, 15NH4 assimilation, and remote sensing of ocean color. Some of the above techniques can help in measuring the new and regenerated production and have varying time scales from hours to days. Traditionally, BoB is considered to be less productive than the adjacent basin (Arabian Sea), although located in the similar latitudes (Qasim 1977; Sen Gupta et al. 1977; Kumar et al. 2002). The low biological productivity of BOB has been hypothesized due to various reasons, such as narrow shelf, cloud cover during summer monsoon, turbidity resulting from sediment influx, freshwater-induced stratification, and even sunlight (Qasim 1977; Gomes et al. 2000; Kumar et al. 2010). Earlier reports suggest that PP values can range up to 3.0–8.7 g C m^2d^{-1} from the inshore waters of the east coast of India in June–July (Nair et al. 1973). A study by Gomes et al. (2000) suggests that during March–April (spring inter-monsoon), the poleward flow of EICC brings nutrient-laden cooler waters that enrich the inshore waters leading to increase in PP. They observed highest biomass (Chl *a*, 53 mg m^{-2}) and productivity (4.5 g C $m^{-2}d^{-1}$) were located in the region of an eddy-like structure along the coast between 13 and 16°N latitude. This study carefully illustrates the linkages between the physical dynamics and primary productivity. During the summer monsoon of 2001, the PP values along the shallow waters of east coast of India ranged 328–520 mg C m^{-2} d^{-1}, except for a low 40 mg C m^{-2} d^{-1} at 15°N (Madhupratap et al. 2003). A vertical profile of primary productions from coastal and open ocean waters in BoB is shown in Fig. 11.4. PP within the mixed layer was about 54% of the total, while below 80 m it was insignificant (8% in both coastal and open waters). About 26% of the PP was associated with the subsurface chlorophyll maxima. The productivity to chlorophyll *a* ratio was between 10 and 20, although a few higher values (up to 40) occurred at some coastal stations (Madhupratap et al. 2003). Phytoplankton population during summer monsoon of 2001 was dominated by diatoms both in coastal and open ocean waters. Low production in open ocean in BoB in comparison to the coastal waters in BoB is attributed to nitrate deficit due to arising from strong stratification and weaker mixing in comparison to Arabian Sea (Prassana Kumar et al. 2002). During a revisit, Prasanna Kumar et al. (2010) occupied several stations along the western boundary of the BoB during three (summer, fall, and inter) monsoon to understand the influence of light limitation on PP. Along the western boundary of the BOB, the highest PP of 502.01 mg C m^{-2} d^{-1} was in the south at 12°N, and the second-highest value of

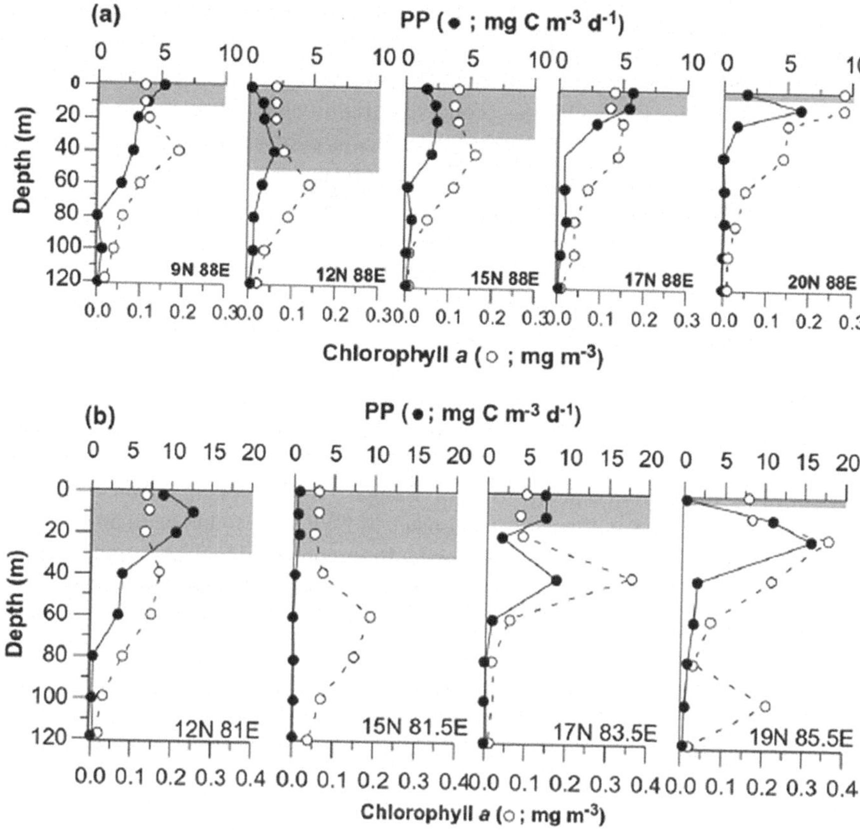

Fig. 11.4 Vertical profiles of primary production (solid circles) and chlorophyll a (open circles) in (**a**) oceanic and (**b**) coastal stations (shown by latitudes and longitudes). Mixed-layer depths are shown as shaded areas. (Adapted from Madhupratap et al. (2003) with permission from Elsevier)

433.8 mg C m^{-2} d^{-1} was in the north at 19°N during summer monsoon (July–August). Their analysis showed that along the western boundary the highest PP in the northern bay was not associated with summer or in the fall inter-monsoon, but occurred in the spring inter-monsoon. This is due to that fact that riverine flux reduced the downward penetration of solar radiation at both times, acting in conjunction with prevailing cloud cover which possibly affected the PP. Recently the role of eddies in influencing the total and size-fractionated PP in BoB has also been illustrated (Sarma et al. 2020). Two cyclonic (CE), one anticyclonic (ACE), and no-eddy (NE) regions were sampled in the BoB during pre-summer monsoon (June 2019). Sarma et al. (2020) illustrated that photic zone integrated total primary production was higher in the CE and NE than ACE regions associated with higher nutrients in the former than latter region. Due to the availability of nutrients within the CE microphytoplankton was more dominant and contributed significant amount to the size-fractionated PP. Here the PP ranged within CE (164 ± 16 mgC m^{-2} d^{-1})

than ACE (60 ± 26 mgC m^{-2} d^{-1}). Higher picophytoplankton production was observed at depth below 10 m from surface (10–80%) than nano- and microphytoplankton (1–30%). This study is important as it suggest cyclonic eddies can enhance PP in BoB otherwise known be less productive traditionally. Further role of eddies and its implications on export fluxes in BoB remains poorly understood.

A localized study from estuarine waters of Hugli suggests that maximum productivity occurred during winter when cell counts showed highest values as well (136 cells ml^{-1}). On the contrary, minimum productivity with respect to both carbon equivalents as well as total phytoplankton cell counts was recorded in the monsoon months (cells ml^{-1}). Phytoplankton productivity (GPP) at the estuarine station was maximum in December 2010 (227.77 mg C m^{-3} h^{-1}) when total phytoplankton cell count was maximum as well (203 cells ml^{-1}) and minimum in August 2010 (58.95 mg C m^{-3} h^{-1}) (Choudhury and Pal 2012). Further, recent satellite-based study suggests a strong link between tropical hilsa shad (*Tenualosa ilisha*) in the estuarine region of Hugli and Bangladesh. This study describes spatial and temporal variability of productivity in the Bay of Bengal (BoB) relating to hilsa fishery (Hossain et al. 2020). Satellite-based estimates of NPP were found be in an order of >2000 mg C m^{-2} day^{-1} within in the Ganges–Brahmaputra–Meghna region covarying with higher fish catch. Their study suggests variations in seasonal productivity linked with nutrients and phytoplankton abundance are important factors for predicting hilsa habitat and their migration patterns in the deltaic regions and shelf waters of BoB. Historical data suggests that most of the primary production is supported by larger phytoplankton called the diatoms although contribution by other classes is also being recognized recently (Madhupratap et al. 2003; Sarma et al. 2020).

11.4 Phytoplankton Community Structure from Estuarine Waters of East Coast of India

The information on phytoplankton community structure from east coast of India is available from 1942 (Chacko 1942), probably the first report that explain the phytoplankton bloom from east coast of India. It is evident that the information on phytoplankton community have been studied and updated on regular intervals by various researchers (Roy 1955; Shetty et al. 1961; Subba Rao 1969; Gopalakrishnan 1971; Santra et al. 1991; Banerjee and Santra 2001; Mukhopadhyay and Pal 2002; Sarkar and Naskar 2002; Choudhury and Pal 2012; Manna et al. 2010; Akhand et al. 2012). The overall results indicate that the phytoplankton community at east coast of India is dominated by diatom group followed by dinoflagellate, blue-green algae, and chlorophyceae. Diatom dominance holds true with different estuarine complex along the east coast of India, such as Hooghly estuary, where diatom contribute more than 50% of the total phytoplankton taxa with 195 species followed by 82 species of green algae, 59 species of cyanophyceae, and 29 species of dinoflagellates (Roshith et al. 2018). Among the dinoflagellate community, further classification based on their mode of nutrition (autotrophic, mixotrophic, heterotrophic), it was noted that that the heterotrophic dinoflagellate were the dominant forms in

Hooghly estuary irrespective of the different seasons (Naik et al. 2011). Mahanadi estuarine regime also exhibited the similar pattern of diatom contributing more than 50% of the total phytoplankton taxa identified, followed by cyanobacteria, chrysophyceae, and dinoflagellates (Mishra et al. 2018). While the phytoplankton communities patter from further south estuarine ecosystems (Krishna and Godavari) along the east coast of India, they were not different than the Hugli and Mahanadi estuarine complex, where the diatom contributed more than 80% to the total phytoplankton biomass followed by dinoflagellates, green algae, cyanobacteria, and silicoflagellates (Baliarsingh et al. 2016). Similarly, Bharathi et al. (2018) studied the phytoplankton community dynamics along the northwest and southwest regions of the Bay of Bengal along the east coast and found the diatom being most abundant group of phytoplankton. Studies carried out based on phytoplankton pigments also indicated the dominance of fucoxanthin (marker pigment for diatom) along the coastal Bay of Bengal (Bandyopadhyay et al. 2017). Studies carried out along the Krishna north, Krishna south, Godavari north, and Godavari south estuarine transects indicated the dominance of diatom, with more than 65% contribution to the total phytoplankton biomass followed by other groups (Bharathi et al. 2018).

11.4.1 Phytoplankton Bloom Dynamics Along the East Coast of India

Among the bloom-forming groups of phytoplankton along the east coast of India, blooms caused by diatoms tops the table followed by dinoflagellate and cyanobacteria (D'Silva et al. 2012). Among the diatom, *Asterionella japonica* (=*Asterionella glacialis*) was the most common bloom-forming species; *Noctiluca scintillans* and *Trichodesmium erythraeum* were the common bloom-forming species belong to dinoflagellate and cyanobacteria group, respectively. D'Silva et al. (2012) in their comprehensive review on algal blooms from Indian waters found that the occurrence of algal bloom along the east coast is throughout the year with the exception of January and November months. While March to May are favorable for diatom dominance, April to August are better for dinoflagellate and cyanobacterial blooms, mostly during month of March.

11.4.2 Causative Factors for Higher Growth of Phytoplankton

Phytoplankton community structure from the east coast of India, specifically from major estuarine habitats, exhibit the more or less similar pattern with dominance of diatom. However, the total biomass and the secondary dominating groups are varying from one region to another. The causative factors explained such as change in salinity due to high influx of freshwater through the rivers, stratification, local upwelling, and also the storms and cyclones. Diatom blooms of *A. japonica* were related to high nutrient, low temperature conditions due to local upwelling (Rao

1969), whereas recent studies show that the dominance of diatoms along the estuarine complex was related to higher silicate concentrations (Bharathi et al. 2018). Higher abundance of *T. erythraeum* during the pre-monsoon period was attributed to the prevalence of stratified condition in the Bay of Bengal (Hegde et al. 2008). Among the higher abundance of dinoflagellates, *Noctiluca scintillans* was associated with nutrient-enriched water during the plankton succession (D'Silva et al. 2012). Physical mechanisms such as eddies and cyclones and their potential role are also highlighted in bloom dynamics of *N. scintillans* (Naik et al. 2011).

11.5 Summary

The biogeochemical dynamics at nearshore waters of the east coast of India is complex and dynamic, as local circulations and mixing affect the chemistry and deposition of organic matter thereby controlling the estuarine–coastal nutrient budget. The fluvial inputs are major sources of nutrients to the Bay of Bengal (BoB) which also regulates the phytoplankton dynamics in both estuaries and coastal waters. The river-borne nutrients (~1.74 ± 0.43 Tg $yr.^{-1}$) support 17% of total PP in the BoB (upper 10 m). Although earlier studies have shown that seasonal dynamics in PP are strongly controlled by riverine inputs and stratification due to freshwater fluxes and low light, however recent analysis suggest t that formation of eddies can be an important source of nutrient enhancement in BoB which remains poorly quantified. Therefore future studies should also address the impact of these mesoscale features on regional productivity and carbon cycling from this region along with quantification of size-fractionated PP budget. This will help us to understand carbon cycling across the phytoplankton groups and its importance across the different oceanic environments.

Acknowledgement RR also acknowledges, Chairman ISRO, Director NRSC, and and Deputy Director ECSA, for their support to carry out these investigations under the umbrella of ISRO–GBP–coastal carbon dynamics project. RN would like to acknowledge Director, NCPOR-MoES for his constant encouragement. PMD acknowledges the Dean, School of Earth, Ocean and Atmospheric Sciences; and the Head, Department of Microbiology, for their support. This is NCPOR contribution number B-12/2020-21.

References

Akhand A, Maity S, Mukhopadhyay A et al (2012) Dinoflagellate Ceratium symmetricum pavillard (Gonyaulacales: ceratiaceae): its occurrence in the Hooghly-Matla estuary and offshore of Indian Sundarban and its significance. J Threatened Taxa 26:2693–2698

Baliarsingh SK, Lotliker AA, Trainer VL et al (2016) Environmental dynamics of red Noctiluca scintillans bloom in tropical coastal waters. Mar Pollut Bull 111:277–286.

Bandyopadhyay D, Biswas H, Sarma VV (2017) Impacts of SW monsoon on phytoplankton community structure along the western coastal BOB: an HPLC approach. Estuar Coasts 40:1066–1081

Banerjee A, Santra SC (2001) Phytoplankton of the rivers of Indian Sundarban mangrove estuary. Indian Biologist 33:67–71

Bharathi MD, Sarma VV, Ramaneswari K et al (2018) Influence of river discharge on abundance and composition of phytoplankton in the western coastal Bay of Bengal during peak discharge period. Mar Pollut Bull 133:671–683

Biswas H, Mukhopadhyay SK, De TK et al (2004) Biogenic controls on the air—water carbon dioxide exchange in the Sundarban mangrove environment, northeast coast of Bay of Bengal, India. Limnol Oceanogr 49:95–101

Bristow LA, Callbeck CM, Larsen M et al (2017) N 2 production rates limited by nitrite availability in the Bay of Bengal oxygen minimum zone. Nat Geosci 10:24–29

Chacko PI (1942) An unusual incidence of mortality of marine fauna. Curr Sci 11:404

Choudhury AK, Pal R (2012) Understanding the seasonal dynamics of primary productivity in relation to phytoplankton populations from the Bhagirathi–Hooghly estuary, eastern Indian coast. J. Algal Biomass Utln 3:80–88

CIFRI (2012) Present status of Hilsa in Hugli-Bhagirathi river, Central Inland Fisheries Research Institute, [Available at www.cifri.ernet.in.179.pdf].

D'Silva MS, Anil AC, Naik RK (2012) Algal blooms: a perspective from the coasts of India. Nat Hazards 63:1225–1253

Das S, Chanda A, Giri S et al (2015) Characterizing the influence of tide on the physico-chemical parameters and nutrient variability in the coastal surface water of the northern Bay of Bengal during the winter season. Acta Oceanol Sin 34:102–111

Das S, Giri S, Das I et al (2017) Nutrient dynamics of northern Bay of Bengal (nBoB)—emphasizing the role of tides. Reg Stud Mar Sci 10:116–134

Dugdale R C, Goering. J J (1967) Uptake of new and regenerated forms of nitrogen in primary productivity. Limnol. Oceanogr. 12:196–206

Fernandes V, Ramaiah N, Paul JT et al (2008) Strong variability in bacterioplankton abundance and production in central and western Bay of Bengal. Mar Biol 153:975–85

Field CB, Behrenfeld MJ, Randerson JT et al (1998) Primary production of the biosphere: integrating terrestrial and oceanic components. Science 281:237–240

Gomes HR, Goes JI, Saino T (2000) Influence of physical processes and freshwater discharge on the seasonality of phytoplankton regime in the Bay of Bengal. Conti Shelf Res 20:313–330

Gopalakrishnan V (1971) The biology of the Hooghly-Matlah estuarine system (West Bengal, India) with special reference to its fisheries. J Mar Biol Ass India 13:182–194

Gupta RS, De Sousa SN, Joseph T (1977) On nitrogen and phosphorus in the western Bay of Bengal Indian. J Mar Sci 6:107–110

Hegde S, Anil AC, Patil JS et al (2008) Influence of environmental settings on the prevalence of Trichodesmium spp. in the Bay of Bengal. Marine Ecol Progr Ser 18:93–101

Hossain MS, Sarker S, Sharifuzzaman SM et al (2020) Primary productivity connects Hilsa fishery in the Bay of Bengal. Sci Rep 10:1–6

Ittekkot V, Nair RR, Honjo S et al (1991) Enhanced particle fluxes in Bay of Bengal induced by injection of fresh water. Nature 351:385–387

Jayappa KS, Mitra D, Mishra AK (2006) Coastal geomorphological and land-use and land-cover study of Sagar Island, Bay of Bengal (India) using remotely sensed data. Int J Remote Sens 27:3671–3682

Konhauser KO, Powell MA, Fyfe WS et al (1997) Trace element geochemistry of river sediment, Orissa State, India. J Hydrol 193:258–269

Krishna MS, Prasad MH, Rao DB et al (2016) Export of dissolved inorganic nutrients to the northern Indian Ocean from the Indian monsoonal rivers during discharge period. Geochim Cosmochim Acta 172:430–443

Kumar SP, Muraleedharan PM, Prasad TG et al (2002) Why is the Bay of Bengal less productive during summer monsoon compared to the Arabian Sea? Geophys Res Lett 29:88–81

Kumar SP, Narvekar J, Nuncio M et al (2010) Is the biological productivity in the Bay of Bengal light limited? Curr Sci 25:1331–1339

Kumari VR, Rao IM (2010) Suspended sediment dynamics in Krishna estuary, east coast of India. Indian J Mar Sci 39:248–256

Madhupratap M, Gauns M, Ramaiah N et al (2003) Biogeochemistry of the Bay of Bengal: physical, chemical and primary productivity characteristics of the central and western Bay of Bengal during summer monsoon 2001. Deep Sea Res Part II Topi Stud Oceanograp 50:881–896

Manna S, Chaudhuri K, Bhattacharyya S et al (2010) Dynamics of Sundarban estuarine ecosystem: eutrophication induced threat to mangroves. Saline Syst 6:8

Mukhopadhyay A, Pal R (2002) A report on biodiversity of algae from coastal West Bengal (South & North 24-parganas) and their cultural behaviour in relation to mass cultivation programme. Indian Hydrobio 5:97–107

Mukhopadhyay SK, Biswas HD, De TK et al (2006) Fluxes of nutrients from the tropical river Hooghly at the land–ocean boundary of Sundarbans, NE Coast of Bay of Bengal, India. J Marine Syst 62:9–21

Naik RK, Hegde S, Anil AC (2011) Dinoflagellate community structure from the stratified environment of the Bay of Bengal, with special emphasis on harmful algal bloom species. Environ Monit Assess 182:15–30

Nair, P.V.R., Samuel, S., Joseph, K.J., Balachandran, VK (1973) Primary production and potential fishery resources in the seas around India. In: Proceedings of the symposium on Living Resources of the Seas Around India, 1968, Special publication. Central Marine Fisheries Research Institute, Cochin, pp. 184–198. Google Scholar

Qasim SZ (1977) Biological productivity of the Indian Ocean. Indian J Mar Sci 6:U1·137

Radhakrishna I (2001) Saline fresh water interface structure in Mahanadi delta region, Orissa, India. Environ Geol 40:369–380

Ramesh R, Robin RS, Purvaja R (2015) An inventory on the phosphorus flux of major Indian rivers. Curr Sci 10:1294–1299

Rao DS (1969) *Asterionella Japonica* bloom and discoloration off Waltair, Bay of Bengal. Limnol Oceanogr 14:632–634

Ray R, Rixen T, Baum A, Malik A et al (2015) Distribution, sources and biogeochemistry of organic matter in a mangrove dominated estuarine system (Indian Sundarbans) during the pre-monsoon. Estuar Coast Shelf Sci 167:404–413

Rengarajan R, Sarma VV (2015) Submarine groundwater discharge and nutrient addition to the coastal zone of the Godavari estuary. Mar Chem 20:57–69

Roshith CM, Meena DK, Manna RK et al (2018) Phytoplankton community structure of the Gangetic (Hooghly-Matla) estuary: status and ecological implications in relation to eco-climatic variability. Flora 240:133–143

Roy HK (1955) Plankton ecology of the river Hooghly at Palta, West Bengal. Ecology 36:169–175

Sadhuram Y, Sarma VV, Murthy TR et al (2005) Seasonal variability of physico-chemical characteristics of the Haldia channel of Hooghly estuary, India. J Earth Syst Sci 114:37–49

Santra SC, Pal UC, Choudhury A (1991) Marine phytoplankton of the mangrove delta region of West Bengal. India J Mar Biol Ass India 33:292–307

Sarin MM, Krishnaswami S, Dilli K et al (1989) Major ion chemistry of the Ganga-Brahmaputra river system: weathering processes and fluxes to the Bay of Bengal. Geochim Cosmochim Acta 53:997–1009

Sarkar SK, Frančišković-Bilinski S, Bhattacharya A et al (2004) Levels of elements in the surficial estuarine sediments of the Hugli River, Northeast India and their environmental implications. Environ Int 30:1089–1098

Sarkar NS, Naskar KR (2002) Taxonomy of the diatoms flora of the Sundarban mangals in West Bengal, India. J Interacad 14:81-108

Sarma VV, Gupta SN, Babu PV et al (2009) Influence of river discharge on plankton metabolic rates in the tropical monsoon driven Godavari estuary, India. Estuar Coast Shelf Sci 85:515–524

Sarma VV, Prasad VR, Kumar BS, et al (2010) Intra-annual variability in nutrients in the Godavari estuary, India. Cont Shelf Res 30:2005–14.

Sarma VV, Krishna MS, Prasad VR et al (2014) Distribution and sources of particulate organic matter in the Indian monsoonal estuaries during monsoon. J Geophys Res Biogeosci 119:2095–2111

Sarma VV, Chopra M, Rao DN et al (2020) Role of eddies on controlling total and size-fractionated primary production in the Bay of Bengal. Conti Shelf Res 24:104186

Schott FA, McCreary JP Jr (2001) The monsoon circulation of the Indian Ocean. Prog Oceanogr 51:1–23
Sridevi B (2013) A comprehensive study on physical processes and their impact on biogeochemistry of Godavari estuary, India; PhD thesis, Andhra University, Visakhapatnam
Sridevi B, Sarma VV, Murty TV et al (2015) Variability in stratification and flushing times of the Gautami–Godavari estuary, India. J Earth Syst Sci 124:993–1003
Srinivas B, Rastogi N, Sarin MM et al (2016) Mass absorption efficiency of light absorbing organic aerosols from source region of paddy-residue burning emissions in the Indo-Gangetic plain. Atmos Environ 125:360–370
Sundaray SK, Panda UC, Nayak BB et al (2006) Multivariate statistical techniques for the evaluation of spatial and temporal variations in water quality of the Mahanadi river–estuarine system (India)–a case study. Environ Geochem Health 28:317–330
Vijith V, Sundar D, Shetye SR (2009) Time-dependence of salinity in monsoonal estuaries. Estuar Coast Shelf Sci 85:601–608

Chapter 12
Microzooplankton in Estuaries, Mangroves, and Lagoons of East Coast of India

Biraja Kumar Sahu and Sourav Das

Abstract The east coast of India is endowed with major estuaries, luxuriant mangroves, and the largest lagoons of India. This study compiled the microzooplankton research works undertaken in estuaries, mangroves, and lagoons of the east coast of India. Regionally, the microzooplankton study is reported from five estuaries, three mangrove areas, and two lagoon studies. The taxonomic groups in microzooplankton comprised ciliates, rotifers, heterotrophic dinoflagellates (HDFs), crustacean nauplii, molluscan nauplii, foraminiferans, radiolarians, etc. The ciliates formed the dominant group in all the ecoregions. A total of 120 ciliate species were listed that are classified into 29 families and 39 genera. The genera *Tintinnopsis*, *Strombidium*, and *Eutintinnus* contained the highest number of species. Among the ciliates, 91 species are loricate (shelled) and 29 are loricate (naked) forms. Among 91 loricate species, 57 are agglomerated and 34 are non-agglomerated (hyaline). Most studies focused on tintinnids only. Tintinnid biomass and production rate are available for four regions. Rotifers comprised 59 species covering 14 families and 25 genera. The families Brachionidae and Lecanidae dominated the species list. *Brachionus* genera covered maximum species (13 species) followed by *Lecane* (10 species). The highest number of rotifer species (35 species) are recorded in Chilika lagoon and Pichavaram mangroves. A few studies of HDFs in microzooplankton are available. A big research gap is there in microzooplankton studies including taxonomy and process studies in the tropical coastal environments of India.

Keywords East coast of India · Bay of Bengal · Estuarine · Plankton · Microzooplankton

B. K. Sahu
Atal Centre for Ocean Science & Technology for Islands, National Institute of Ocean Technology, Port Blair, India

S. Das (✉)
School of Oceanographic Studies, Jadavpur University, Kolkata, West Bengal, India

S. Das, T. Ghosh (eds.), *Estuarine Biogeochemical Dynamics of the East Coast of India*, https://doi.org/10.1007/978-3-030-68980-3_12

12.1 Introduction

Microzooplankton represents a significant constituent of the plankton community. Their size ranges from 20 to 200 µm. They form an important link between small cells (like pico-, nano-, bacterioplankton) and higher trophic levels (like mesozooplankton and fishes) (Mukherjee et al. 2015). It consists of planktonic ciliates, rotifers, heterotrophic dinoflagellates, crustacean nauplii, molluscan nauplii, foraminiferans, etc. They play a significant role in carbon transfer in the food web by linking higher trophic levels from primary production of small organisms (Hopkins 1987). According to Gauns et al. (2005), "it is believed that due to the close coupling between microbial and microzooplankton components of the aquatic food webs, less organic carbon leaves the euphotic zone especially where microzooplankton form the mediator route for the uptake of organic carbon thereby influencing biogeochemical cycles" (Gauns et al. 2005). Only a small portion of the organic matter produced by autotrophs takes the "fast lane" to upper trophic levels, to be grazed directly by large metazoans (e.g., copepods) and most primary production circulates through different trophic levels, including microzooplankton, and is eventually respired within the microbial loop (Azam et al. 1983; Calbet 2008). According to Bruggeman and Kooijman (2007), if the role of each group of the microzooplankton is depicted, then the dynamics of the food web will be better understood, and more precise plankton models will be built.

The east coast of India is endowed with giant river estuaries, luxuriant mangroves, and the largest lagoons of India. Major rivers of India drain into the Bay of Bengal (BoB) on the east coast of India, such as the Ganges, Brahmaputra, Godavari, Mahanadi, etc. Estuaries are semi-enclosed bodies of water where freshwater from rivers and a coastal stream merges with the ocean, and it mediates material flux between terrestrial and marine ecosystems (Sooria et al. 2015). The mixing of freshwater and seawater generates salinity gradient and variable physical characteristics that create multiple unique habitats supporting high diverse communities and it provides vital links to nearby ecosystems (McLusky and Elliott 2004; Sun et al. 2012). The estuaries can be microtidal (<2 m), mesotidal (2–4 m), macrotidal (4–6 m), and hypertidal (>6 m) based on a tidal range (McLusky and Elliott 2004). On the Indian east coast, the tidal range increases from the south (0.6 m in Nagapattinam) toward the north (5.5 m in Sunderban) (Misra et al. 2013; Das 2015). Estuaries receive considerable amounts of freshwater, nutrients, dissolved and particulate organic matter, suspended matter, and also contaminants from land, and exchange materials and energy with the open ocean (Sarma et al. 2010). The freshwater flow into the estuaries varies largely with seasons; the highest flow is in monsoon and the lowest in summer seasons. The northern part of the east coast sees high river flow in the southwest monsoon (June to September) while the southern part (mostly Tamil Nadu) sees the high flow in the northeast monsoon (November to December). The mangrove regions which thrive adjacent to these estuarine regions also face similar freshwater flow. Most of the lagoons on the east coast of India also demonstrate estuarine characteristics with many rivers flowing into them.

Microzooplankton population dynamics is a prerequisite to fully understand and quantitatively model carbon cycling pathways and other biogenic elements on a global scale (Godhantaraman 2004). The present study collected all the information on microzooplankton studies carried out in estuaries, mangroves, and lagoons of the east coast of India. It compiles all the data and aims to give a holistic view of the microzooplankton of the east coast of India.

12.2 Methodology

The microzooplankton studies were collected from the published articles, book chapters, and books. The literature collection retrieved the studies from Hooghly estuary, Sunderban mangroves, Chilika lagoon, Bahuda estuary, Godavari–Coringa mangroves, Pulicat lagoon, Vellar estuary, Pichavaram mangroves, Parangipettai backwater, Kaduviyar estuary, Muthupet lagoon, and Pullavazhi estuary (Fig. 12.1). It covered the studies carried out from 1973 to 2020. Before 1973, literature is there

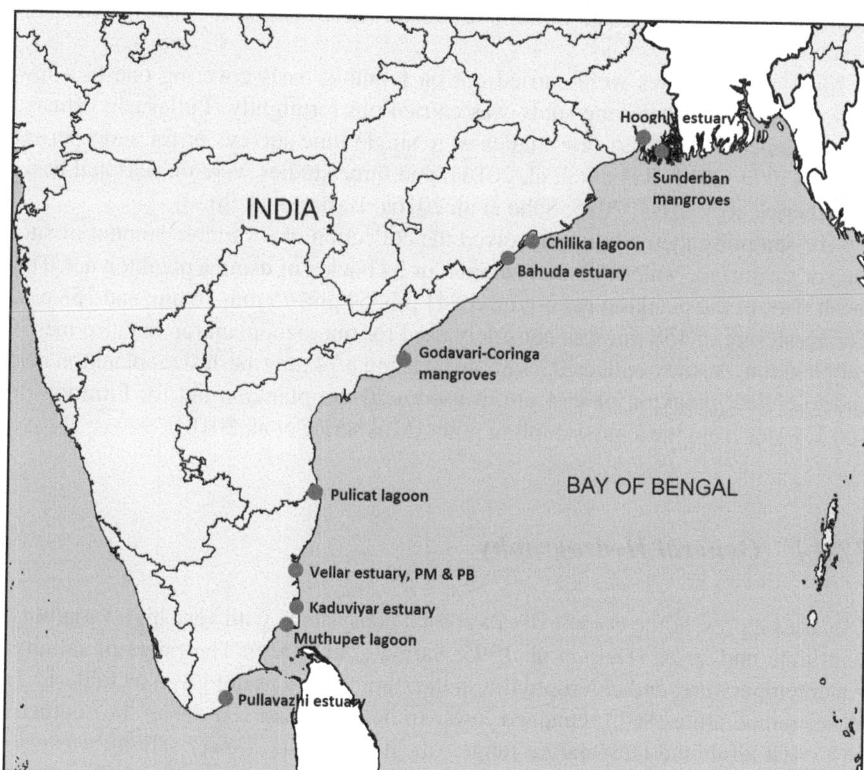

Fig. 12.1 Map showing microzooplankton studied sites on east coast of India (PM: Pichavaram mangroves, PB: Parangipettai backwater). (Source: Google Earth)

on taxa coming under microzooplankton; however they are exempted from this study due to their inaccessibility. Basic hydrographic parameters such as salinity, temperature and chlorophyll-*a*, and microzooplankton species list was prepared.

12.3 Results and Discussion

The microzooplankton study literature in the eastern Indian estuaries/mangroves/lagoons included Hooghly estuary (Biswas et al. 2013, 2014; Rakshit et al. 2015, 2016a), Sunderban mangroves (Rakshit et al. 2016b, 2017), Chilika lagoon (Pattnaik 1973; Kalavati and Raman 2008; Sahu et al. 2016a; Mukherjee et al. 2014, 2015, 2018), Bahuda estuary (Mishra and Panigrahy 1999), Godavari–Coringa mangroves (Kalavati and Raman 2008), Pulicat lagoon (Basuri et al. 2020), Vellar estuary (Krishnamurthy and Santhanam 1975; Godhantaraman 2001, 2002; Prabu et al. 2005), Pichavaram mangroves (Krishnamurthy and Santhanam 1975; Godhantaraman 1994, 2001, 2002), Parangipettai backwater (Krishnamurthy and Santhanam 1975), Kaduviyar estuary (Perumal et al. 2009), Muthupet lagoon (Santhanam et al. 2019), and Pullavazhi estuary (Srinivasan et al. 1988) from north to south.

Most of the studies were carried out on monthly basis covering one to a few years, and among them one study was carried out fortnightly (Pullavazhi estuary, Srinivasan et al. 1988). A few studies were single time surveys or for short period (Biswas et al. 2014; Rakshit et al. 2016a) and three studies were on seasonal basis (Mukherjee et al. 2014, 2015; Sahu et al. 2016a; Basuri et al. 2020).

The sampling methodology involved the collection of a variable amount of surface or subsurface water by water sampler or by bucket or using a plankton net. The mesh sizes of the plankton net used were 41 μm, 54 μm, 75 μm, 76 μm, and 158 μm. The mesh size of 158 μm was not solely used for microzooplankton but also mesozooplankton. A study collected a sample by using a 74 μm (mesh size) plankton net, and to collect plankton of <74 μm, it used a 20 μm plankton net for filtration of 100 L water from the same sampling point (Mukherjee et al. 2018).

12.3.1 General Hydrography

Estuaries represent the complexity of coastal ecosystems with very high variability with time and space (Heip et al. 1995; Sarma et al. 2009). The range of salinity, water temperature, and chlorophyll-*a* in the studied regions are given in Table 12.1. Water temperature (WT) remained lower in northern east coast than the southern east coast while the temperature range was just opposite. Lower salinity was also observed in northern parts due to prevalence of more freshwater. Sunderban mangrove area showed higher salinity and the range was also high as compared to nearby Hooghly estuary. The salinity upper value (36.3 PSU) of Bahuda estuary is

Table 12.1 Hydrographical parameters, microzooplankton (ciliate) species number, and population range

	Salinity (PSU)	WT (°C)	Chl-*a* (mg/m³)	Species (Nos.)	Population[a] (Nos./L)	Collection method	References
Sunderban mangroves	5.4–30.8	19.0–34.8	0.67–5.67	35	0–2747	Water sample[b]	Biswas et al. (2013, 2014), Rakshit et al. (2015, 2016a)
Hooghly estuary	0.21–22.28	19–34	1.10–2.71	32	52–9113	Water sample	Rakshit et al. (2016b, 2017)
Chilika lagoon	0–33.6	22.2–33.4	2.1–49.6	48	0–5244	Water sample and by net [20μm and 74μm, Mukherjee et al. 2015; 75μm, Pattnaik 1973]	Pattnaik (1973), Kalavati and Raman (2008), Mukherjee et al. (2014, 2015, 2018), Sahu et al. (2016a)
Bahuda estuary	2.7–36.3	24.0–32.5	–	24	1–16	By net (76μm)	Mishra and Panigrahy (1999)
Godavari–Coringa mangroves	0–34.36	–	–	5	–	Water sample	Kalavati and Raman (2008)
Pulicat lagoon	12.5–61.0	25.2–32.8	0.6–106.8	29	96–25,546	Water sample	Basuri et al. (2020)
Vellar estuary	2.3–35.5	22.5–34	~1–~23	53	1–420	Water sample and by net [54μm, 75μm, Prabu et al. 2005; Krishnamurthy and Santhanam 1975]	Krishnamurthy and Santhanam (1975), Godhantaraman (2001, 2002), Prabu et al. (2005)
Parangipettai backwater	5.2–35.5	23–34	–	7	9–79	By net (75μm)	Krishnamurthy and Santhanam (1975)
Pichavaram mangroves	2.6–35.5	22.5–34	~1–~24	37	1–290	Water sample and by net [75μm, Krishnamurthy and Santhanam 1975]	Krishnamurthy and Santhanam (1975), Godhantaraman (1994, 2001, 2002)
Kaduviyar estuary	0–34	24–31	–	11		By net (158μm)	Perumal et al. (2009)
Muthupet lagoon	0.9–39	26–34	–	11		By net (158μm)	Santhanam et al. (2019)

(continued)

Table 12.1 (continued)

	Salinity (PSU)	WT (°C)	Chl-*a* (mg/m³)	Species (Nos.)	Population[a] (Nos./L)	Collection method	References
Pullavazhi estuary	25.6–35.5	24.5–29.5	–	17	0–148	By net (41μm)	Srinivasan et al. (1988)

[a]Range is not representative of all studies given. It is given wherever it is available. Some populations are only of tintinnids

[b]Here water sample means collection by water sampler or by bucket

high as compared to the coastal waters and it is so with Godavari–Coringa mangroves (34.36 PSU). Pulicat lagoon is a hyper-saline environment with the recorded salinity range of 12.5–61.0 PSU. Vellar estuary, Pichavaram mangroves, and Parangipettai backwater showed similar salinity range with somewhat higher value in backwater. In Muthupet lagoon, it ranged 0.9–39.0 PSU. Pullavazhi estuary seemed to be a seawater-dominated estuary with salinity range of 25.6–35.5 PSU. Chlorophyll-*a* (proxy of primary production) was found low in Sunderban and Hooghly estuary as compared to other reported estuaries and lagoons. Highest chlorophyll-*a* (Chl-*a*) concentrations were in lagoons.

12.3.2 Ciliates (Phylum: Ciliophora)

These unicellular eukaryotes are widely distributed and diverse group of protists. Presence of nuclear dualism, short generation time, species-specific ciliature, and specialized oral apparatus to grasp food are some of the unique characteristics which distinguish ciliates from other protists (Corliss 1979; Fukang et al. 2002; Kaur et al. 2019). Ciliates thrive in different habitats – planktonic, epiphytic, psammophilic/in sediments and as parasite. Microzooplankton include the planktonic forms only. They are an important component of all aquatic ecosystems on earth, from freshwater to almost saturated brines (Pandey et al. 2008). Tintinnids are the most recognizable among the numerous species of ciliates, and they are beautiful due to their distinctive tunics called *loricae*, a name borrowed from the armor of Roman soldiers (Sardet 2015).

Ciliates are mostly heterotrophs, and many of them feed on bacteria, diatoms, and algae (Finlay et al. 1996). Some ciliates are mixotrophs that have characters like acquired phototrophy and karyoklepty (Sahu et al. 2016b). So, ciliates can be found at various trophic levels in food webs, which make them important for the proper functioning of an ecosystem (Verni and Rosati 2011). These are source of protein with a low C:N ratio as compared to phytoplankton, and thus considered as high-quality food for higher trophic level (Stoecker and Capuzzo 1990; Gifford 1991). The ciliate community structure is highly complex and most likely an important driver for structuring the phytoplankton community (Haraguchi et al. 2018).

In the east coast of India, most of the microzooplankton studies are focused on ciliates, more specifically tintinnids (Class: Spirotrichea, Subclass: Choreotrichia, Order: Tintinnida). A total of 120 ciliate species are recorded under 29 families and 39 genera. The genera *Tintinnopsis*, *Strombidium*, and *Eutintinnus* contained the highest number of species. Among all, 91 species are loricate and 29 are aloricate. Among 91 loricate species, 57 are agglomerated and 34 are non-agglomerated (hyaline). Generally, the two lorica types approximately correspond to different habitats where they live (Dolan 2013). The agglomerated ciliates mostly dominate in coastal regions while non-agglomerated ones in open ocean (Dolan 2013).

The ciliate community of Pulicat lagoon largely differed from other regions (Table 12.2). In Hooghly estuary and Sunderban mangroves, the species like *Tintinnopsis beroidea*, *Tintinnidium primitivum*, and *Leprotintinnus simplex* form the core species of the environment. Rest of the species of this region are observed seasonally or occasionally. If the mangrove regions are compared, Pichavaram mangrove region has shown higher number of species (37). Among all, Vellar estuary has shown highest number of species (53). It was observed in this case that more investigations lead to more number of species records (Table 12.1).

The species that found in most regions are *Tintinnopsis beroidea*, *Tintinnopsis cylindrica*, *Tintinnopsis tocantinensis*, *Tintinnopsis tubulosa*, *Tintinnopsis directa*, *Leprotintinnus nordqvistii*, *Favella philippinensis* (7 species). When the species list was compared, it was observed that some species remained confined to specific study regions. It is so because some studies focused on only loricates while others focused on both. Maximum number of species remained constrained in Pulicat lagoon – *Rimostrombidium* sp., *Strombidium tintinnodes* (=*Strombidium oculatum*), *Strombidium spiralis*, *Strombidium acutum*, *Strombidium conicum*, *Strombidium elongatum*, *Strombidium* sp1., *Strombidium* sp2., *Strobilidium* sp., *Euplotes* sp., *Euplotes vannus*, *Euplotes crassus*, *Fabrea salina*, *Uronema* sp., *Frontonia* sp., *Coleps* sp., *Mesodinium* sp., *Prorodon* sp., *Holosticha* sp., *Epistylis* sp., and *Amphirellopsis* sp. (21 species). *Tintinnopsis brasiliensis*, *Tintinnopsis amoyensis*, *Tintinnopsis orientalis*, *Codonellopsis lusitanica*, *Tintinnidium mucicola*, *Amphorellopsis tetragona*, and *Amphorellopsis acuta* (7 species) were confined to Sunderban mangroves. *Wangiella dicollaria*, *Tintinnopsis urnula*, *Tintinnopsis tentaculata*, *Tintinnopsis turbo*, *Tintinnopsis acuminata*, and *Helicostomella* sp. (6 species) confined to Hooghly estuary. *Lacrymaria coronata*, *Loxophyllum setigerum*, *Nassula citrea*, *Oxytricha oxymarina* (=*Oxytricha marina*), *Holophrya simplex*, *Tintinnopsis compressa*, *Tintinnopsis failakkaensis*, *Tintinnopsis spiralis*, *Tintinnopsis fimbriata*, *Eutintinnus fraknoi*, *Eutintinnus elongatus*, *Metacylis tropica*, *Favella adriatica*, *Favella campanula* (14 species) were confined to Chilika lagoon. *Prorodon discolor* and *Tintinnidium fluviatile* (2 species) were confined to Godavari–Coringa mangroves. *Didinium* sp., *Metacylis corbula*, *Helicostomella fusiformis*, *Epiplocylis undella*, and *Rhabdonella spiralis* (5 species) were confined to Vellar estuary. *Tintinnopsis amphora* (1 species) is confined to Pichavaram mangroves. There may be two reasons for this confinement in this coast:

Table 12.2 List of ciliates in estuaries, mangroves, and lagoons of east coast of India (P: present)

Species name	Sunderban mangroves	Hooghly estuary	Chilika lagoon	Bahuda estuary	Godavari–Coringa mangroves	Pulicat lagoon	Vellar estuary	Parangipettai backwater	Pichavaram mangroves	Kaduviyar estuary	Muthupet lagoon	Pullavazhi estuary
Rimostrombidium sp.						P						
Strombidium tintinnodes (=*Strombidium oculatum*)						P						
Strombidium spiralis						P						
Strombidium acutum						P						
Strombidium conicum						P						
Strombidium elongatum						P						
Strombidium sp1						P						
Strombidium sp2						P						
Strobilidium sp.						P						
Euplotes sp.						P						
Euplotes vannus						P						
Euplotes crassus						P						
Fabrea Salina						P						
Uronema sp.						P						
Frontonia sp.						P						
Coleps sp.						P						
Mesodinium sp.						P						

Prorodon discolor					P							
Prorodon sp.						P						
Holosticha sp.						P						
Epistylis sp.						P						
Lacrymaria coronata			P									
Loxophyllum setigerum			P									
Nassula citrea			P									
Tetrahymena pyriformis			P		P							
Halteria chlorelligera			P		P							
Oxytricha oxymarina (=Oxytricha marina)			P									
Holophrya simplex			P									
Didinium sp.							P					
Wangiella dicollaria		P										
Tintinnopsis urnula		P										

(continued)

Table 12.2 (continued)

Species name	Sunderban mangroves	Hooghly estuary	Chilika lagoon	Bahuda estuary	Godavari–Coringa mangroves	Pulicat lagoon	Vellar estuary	Parangipettai backwater	Pichavaram mangroves	Kaduviyar estuary	Muthupet lagoon	Pullavazhi estuary
Tintinnopsis tentaculata		P										
Tintinnopsis parvula		P	P				P					
Tintinnopsis turbo		P										
Tintinnopsis nana		P	P				P		P			
Tintinnopsis amphora									P			
Tintinnopsis glans							P		P			
Tintinnopsis mortensenii				P			P	P	P	P		P
Tintinnopsis kofoidi							P		P	P		
Tintinnopsis tenuis							P		P			
Tintinnopsis sacculus			P	P				P	P			P
Tintinnopsis compressa			P									
Tintinnopsis failakkaensis			P									
Tintinnopsis spiralis			P									
Tintinnopsis acuminata		P										

Tintinnopsis rotundata		P	P									
Tintinnopsis fimbriata			P									
Tintinnopsis bermudensis		P	P	P			P			P	P	P
Tintinnopsis beroidea	P	P	P	P	P	P	P		P	P		P
Tintinnopsis minuta	P	P					P		P	P	P	
Tintinnopsis lohmanni	P	P	P	P			P		P			P
Tintinnopsis cylindrica	P	P	P	P		P	P	P	P	P	P	P
Tintinnopsis lobiancoi	P	P				P						
Tintinnopsis tocantinensis	P		P	P		P	P	P	P	P	P	P
Tintinnopsis uruguayensis	P		P	P			P					P
Tintinnopsis tubulosa	P	P	P	P			P	P	P	P	P	P
Tintinnopsis nucula	P	P	P				P		P			

(continued)

Table 12.2 (continued)

Species name	Sunderban mangroves	Hooghly estuary	Chilika lagoon	Bahuda estuary	Godavari–Coringa mangroves	Pulicat lagoon	Vellar estuary	Parangipettai backwater	Pichavaram mangroves	Kaduviyar estuary	Muthupet lagoon	Pullavazhi estuary
Tintinnopsis parva	P	P										
Tintinnopsis buetschlii	P		P	P			P			P	P	P
Tintinnopsis directa	P	P	P	P			P		P		P	P
Tintinnopsis brasiliensis	P											
Tintinnopsis amoyensis	P											
Tintinnopsis radix	P	P	P				P		P			
Tintinnopsis orientalis	P											
Tintinnopsis karajacensis	P	P	P				P		P			
Tintinnopsis gracilis	P	P	P				P		P			
Tintinnopsis sp1	P											
Tintinnopsis sp2	P											
Tintinnopsis sp.			P									
Codonellopsis schabi	P	P					P		P			
Codonellopsis lusitanica	P											
Codonellopsis ostenfeldi			P	P			P	P	P			

Codonellopsis orthoceras							P		P			
Codonellopsis tessellata							P		P			
Codonellopsis sp.							P					
Stenosemella ventricosa	P	P	P				P					
Stenosemella nivalis			P				P					
Stenosemella parvicollis							P		P			
Stenosemella steineii							P		P			
Stenosemella sp.	P		P			P						
Tintinnidium primitivum	P	P					P		P			
Tintinnidium mucicola	P											
Tintinnidium incertum		P		P			P		P			
Tintinnidium fluviatile					P							
Leprotintinnus simplex	P	P	P	P			P		P			

(continued)

Table 12.2 (continued)

Species name	Sunderban mangroves	Hooghly estuary	Chilika lagoon	Bahuda estuary	Godavari–Coringa mangroves	Pulicat lagoon	Vellar estuary	Parangipettai backwater	Pichavaram mangroves	Kaduviyar estuary	Muthupet lagoon	Pullavazhi estuary
Leprotintinnus nordqvistii	P	P	P	P		P	P	P	P			P
Amphorellopsis acuta	P		P	P			P		P			
Eutintinnus apertus	P		P									
Eutintinnus fraknoi			P									
Eutintinnus elongatus			P									
Eutintinnus sp.	P	P				P						
Eutintinnus tenuis				P			P		P		P	P
Eutintinnus lusus-undae							P		P			
Eutintinnus rectus							P		P			
Amphorellopsis tetragona	P											
Amphorellopsis acuta	P											
Amphirellopsis sp.						P						
Dadayiella ganymedes		P					P					
Dadayiella bulbosa			P				P					
Metacylis tropica			P									

Metacylis jorgensenii			P	P								P
Metacylis corbula							P					
Metacylis sp.	P	P										
Coxliella ampla				P			P					
Coxliella annulata							P		P			
Helicostomella longa				P			P		P			
Helicostomella fusiformis							P					
Helicostomella sp.		P										
Favella ehrenbergii	P	P	P			P	P		P			
Favella adriatica			P									
Favella campanula			P									
Favella philippinensis			P	P			P	P	P	P	P	P
Favella brevis				P			P			P	P	P
Favella sp.							P				P	
Cyttarocylis sp.			P									
Dictyocysta sp.			P				P					

(continued)

Table 12.2 (continued)

Species name	Sunderban mangroves	Hooghly estuary	Chilika lagoon	Bahuda estuary	Godavari–Coringa mangroves	Pulicat lagoon	Vellar estuary	Parangipettai backwater	Pichavaram mangroves	Kaduviyar estuary	Muthupet lagoon	Pullavazhi estuary
Dictyocysta seshaiyai				P			P		P			P
Epiplocylis undella							P					
Rhabdonella sp.				P								
Rhabdonella spiralis							P					

(i) It may be due to the prevailed salinity and temperature of the region; as we go from north to south, the salinity of the coastal area increases towards the southern coast.
(ii) The regions may be understudied, as only two to three regions are well-studied and some studies focused on tintinnids only.

Ciliate population varied significantly from 0 to 25,546 Nos./L (Table 12.1). The population was very low in Bahuda estuary (1–16 Nos./L) and Parangipettai backwater (9–79 Nos./L) while high population was found in Sunderban mangroves, Hooghly estuary, Chilika lagoon, and Pulicat lagoon. Sometimes, swarming of the ciliates was also observed in Vellar estuary and Pichavaram mangroves (Krishnamurthy and Naidu 1977; Naidu 1986). The reason for this type of varied population may be due to the following:

(i) In the different types of sampling methods used in these studies, some used higher mesh size net. The 74/76 μm net misses the population of 20–74 μm range.
(ii) The growth of the population normally depends on the food availability and mesozooplankton grazing pressure.

Small-sized tintinnid (lorica length < 76 μm) was found to be dominated over the large-sized, and this might be favored by environmental variables (low Chl-*a*, salinity and high turbidity) as well as biotic interaction (e.g., reduced prey size) (Rakshit and Sarkar 2016). *Tintinnopsis* species use silica grains to build their lorica (Burns 1983), and showed strong relation to silicate as they are more agglomerated with silica or diatoms shells on lorica (Mukherjee et al. 2018). This may be the reason for dominance of agglomerated ciliates in the coastal environments.

Tintinnid biomass and production rate is available in many regions. These were calculated by measuring lorica length and lorica oral diameter, and then calculating the volume and converting it into biomass and production rate by using previously calculated factors, measured salinity, and temperature. In Sunderban mangroves, the biomass ranged from 0.03 to 18.08 μg C L^{-1} and the production rate varied from 0.006 to 7.65 μg C L^{-1} day^{-1} (Biswas et al. 2013, 2014; Rakshit et al. 2016a). In Hooghly estuary, the biomass ranged from 0.004 to 3.32 μg C L^{-1} and the production rate varied from 0.04 to 3.55 μg C L^{-1} day^{-1} (Rakshit et al. 2016b, 2017). In Vellar estuary, the biomass ranged from 0.04 to 3.01 μg C L^{-1} and production rate varied from 0.07 to 2.50 μg C L^{-1} day^{-1} (Godhantaraman 2002). In Pichavaram mangroves, the biomass ranged from 0.02 to 2.18 μg C L^{-1} and production rate varied from 0.02 to 2.01 μg C L^{-1} day^{-1} (Godhantaraman 2002). In Pullavazhi estuary, the biomass (as plasma volume) varied from 0.35 to 55.98 mm^3 m^{-3} (Srinivasan et al. 1988). The overall idea on size range of tintinnids was given afterwards. The size range of tintinnids were LL (lorica length): 13.90–531.9 μm and LOD (lorica oral diameter): 4.3–151.1 μm in Sunderban mangroves (Biswas et al. 2013, 2014; Rakshit et al. 2016a); LL: 22.76–260.8 μm and LOD: 8.45–74.52 μm in Hooghly estuary (Rakshit et al. 2016b, 2017); LL: 22.70–307.97 μm and LOD: 9.60–107.16 μm in Chilika lagoon (Mukherjee et al. 2015); LL: 44–540 μm and LOD: 24–144 μm in Pullavazhi estuary (Srinivasan et al. 1988).

By this compiled study, it was observed that the taxonomic position of some species is uncertain and is to be rechecked. Some species names are not there in species databases (WoRMS – World Register of Marine Species, Marine Species Identification Portal). Some species' taxonomic validity is uncertain or disputed by different experts (*taxon inquirendum*). There are also spelling mistakes/changes in various studies. Care was taken to present the spellings correctly in this article.

Traditionally, tintinnid species are distinguished via lorica features (Dolan et al. 2014). Recently, sequencing has revealed polymorphism, that is, genetically identical individuals with distinct lorica morphologies (Dolan et al. 2014). We should go for closer look at the scaling between morphological and genetic diversity as the two may be inconsistent (Andre et al. 2013; Dolan et al. 2014).

12.3.3 Rotifers (Phylum: Rotifera)

Rotifers, commonly known as "wheel animalcule," are metazoans with ubiquitous presence in freshwater (95%) and limited presence (5% of the community) in the marine environment (Sharma 1987; Sharma and Naik 1996; Anjusha et al. 2018). Rotifers play an essential role in the biodiversity and functioning of zooplankton communities in estuarine ecosystems (Telesh 2004; Gopko and Telesh 2013). Planktonic rotifers are notably dependent on the water body trophic conditions and therefore they may be used as bioindicators (Duggan et al. 2001; Gopko and Telesh 2013). Rotifers have received little attention in studies of estuarine and marine plankton, and while this can be understood as due to their limited distribution and underrepresentation in net samples (Heinbokel et al. 1988; Dolan and Gallegos 1992).

When we study an estuarine environment, these animals play an important role when salinity is low during monsoonal flows. So, they contribute significantly to the biomass in seasonal periods in tidal freshwater and estuarine waters (Park and Marshall 2000). They are also important because of their reproductive rates that are among the fastest of the metazoans due to their parthenogenetic production and short developmental periods (Herzig 1983; Park and Marshall 2000). Accordingly, they can populate vacant niches with extreme rapidity, and convert primary production into a form usable for secondary consumers, producing up to 30% of the total plankton biomass (Nogrady et al. 1993; Park and Marshall 2000).

Five regions were studied for rotifers in estuaries, mangroves, and lagoons of east coast of India (Table 12.3). They are in Chilika lagoon (Mukherjee et al. 2014, 2018; Sahu et al. 2016a), Vellar estuary (Prabu et al. 2005), Pichavaram mangroves (Govindasamy and Kannan 1991), Kaduviyar estuary (Perumal et al. 2009), and Muthupet lagoon (Santhanam et al. 2019).

In total, 59 species of rotifers were recorded covering 14 families and 25 genera. The families Brachionidae and Lecanidae dominated the species list. *Brachionus*

Table 12.3 Rotifers in estuaries, mangroves, and lagoons of east coast of India

Species name	Chilika lagoon	Vellar estuary	Pichavaram mangroves	Kaduviyar estuary	Muthupet lagoon
Brachionus rubens	P		P	P	
Brachionus plicatilis	P	P		P	P
Brachionus angularis	P	P	P		P
Brachionus urceolaris	P	P	P		P
Brachionus bidentata	P		P		P
Brachionus calyciflorus	P		P	P	P
Brachionus caudatus			P	P	P
Brachionus falcatus	P		P	P	P
Brachionus forficula			P		
Brachionus quadricornis					P
Brachionus quadridentata	P		P		P
Brachionus rubens					P
Brachionus sp.	P				
Kellicottia longispina	P				
Anuraeopsis fissa			P		
Keratella cochlearis			P		
Keratella procurva			P		
Keratella quadrata			P		
Keratella tropica	P		P		
Keratella sp.	P			P	
Platyias patulus	P		P		
Platyias quadricornis			P	P	P
Monostyla bulla	P	P	P		
Monostyla sp.	P			P	
Lecane curvicornis			P		

(continued)

Table 12.3 (continued)

Species name	Chilika lagoon	Vellar estuary	Pichavaram mangroves	Kaduviyar estuary	Muthupet lagoon
Lecane leontina	P		P		
Lecane luna	P		P		
Lecane papuana			P		
Lecane ungulata	P		P		
Lecane styrax	P				
Lecane batillifer	P				
Lecane crepida	P				
Lecane inopinata	P			P	
Lecane sp.	P				
Monostyla bulla	P				
Monostyla closterocerca			P		
Monostyla quadridentata			P		P
Monostyla stenroosi			P		
Monostyla unguitata			P		
Dipleuchlanis propatula			P		
Euchlanis dilatata			P		
Euchlanis oropha			P		
Tripleuchlanis plicata			P		
Polyarthra sp.	P				
Ploesoma lenticularia			P		
Trichocerca sp.	P				
Trichotria tetractis			P		
Hexarthra sp.	P				
Lepadella sp.	P				
Cephalodella gibba		P	P		
Asplanchna sp.	P				
Testudinella patina	P				
Pompholyx sulcata					
Testudinella sp.	P				

(continued)

Table 12.3 (continued)

Species name	Chilika lagoon	Vellar estuary	Pichavaram mangroves	Kaduviyar estuary	Muthupet lagoon
Mytilina ventralis			P		
Filinia longiseta	P		P		P
Filinia opoliensis	P				
Filinia sp.	P				
Conochilus dossuarius	P				
References	Mukherjee et al. (2014, 2018) Sahu et al. (2016a)	Prabu et al. (2005)	Govindasamy and Kannan (1991)	Perumal et al. (2009)	Santhanam et al. (2019)

genera covered maximum species (13 species) followed by *Lecane* (10 species). Highest numbers of species (35 species each) were found in Chilika lagoon and Pichavaram mangroves. These species were mostly observed in low salinities like 1.02–7.10 PSU in Pichavaram mangroves (Govindasamy and Kannan 1991). The species number were low in Vellar estuary (5), Kaduviyar estuary (9), and Muthupet lagoon (13). Saline or brackish water species like *Brachionus plicatilis* and *Brachionus falcatus* observed in more saline regions (Mukherjee et al. 2014).) Rotifers can be an important component of the microzooplankton during some specific periods in estuarine habitats but play a definite trophic role by replacing other microzooplankton (Park and Marshall 2000).

The common species recorded in most of the regions were *Brachionus plicatilis*, *Brachionus angularis*, *Brachionus urceolaris*, *Brachionus calyciflorus*, *Brachionus falcatus*, *Brachionus rubens*, *Brachionus bidentata*, *Brachionus caudatus*, *Brachionus quadridentata*, *Platyias quadricornis*, *Monostyla bulla*, and *Filinia longiseta* (12 species). The species confined to Chilika lagoon were *Brachionus* sp., *Kellicottia longispina*, *Lecane styrax*, *Lecane batillifer*, *Lecane crepida*, *Lecane* sp., *Monostyla bulla*, *Polyarthra* sp., *Trichocerca* sp., *Hexarthra* sp., *Lepadella* sp., *Asplanchna* sp., *Testudinella patina*, *Testudinella* sp., *Filinia opoliensis*, *Filinia* sp., and *Conochilus dossuarius* (17 species). The species confined to Vellar and Kaduviyar estuary were zero. The species confined to Pichavaram mangroves were *Brachionus forficula*, *Anuraeopsis fissa*, *Keratella cochlearis*, *Keratella procurva*, *Keratella quadrata*, *Lecane curvicornis*, *Lecane papuana*, *Monostyla closterocerca*, *Monostyla quadridentata*, *Monostyla stenroosi*, *Monostyla unguitata*, *Dipleuchlanis propatula*, *Euchlanis dilatata*, *Euchlanis oropha*, *Tripleuchlanis plicata*, *Ploesoma lenticularia*, *Trichotria tetractis*, and *Mytilina ventralis* (18 species). The species confined to Muthupet lagoon were *Brachionus quadricornis* and *Brachionus rubens* (2 species).

Like ciliates, in rotifers also some species have taxonomic uncertainty. For example, the species *Ploesoma lenticularia* was reported by Govindasamy and Kannan (1991) from Pichavaram mangroves. Any reference to the species was not found on Google Scholar search engine. Therefore, there is a need for taxonomic validation of some species.

12.3.4 Heterotrophic Dinoflagellates (Class: Dinophyceae) and Other Groups

Heterotrophic dinoflagellates (HDFs) are a significant component of microzooplankton biomass and major grazers of diatoms in the sea (Sherr and Sherr 2007). Approximately 50% of dinoflagellate species are heterotrophic and the rest are autotrophic or mixotrophic (Dale 2009). They have the greatest potential to consume diatoms of the major groups of herbivores in pelagic systems (Sherr and Sherr 2007). HDFs can prey on organisms as large as, or larger than, themselves in size, while other categories of phagotrophic protists (heterotrophic flagellates and ciliates) in general feed on smaller-sized prey (Sherr and Sherr 2007; Lan et al. 2009). A mathematical model based on observed rates of grazing by *Gyrodinium* sp. suggested that the heterotrophic dinoflagellates could have grazed down the diatom bloom to ambient abundance within 7 to 8 days (Saito et al. 2006). With their potential fast-growth rate, HDFs respond quickly to blooms and play a role as significant as the mesozooplankton in consuming phytoplankton blooms (Sherr and Sherr 2007; Lan et al. 2009).

Studies on heterotrophic dinoflagellates (HDFs) in microzooplankton are very rare in the estuaries, mangroves, and lagoons of east coast of India. Only one study (Sahu et al. 2016a) in Chilika lagoon has included these HDFs in the microzooplankton study. Two genera of HDF was recorded – *Gymnodinium* sp. and *Protoperidinium* sp. Normally these HDFs dominate the microzooplankton in open ocean in Bay of Bengal (Jyothibabu et al. 2003) (Table 12.4).

Apart from ciliates, rotifers, and HDFs, the other groups that come in microzooplankton are crustacean larvae, molluscan larvae, foraminiferans, radiolarians, etc. These were recorded in Chilika lagoon (Sahu et al. 2016a), Vellar estuary (Prabu et al. 2005), and Kaduviyar estuary (Perumal et al. 2009) (Table 12.4). The crustacean larvae (copepod naupli and cirripede naupli) and molluscan larvae (bivalve veliger and gastropod veliger) were recorded in all three regions. The foraminiferans and radiolarians were recorded in Vellar estuary. The naupliar forms are important intermediaries between microbial and classical food web as they can efficiently graze on autotrophic picophytoplankton and heterotrophic bacteria in diverse aquatic ecosystems that allows picophytoplankton carbon transfer into classical food chain (Richardson and Jackson 2007; Bemal and Anil 2019). The high abundance of crustacean nauplii is linked to the high availability of picoplankton, the

Table 12.4 Heterotrophic dinoflagellates and other groups in estuaries, mangroves, and lagoons of east coast of India (Sahu et al. 2016a; Prabu et al. 2005; Perumal et al. 2009)

Groups/Taxa	Chilika lagoon	Vellar estuary	Kaduviyar estuary
Heterotrophic dinoflagellates			
Gymnodinium sp.	P		
Protoperidinium sp.	P		
Crustacean larvae			
Copepod nauplii	P	P	P
Cirripede nauplii	P	P	P
Molluscan larvae			
Bivalve veliger	P	P	P
Gastropod veliger	P	P	P
Foraminifera			
Globigerina rubescens		P	
Globigerina plicatilis		P	
Globigerina sp.		P	
Radiolaria			
Acantharia sp.		P	
Thalassicolla sp.		P	
Others			
Polychaete larvae		P	
Oikopleura larvae		p	

major preferred food source of invertebrate larval forms (Roff et al. 1995). They are abundant, but their role in food chain is undermined.

12.4 Conclusion

The study described the microzooplankton study of Hooghly estuary, Sunderban mangroves, Chilika lagoon, Bahuda estuary, Godavari–Coringa mangroves, Pulicat lagoon, Vellar estuary, Pichavaram mangroves, Parangipettai backwater, Kaduviyar estuary, Muthupet lagoon, and Pullavazhi estuary of the east coast of India. A total of 120 ciliate species was classified into 29 families and 39 genera. In total, 59 species of rotifers were recorded including 14 families and 25 genera. In these coastal environments, major studies were carried out in mesozooplankton and phytoplankton, but there is a big gap in microzooplankton studies in all the estuaries, mangroves, and lagoons of the east coast of India. We have also look into the microzooplankton for their mixotrophy characteristics to understand the entire food web dynamics and carbon flow. It is also proposed that a better understanding of each group of microzooplankton will lead to better modeling of the food web.

Acknowledgments The first author wishes to thank to SERB, Department of Science and Technology (DST), India for National Postdoctoral Fellowship (PDF/2017/000522). The corresponding author wish to thank to UKRI–Living Delta Research Hub for his Postdoctoral Fellowship.

References

Andre A, Weiner A, Quillévére F, Aurahs R, Morard R, Douady CJ, de Garidel-Thoron T, Escarguel G, de Vargas C, Kucera M (2013) The cryptic and apparent reversed: lack of genetic differentiation within the morphologically diverse plexus of the planktonic foraminifera *Globigerinoides sacculifer*. Paleobiology 39:21–39

Anjusha A, Jyothibabu R, Jagadeesan L, Arunpandi N (2018) Role of rotifers in microzooplankton community in a large monsoonal estuary (Cochin backwaters) along the west coast of India. Environ Monit Assess 190(5):295

Azam F, Fenchel T, Field JG, Gray JS, Meyer-Reil LA, Thingstad F (1983) The ecological role of water-column microbes in the sea. Mar Ecol Prog Series 10:257–263

Basuri CK, Pazhaniyappan E, Munnooru K, Chandrasekaran M, Vinjamuri RR, Karri R, Mallavarapu RV (2020) Composition and distribution of planktonic ciliates with indications to water quality in a shallow hypersaline lagoon (Pulicat Lake, India). Environ Sci Pollut Res 27:18303–18316

Bemal S, Anil AC (2019) Picophytoplankton *Synechococcus* as food for nauplii of *Amphibalanus amphitrite* and *Artemia salina*. Hydrobiologia 835(1):21–36

Biswas SN, Godhantaraman N, Rakshit D, Sarkar SK (2013) Community composition, abundance, biomass and production rates of Tintinnids (Ciliata: Protozoa) in the coastal regions of Sundarban Mangrove wetland, India. Indian J Geo-Marine Sci 42(2):163–173

Biswas SN, Rakshit D, Sarkar SK, Sarangi RK, Satpathy KK (2014) Impact of multispecies diatom bloom on plankton community structure in Sundarban mangrove wetland, India. Mar Pollut Bull 85(1):306–311

Bruggeman J, Kooijman ALM (2007) A biodiversity-inspired approach to aquatic ecosystem modeling. Limnol Oceanogr 52:1533–1544

Burns DA (1983) The distribution and morphology of tintinnids (ciliate protozoans) from the coastal waters around New Zealand. N Z J Mar Freshw Res 17(4):387–406

Calbet A (2008) The trophic roles of microzooplankton in marine systems. ICES J Mar Sci 65(3):325–331

Corliss JO (1979) The Ciliated Protozoa, 2nd edn. Pergamon Press, Oxford

Dale B (2009) Eutrophication signals in the sedimentary record of dinoflagellate cysts in coastal waters. J Sea Res 61:103–113

Das GK (2015) Sunderbans: physical aspects and configurations. In: Estuarine Morphodynamics of the Sunderbans. Springer, Cham, pp 1–21

Dolan JR (2013) Introduction to tintinnids. In: The biology and ecology of tintinnid ciliates: models for marine plankton, vol 1. Wiley, USA, pp 1–16

Dolan JR, Gallegos CC (1992) Trophic role of planktonic rotifers in the Rhode River Estuary, spring–summer 1991. Mar Ecol Prog Ser 85(1):187–199

Dolan JR, Pierce RW, Bachy C (2014) *Cyttarocylis ampulla*, a polymorphic tintinnid ciliate of the marine plankton. Protist 165(1):66–80

Duggan IC, Green JD, Shiel RJ (2001) Distribution of rotifers in North Island, New Zealand, and their potential use as bioindicators of lake trophic state. Hydrobiologia 446(447):155–164

Finlay BJ, Corliss JO, Esteban G, Fenchel T (1996) Biodiversity at the microbial level: the number of free-living ciliates in the biosphere. Q Rev Biol 71:221–237

Fukang G, Bing N, Zhenyun Y, Baojian D (2002) Ultrastructure of the vegetative cell and resting cyst in Pseudourostyla cristata (Ciliophora, Hypotrichida). Chinese J Zool 48:251–257
Gauns M, Madhupratap M, Ramaiah N, Jyothibabu R, Fernandes V, Paul JT, Kumar SP (2005) Comparative accounts of biological productivity characteristics and estimates of carbon fluxes in the Arabian Sea and the Bay of Bengal. Deep-Sea Res II Top Stud Oceanogr 52:2003–2017
Gifford DJ (1991) The protozoan–metazoan trophic link in pelagic ecosystems. J Protozool 38:81–86
Godhantaraman N (1994) Species composition and abundance of tintinnids and copepods in the Pichavaram mangroves (South India). Cienc Mar 20(3):371–391
Godhantaraman N (2001) Seasonal variations in taxonomic composition, abundance and food web relationship of microzooplankton in estuarine and mangrove waters, Parangipettai region, southeast coast of India. Indian J Marine Sci 30:151–160
Godhantaraman N (2002) Seasonal variations in species composition, abundance, biomass and estimated production rates of tintinnids at tropical estuarine and mangrove waters, Parangipettai, southeast coast of India. J Mar Syst 36(3–4):161–171
Godhantaraman N (2004) Ecological significance of marine Microzooplankton. In: Ramaiah N (ed) Marine microbiology: facets & opportunities, NIO, Goa, pp 181–188
Gopko M, Telesh IV (2013) Estuarine trophic state assessment: new plankton index based on morphology of Keratella rotifers. Estuar Coast Shelf Sci 130:222–230
Govindasamy C, Kannan L (1991) Rotifers of the Pitchavaram mangroves (Southeast Coast of India): a hydrobiological approach. Mahasagar 24(1):39–45
Haraguchi L, Jakobsen HH, Lundholm N, Carstensen J (2018) Phytoplankton community dynamic: a driver for ciliate trophic strategies. Front Mar Sci 5:272
Heinbokel JF, Coats DW, Henderson KW, Tyler MA (1988) Reproduction rates and secondary production of three species of the rotifer genus Synchaeta in the estuarine Potomac River. J Plankton Res 10:659–674
Heip CHR, Goosen NK, Herman PMJ, Kromkamp J et al (1995) Production and consumption of biological particles in temperate tidal estuaries. Oceanogr Mar Biol Ann Rev 33:1–149
Herzig A (1983) Comparative studies on the relationship between temperature and duration of embryonic development of rotifers. Hydrobiologia 104:237–246
Hopkins TL (1987) Midwater food web in Mc Murdo sound, Ross Sea, Antarctica. Mar Biol 96:93–106
Jyothibabu R, Madhu NV, Maheswaran PA, Nair KKC, Venugopal P, Balasubramanian T (2003) Dominance of dinoflagellates in micro-zooplankton community in the oceanic regions of the Bay of Bengal and the Andaman Sea. Curr Sci 87(6):783–791
Kalavati C, Raman AV (2008) Taxonomy and ecology of ciliated Protozoa from marginal marine environments of East Coast of India. Rec Zoolog Surv India Occ Paper No 282:1–136
Kaur H, Iqbal S, Inga E, Yawe D (2019) Encystment and excystment in ciliated protists: multidimensional approach. Curr Sci 117(198):198–203
Krishnamurthy K, Naidu WD (1977) Swarming of the tintinnids (Protozoa: Ciliata) in the Vellar estuary. Curr Sci 46(11):384
Krishnamurthy K, Santhanam R (1975) Ecology of tintinnids (Protozoa: Ciliata) in Porto Novo region. Indian J Mar Sci 4:181–184
Lan W, Huang B, Dai M, Ning X, Huang L, Hong H (2009) Dynamics of heterotrophic dinoflagellates off the Pearl River estuary, northern South China Sea. Estuar Coast Shelf Sci 85(3):422–430
McLusky DS, Elliott M (2004) The estuarine ecosystem: ecology, threats and management, 3rd edn. Oxford University Press Inc., New York
Mishra S, Panigrahy RC (1999) The tintinnids (Protozoa: Ciliata of the Bahuda estuary, east coast of India). Indian J Mar Sci 28:219–221
Misra SK, Chandramohan P, Murty AS, Panigrahi JK, Mahadevan R (2013) Nature of the tide induced flow field along the East Coast of India. Int J Oceans Oceanogr 7(1):57–71

Mukherjee M, Banik SK, Suresh VR, Manna RK, Panda D, Sharma AP (2014) Rotifers, their distribution, abundance and seasonal variation in Chilika lagoon. J Inland Fisheries Soc India 46(1):29–37

Mukherjee M, Banik SK, Pradhan SK, Sharma AP, Suresh VR, Manna RK, Panda D, Roshith CM, Mandal S (2015) Diversity and distribution of tintinnids in Chilika Lagoon with description of new records. Indian J Fisheries 62(1):25–32

Mukherjee M, Suresh VR, Manna RK (2018) Microplankton dynamics of a coastal lagoon, Chilika: interactive effect of environmental parameters on microplankton groups. Environ Monit Assess 190(11):689

Naidu WD (1986) Tintinnid swarms of Porto Novo waters. Mahasagar 19(1):23–27

Nogrady T, Wallace RL, Snell TW (eds) (1993) Rotifera—guides to the identification of the micro-invertebrates of the continental waters of the world, 4. SPB Academic Publishing, The Hague. 142pp

Pandey BD, Yeragi SG, Reddy AK, Hagiwara A (2008) Life strategies and aquacultural usability of a hypersaline ciliate, Fabrea salina. Curr Sci 94(3):307–309

Park GS, Marshall HG (2000) The trophic contributions of rotifers in tidal freshwater and estuarine habitats. Estuar Coast Shelf Sci 51(6):729–742

Pattnaik S (1973) Observation on seasonal fluctuations of plankton in the Chilika lagoon. Indian J Fisheries 19:43–55

Perumal NV, Rajkumar M, Perumal P, Rajasekar KT (2009) Seasonal variations of plankton diversity in the Kaduviyar estuary, Nagapattinam, southeast coast of India. J Environ Biol 30(6):1035–1046

Prabu VA, Perumal P, Rajkumar M (2005) Diversity sf microzooplankton in Parangipettai coastal waters, southeast coast of India. J Mar Biol Assoc India 47(1):14–19

Rakshit D, Sarkar SK (2016) Diversity, distribution and polymorphism of loricate ciliate tintinnids along Hooghly estuary, India. J Mar Biol Assoc India 58(2):61–68

Rakshit D, Sarkar SK, Bhattacharya BD, Jonathan MP, Biswas JK, Mondal P, Mitra S (2015) Human-induced ecological changes in western part of Indian Sundarban megadelta: a threat to ecosystem stability. Mar Pollut Bull 99(1–2):186–194

Rakshit D, Sarkar SK, Satpathy KK, Ganesh PS, Godhantaraman N, Biswas JK (2016a) Diversity and distribution of microzooplankton tintinnid (Ciliata: Protozoa) in the core region of Indian Sundarban wetland. CLEAN–Soil, Air, Water 44(10):1278–1286

Rakshit D, Ganesh PS, Sarkar SK (2016b) Choreotrich ciliate tintinnid (Protozoa: Ciliophora) in a tropical meso–macrotidal estuary, eastern part of India. Reg Stud Mar Sci 3:89–100

Rakshit D, Murugan K, Biswas JK, Satpathy KK, Ganesh PS, Sarkar SK (2017) Environmental impact on diversity and distribution of tintinnid (Ciliata: Protozoa) along Hooghly Estuary, India: a multivariate approach. Reg Stud Mar Sci 12:1–10

Richardson TL, Jackson GA (2007) Small phytoplankton and carbon export from the surface ocean. Science 315:838–840

Roff JC, Turner TJ, Webber MK, Russell R, Hopcroft RR (1995) Bacterivory by tropical copepod nauplii: extent and possible significance. Aquat Microb Ecol 9:165–175

Sahu BK, Srichandan S, Panigrahy RC (2016a) A preliminary study on the microzooplankton of Chilika Lake, a brackish water lagoon on the east coast of India. Environ Monit Assess 188(1):69

Sahu BK, Panigrahy RC, Baliarsingh SK, Parida C, Sahu KC, Lotliker AA (2016b) Red-tide of *Mesodinium rubrum* (Lohmann, 1908) in Indian waters. Curr Sci 110(6):982–983

Saito H, Ota T, Suzuki K, Nishioka J, Tsuda A (2006) Role of heterotrophic dinoflagellate *Gyrodinium* sp. in the fate of an iron induced diatom bloom. Geophys Res Lett 33:L09602

Santhanam P, Ananth S, Kumar SD, Sasirekha R, Premkumar C, Jeyanthi S, Devi AS (2019) An intensive culture techniques of marine copepod Oithona rigida (Dioithona rigida) Giesbrecht. In: Basic and applied zooplankton biology. Springer, Singapore, pp 367–394

Sardet C (2015) Plankton: wonders of the drifting world. University of Chicago Press, Chicago, USA, pp 222

Sarma VVSS, Gupta SNM, Babu PVR, Acharya T, Harikrishnachari N, Vishnuvardhan K et al (2009) Influence of river discharge on plankton metabolic rates in the tropical monsoon driven Godavari estuary, India. Estuar Coast Shelf Sci 85(4):515–524
Sarma VVSS, Prasad VR, Kumar BSK, Rajeev K, Devi BMM, Reddy NPC et al (2010) Intra-annual variability in nutrients in the Godavari estuary, India. Cont Shelf Res 30(19):2005–2014
Sharma BK (1987) Indian Brachionidae (Eurotatoria: Monogononta) and their distribution. Hydrobiologia 144:269–275
Sharma BK, Naik LP (1996) Results on planktonic rotifers in the Narmada River (Madhya Pradesh, Central India). In: Schiemer F, Boland KT (eds) Perspectives in tropical limnology, SPB Academic Publishing, Amsterdam, The Netherlands, pp 189–198
Sherr EB, Sherr BF (2007) Heterotrophic dinoflagellates: a significant component of microzooplankton biomass and major grazers of diatoms in the sea. Mar Ecol Prog Ser 352:187–197
Sooria PM, Jyothibabu R, Anjusha A, Vineetha G, Vinita J, Lallu KR, Paul M, Jagadeesan L (2015) Plankton food web and its seasonal dynamics in a large monsoonal estuary (Cochin backwaters, India) – significance of mesohaline region. Environ Monit Assess 187(7):427
Srinivasan A, Santhanam R, Jegatheesan G (1988) Biomass and seasonal distribution of planktonic tintinnids of Pullavazhi Estuary, southeast coast of India. Indian J Mar Sci 17:131–133
Stoecker DK, Capuzzo JM (1990) Predation on protozoa: its importance to zooplankton. J Plankton Res 12:891–908
Sun J, Wang MH, Ho YS (2012) A historical review and bibliometric analysis of research on estuary pollution. Mar Pollut Bull 64(1):13–21
Telesh IV (2004) Plankton of the Baltic estuarine ecosystems with emphasis on Neva Estuary: a review of present knowledge and research perspectives. Mar Pollut Bull 49:206–219
Verni F, Rosati G (2011) Resting cysts: a survival strategy in Protozoa Ciliophora. Italian J Zool 78:134–145

Chapter 13
Influence of Physical Processes on Nutrient Dynamics and Phytoplankton in the Coastal Bay of Bengal

Madhusmita Dash, Chandanlal Parida, Biraja Kumar Sahu, Kali Charan Sahu, and Sourav Das

Abstract This review chapter addresses the impact of physical forcing mechanisms on spatiotemporal variation of biological productivity aided by nutrient dynamics in the nearshore waters of the Bay of Bengal (BoB). The BoB has a unique physical forcing mechanism, viz., reversal pattern of monsoon wind, large freshwater influx, and remote forcing from the equator. The substantial freshwater influx into the BoB leads to strong stratification and a strong halocline below the mixed layer. Nonetheless, the strong southwest monsoon wind is inadequate to break this stratification and hence obstructs the vertical transfer of nutrients from the nutricline. This phenomenon makes the BoB oligotrophic. In contrast to this, several physical processes increase productivity by bringing nutrients from the nutricline to euphotic depths. The southwest monsoon-induced wind-driven upwelling and the ocean Ekman pumping by positive cyclonic wind stress circulation triggered by the northeast monsoon enhance the primary production in the BoB. Before the onset of the southwest monsoon, the water current along India's east coast brings more saline, cooler water to the surface by eddy-like structures, and this process causes surplus productivity in the BoB. Moreover, tropical cyclones moving over the BoB are adequate to break the stratification and thereby injecting nutrients, subject to phytoplankton blooms.

Keywords Physical forcing · Nutrient dynamics · Chlorophyll · Bay of Bengal

---

M. Dash
National Institute of Ocean Technology, MoES, Chennai, India

C. Parida (✉)
Centre for Atmospheric and Oceanic Sciences, Indian Institute of Science, Bangalore, India

B. K. Sahu
Atal Centre for Ocean Science & Technology for Islands, National Institute of Ocean Technology, MoES, Port Blair, India

K. C. Sahu
Department of Marine Sciences, Berhampur University, Berhampur, India

S. Das
School of Oceanographic Studies, Jadavpur University, Kolkata, India

S. Das, T. Ghosh (eds.), *Estuarine Biogeochemical Dynamics of the East Coast of India*, https://doi.org/10.1007/978-3-030-68980-3_13

13.1 Introduction

The northeastern part of the Indian Ocean encompasses a semi-enclosed ocean basin, which experiences a typical tropical climate, and we refer to this region as the Bay of Bengal (BoB). The BoB is a unique area in the perspective of the physical oceanographic processes. A typical monsoon climate characterizes the BoB, where the wind reverses twice in a year, and therefore the bay experiences a seasonal change in circulation (Shetye et al. 1991). Wind over the BoB blows from the southwest from May to September and northeast from November to January. In addition to the seasonal reversal wind pattern, the BoB receives a large volume of freshwater in the form of precipitation and river influx (Girishkumar et al. 2012). The BoB receives freshwater plumes of about 1.5×10^{12} m^3/year (Martin et al. 1981; UNESCO 1988) from several perennial rivers such as Ganges, Mahanadi, Godavari, and Brahmaputra (Gomes et al. 2000) and ~2 m/year in the form of precipitation (Prasad 1997). The riverine influx in the BoB is substantially higher than that in the Arabian Sea. The freshwater of BoB and the high saline water from the Arabian Sea (AS) through Summer Monsoon Current (SMC) create two different layers of water masses.

The freshwater on the top with subducted high saline water reduces the salinity in the upper surface layer and stratifies the water column in the photic region (Prasanna Kumar et al. 2002). It also leads to a strong halocline with a shallow mixed layer (Shetye et al. 1996; Vinayachandran et al. 2002). Such stratification enables the formation of a strong barrier layer (BL) between mixed layer and thermocline (Vinayachandran et al. 2002; Girishkumar et al. 2012). The formation and dissipation of BL play a crucial role in monsoon variation and the genesis of the tropical cyclones (George et al. 2019). The BL acts as an obstruction towards stratification and upwelling of nutrient-enriched water, and hence forms an oligotrophic condition (Vidya and Das 2017; Prasanna Kumar et al. 2002). Apart from the freshwater influx and seasonal wind pattern, the proximity to the equatorial Indian Ocean leads to a third forcing mechanism in the BoB. The Kelvin waves that occur in the coastal regions of the Bay merge with the equatorial Rossby waves along the eastern border of the BoB, which, in turn leads to wind forcing in the entire coastal periphery of the equatorial Indian Ocean (McCreary et al. 1993; Shankar et al. 1996). The combination of all the three factors, viz., reversal monsoon wind, distant forcing from the equator, and freshwater influx, regulate the circulation pattern in the BoB. The entire physical phenomenon that takes place in the BoB has a significant impact on nutrient dynamics and biological productivity.

The biological productivity is comparatively much lower in the BoB than that observed in the Arabian Sea (Radhakrishna et al. 1978; Prasanna Kumar et al. 2002, 2004; Vinayachandran 2009). Despite the fact that the nutrient input in the BoB is much higher than the Arabian Sea, the net primary production in the BoB is significantly less than that of Arabian Sea (Madhupratap et al. 2003). This point has been reiterated by several shipboard observational studies (Gomes et al. 2000; Prasanna Kumar et al. 2002; Madhupratap et al. 2003; Jyothibabu et al. 2004; Vinayachandran

2009). The freshwater influence acts as a major reason for low biological productivity in the BoB. The strong stratification due to freshwater caps in the upper layer of the BoB prevents the nutrients upwelling from the mixed layer to the euphotic depths (McCreary et al. 2009; Vinayachandran 2009; Amol et al. 2019). Even the strong winds accompanied by the southwest monsoon are inadequate to disrupt the stratification (Shenoi et al. 2002) that could enable the transport of nutrients from the nutricline to the euphotic zone, and this limits the productivity in the BoB (Gomes et al. 2000). Besides, higher concentrations of suspended sediments, intense cloud cover, and shallow continental shelf are also possible reasons behind the low productivity in the BoB (Muraleedharan et al. 2007; Vinayachandran 2009).

Several physical processes promote productivity by bringing nutrients from the subsurface to the euphotic zone, leading to phytoplankton blooms. Shetye et al. (1991) observed that the southwest monsoon facilitates the local alongshore wind-driven upwelling, and enhances the productivity along the southern part of the Indian coast (Gomes et al. 2000; Madhupratap et al. 2003; Madhu et al. 2006). The coastal upwelling process (Vinayachandran et al. 2004; Girishkumar et al. 2012) causes the phytoplankton bloom in the south of Sri Lanka. During northeast monsoon, the open ocean Ekman pumping acts as a principal factor for the phytoplankton bloom in the southwestern BoB (Vinayachandran and Mathew 2003). The circulation process during northeast monsoon consists of a cyclonic gyre in the southwestern BoB (Vinayachandran and Yamagata 1998). The Ekman pumping caused by the positive wind stress curl augments the chlorophyll concentration in the southwestern part of the BoB (Vinayachandran et al. 2005) and also enhances the chlorophyll concentration within the Sri Lanka dome during the southwest monsoon (Vinayachandran and Yamagata 1998; Vinayachandran et al. 2004). In the BoB, the cyclonic cold-core eddies are also promoting biological productivity (Falkowski et al. 1991; Prasanna Kumar et al. 2002, 2004; Sarangi et al. 2008; McGillicuddy et al. 2003; Sarma et al. 2013). During northeast monsoon, the cooling of sea surface takes place due to the heat loss from the sea. The high wind speed is capable of mixing and churning the water column as the stability remains comparatively low during this time of the year. The cold-core eddies are capable of supplying nutrients under such lowered stability conditions. This, in turn, increases the biological productivity in the BoB (Gomes et al. 2000; Muraleedharan et al. 2007). The tropical cyclones that originate in the BoB can also de-stratify the water column, which facilitates nutrient injection from the subsurface to the euphotic zone (Vinayachandran and Mathew 2003; Madhu et al. 2003, 2006; Rao et al. 2006; Smitha et al. 2006; McCreary et al. 2009; Vinayachandran 2009; Baliarsingh et al. 2015; Sarangi et al. 2014; Amol et al. 2019; Jayaram et al. 2019). Before the onset of the southwest monsoon, the East India coastal current (EICC) brings more saline, cooler water to the surface by eddy-like structures, and this process also causes surplus productivity in the BoB. Along with the coastal areas of the BoB, the rivers play a crucial role in supplying nutrients and thereby increasing the biological production (Muraleedharan et al. 2007; Choudhury and Pal 2010). The principal objective of this review is to characterize the role of the physical processes on phytoplankton and nutrient variability of coastal BoB.

13.2 Physical Processes

The growth of phytoplankton primarily depends upon the availability of sunlight and nutrients, and consequently, the scarcity of either one limits the productivity and growth of phytoplankton. In the tropical oceans, availability of light throughout the year usually never delimits the productivity; however the nutrient concentrations often remain significantly low to limit the biological growth of the autotrophs (Prasanna Kumar et al. 2002). However, during the southwest monsoon, the sky over the BoB remains cloudy for a longer period. This lack of an adequate amount of light might be a possible reason for low productivity in the BoB (Radhakrishna et al. 1978; Gomes et al. 2000; Muraleedharan et al. 2007). In tropical oceans, the significant limiting factor is the nutrient availability. Enrichment of nutrients to the euphotic layers by the process of coastal and open ocean upwelling and entrainment increases the chlorophyll. The augmentation of phytoplankton due to these physical processes, along with strong monsoonal wind, is highly familiar in the Arabian Sea (Lévy et al. 2007). Indian monsoon impacts both the western and eastern tropical Indian Ocean basins, that is, the Arabian Sea and the BoB. However, the response of the BoB to the monsoon is different from that of the Arabian Sea. The mixing phenomena and the entrainment are absent in the BoB and are mainly due to the (a) weak monsoonal wind over the BoB than those over Somalia and Oman in the Arabian Sea, (b) remotely forced thermocline deepens around the perimeter of the BoB, and (c) substantial freshwater influx into the BoB strongly stratifies the surface layer of the BoB, which needs strong energy to break the stratification (McCreary et al. 2009; Vinayachandran 2009). Nevertheless, several physical processes in different spatiotemporal scale help to break the barrier layer, cause upwelling of nutrients from subsurface to the near-surface zone, and allow an increase in the primary productivity in the BoB. The details of the physical processes are described in the next section. Figure 13.1 indicates the Ocean Surface Current Analysis Real-time (OSCAR) current vectors (averages for the period of pre-monsoon, monsoon, and post-monsoon, 2016), shown as arrows.

13.2.1 *Coastal Upwelling*

The winds over the BoB are weak during the initial months (March and April) of the southwest monsoon, but during this period the anticyclonic curls are developed by the wind, and its center is located in the BoB. The southwesterlies cover the entire BoB during May and remains so till September. During the transition period between the southwesterlies and northeasterlies, the tropical cyclones originate in the western BoB, and the southwesterlies move to the south. The northeasterlies are strong and occur from November to February. The wind during the southwest monsoon blows toward Sri Lankan's west coast and away from its east coast. However, the wind blows parallel to the south of Sri Lankan coast, which is conducive for upwelling. This upwelling enhances the primary production (Vinayachandran et al.

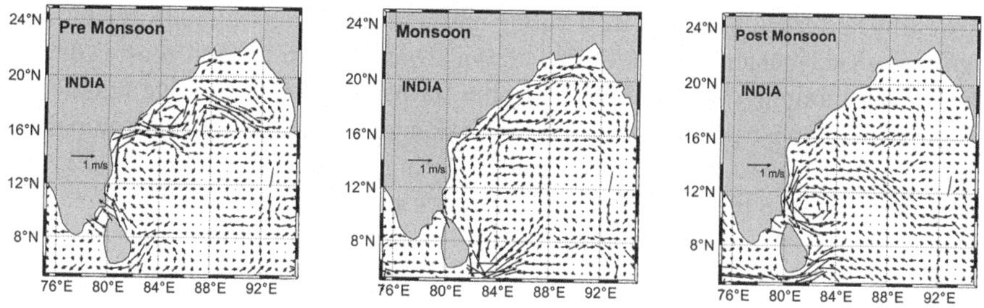

Fig. 13.1 Map of the Bay of Bengal, arrows indicate the Ocean Surface Current Analysis Real-time (OSCAR) current vectors shown in this figure are averages for period for pre-monsoon, monsoon, and post-monsoon for the year 2016

2004). The wind flow across the east coast of the BoB during the southwest monsoon favors upwelling and enhance the biological productivity (Shetye et al. 1991; Gomes et al. 2000; Madhupratap et al. 2003; Thushara and Vinayachandran 2016; Lotliker et al. 2020). In the BoB, during the transition period of the monsoon, the westerlies over the equatorial Indian Ocean generate coastal Kelvin waves and oceanic Rossby waves (Shankar et al. 2002). The Rossby wave propagation acts as a driving mechanism for the offshore spread of chlorophyll. The upwelling on the east coast of the bay is not common because of the advection of the equatorial river plume and propagation of the downwelling coastal Kelvin wave (Vinayachandran et al. 1996). This indicates localized blooming of phytoplankton during pre-southwest monsoon. And the time lag between river influence and local upwelling also shows an increase in chlorophyll in the early summer monsoon period, and this could be due to the coastal upwelling (Vinayachandran 2009). During the winter monsoon, the wind over the BoB is favorable for upwelling in the northeastern part of the Bay (McCreary et al. 1996).

13.2.2 Ekman Pumping

The local wind stress-driven Ekman pumping is an important forcing factor of the Bay. During pre-summer monsoon months, the downward Ekman pumping is the major driving mechanism of the anticyclonic gyre (Shankar et al. 1996). However, the upward Ekman pumping during the southwest monsoon in the Sri Lankan east coast leads to the formation of the Sri Lankan dome. A cyclonic gyre formation due to the upward Ekman pumping takes place in the southwestern part of the bay during the northeast monsoon (Vinayachandran and Yamagata 1998). During northeast monsoon, enhancement of phytoplankton blooms in the southwestern BoB takes place due to the nutrient injection into the upper surface led by Ekman pumping (Vinayachandran and Mathew 2003). The downward pumping makes the bay less productive, whereas upward Ekman pumping results in higher chlorophyll concentrations. A numerical model study by Vinayachandran et al. (2005) using a coupled physical–biological model has also stated that the physical process, that is, upwelling in the southwestern BoB during northeast monsoon caused by the Ekman pumping (Vinayachandran 2009). The bloom around the Sri Lankan coast is due to the Ekman pumping. It usually leads to the formation of two separate regions of bloom. One of these patches cover the Sri Lankan dome and the other encompasses the southeastern coast of Sri Lanka. Besides, during the southwest monsoon, Ekman pumping facilitates the upwelling in the northwestern BoB.

13.2.3 Tropical Cyclone

Over the tropical oceanic region, a weather system with anticyclonic wind forms in the northern hemisphere and is known as tropical cyclones. The wind speeds associated with cyclones are very high, and sometimes they can reach as high as 85 m/s (Emanuel 2003), which abruptly changes the circulation pattern in the ocean. The tropical cyclones cause the cooling of SST and mixed-layer deepening. The forcing by a storm can simulate oceanic productivity. The periodic churning process by cyclonic storms entrains nutrients into the near-surface layer and thereby increasing the chlorophyll concentration. Five percent of all the tropical cyclones that occur throughout the world occurs in the BoB (Neetu et al. 2012; Jayaram et al. 2019). The cyclonic storms in the BoB generally originate from May to October (Madhu et al. 2006). In the BoB, cyclones affect the biological as well as physical characteristics by decreasing the sea surface temperature, increasing the dissolved oxygen, and enhancing productivity (Chacko 2017). A significant increase in chlorophyll concentration took place in the northern BoB after the 1999 super-cyclonic storm (Patra et al. 2007) and increased the productivity in the southwestern BoB (Madhu et al. 2006). The passage of a cyclone during November 2000 over the BoB led to the extensive growth of surface chlorophyll and primary productivity (Rao et al. 2006). During northeast monsoon, cyclones over the BoB lead to intense localized bloom (Vinayachandran and Mathew 2003). Another tropical cyclone developed over BoB during May 2003 and has influenced the increase in chlorophyll and primary productivity in the southern BoB (Smitha et al. 2006). Similarly, the passages of other tropical cyclones, viz., Phailin, HudHud, Helen, Leher, and Madi, etc., over the BoB have also influenced an increase in the chlorophyll concentration and thereby in the biological productivity over the bay (Sarangi 2011; Sarangi et al. 2014; Baliarsingh et al. 2015; Jayaram et al. 2019).

13.2.4 Cyclonic Eddies

Mesoscale processes like cyclonic eddies are accompanied by upward movement of the thermocline, which can upwell nutrients into the euphotic zone. The cyclonic eddies also play a pivotal role in supplying the nutrients to the photic zone (Falkowski et al. 1991). In the BoB, cyclonic eddies enhance the autotrophic production by bringing the nutrients into the upper surface oligotrophic waters during the southwest monsoon. During this time, a baroclinic instability at the junction of two opposing currents, that is, the Western Boundary Current (WBC) and the freshwater discharge from the rivers occurs along the east coast of India. This instability can effectively generate the cyclonic eddies and the wind stress curl over these regions aids in the formation of cold-core eddy (Muraleedharan et al. 2007). Though the upper layers of the BoB are highly stratified during the southwest monsoon, cold-core eddy pumping acts as the possible mechanism for transferring nutrients into the photic zone and increasing biological production (Gomes et al. 2000; Prasanna Kumar et al. 2004; Muraleedharan et al. 2007). The cold-core eddy enhances biological productivity by more than double in the BoB (Prasanna Kumar et al. 2004). Sardessai et al. (2007) also reported that the upliftment of water mass to shallower depth by the cyclonic eddies was the prime mechanism supplying nutrients to the surface waters in the BoB. From October to December the sea level decreases because of the seasonal cyclonic gyre in the southwestern BoB and the eddies could be distinguished. The cold-core eddies fewer its number and much weaker in the eastern BoB as compared to the west. And a cyclonic eddy is defined to occur negative sea level anomaly becomes less than 10 cm (Vinayachandran 2009).

13.2.5 Circulation and Advection

The seasonal circulation in the BoB does not take place according to the monsoon wind reversal due to the combination of remote and local forcing that governs the seasonality of the surface circulation of the BoB (Vinayachandran et al. 1996; MacCreary et al. 1996). In the BoB during the initial period of the pre-monsoon season, that is, March to April, a well-defined anticyclonic gyre with poleward East Indian Coastal Current (EICC) along the east coast of India (Shetye et al. 1993; Sanilkumar et al. 1997). The EICC flows through an eastward direction north of 19°N. In the western part of Andaman Islands, the gyre starts flowing southward and thereafter this anticyclonic gyre meets the westward winter monsoon current (Shankar et al. 2002). The disintegration of this gyre commences in May, and by the month of June its complete disappearance takes place. During the southwest monsoon, the intense current in the east and south of Sri Lanka, where the Southwest Monsoon Current (SMC) turns around Sri Lanka and flows into the BoB (Vinayachandran et al. 1999). And during this period, there is an intense coastal upwelling along the southern coast, and the SMC transports cold upwelled nutrient-rich water along its path into the BoB (Vinayachandran et al. 2004; Jyothibabu et al. 2015; Thushara et al. 2019), which led to the formation of a cold pool and increased the biological productivity. The summertime coastal current, EICC flows northward up to 15°N, where it turns eastward. During the northeast monsoon months, the EICC flows downward and carries low saline water along its path (Shetye et al. 1996). In this period, the southwestern BoB consists of a cyclonic gyre (Vinayachandran and Yamagata 1998). The SMC advects high concentration chlorophyll from the west coast to Sri Lanka and from there into the BoB. During northeast monsoon, the EICC flows equatorward and advects nutrients-rich water along its route (Vinayachandran 2009).

13.2.6 River Influx

During the summer monsoon period, the freshwater discharge from peninsular rivers is at its peak. Therefore, in the BoB, elevated chlorophyll was also observed at the mouths of several rivers dumping into the BoB. The discharge from these rivers can often penetrate several hundred kilometers offshore. However, the direction of flow and extent depends upon the volume, wind, and current patterns (Vinayachandran 2009; Lotliker et al. 2020). The river discharge acts as a potential source of nutrients for enhancing the biological productivity in the BoB (Narvekar and Kumar 2006; Sarma et al. 2013). Specifically, the coastal regions receive inputs of nutrients via riverine outflows and runoff from the land inputs, which sustain new production. Hence, the coastal region can respond rapidly to episodic events like upwelling (Muraleedharan et al. 2007). During spring inter-monsoon seasons, eddies and recirculation zones in the coastal regions forms due to the Western Boundary Currents (WBC) fund to enhance the chlorophyll concentration (Gomes et al. 2000; Prasanna Kumar et al. 2004; Madhu et al. 2006). However, during the summer monsoon season, the nutrient input from river runoff triggers the primary production of the coastal waters of the BoB (Madhupratap et al. 2003; Madhu et al. 2006). Vinayachandran (2009) reported that there is a large influx of fresh water from the River Irrawaddy into the BoB during the southwest monsoon, and during October to December a large patch of chlorophyll flows directly outward, and the current pattern prevails over the BoB advect this plume in a northwestward direction. They also mentioned similar observations off the mouth of the Ganges and Brahmaputra rivers. After the withdrawal of southwest monsoon, when EICC flows equatorward, the chlorophyll regions cling to the coast, and when the current reverse, the chlorophyll plumes flow offshore directly, away from the coast to the open ocean. During northeast monsoon, the river plumes from the mouth of the River Mahanadi and flows southward of the Indian coast and then toward the offshore.

13.3 Conclusion

The overall primary productivity in the Bay of Bengal is substantially less than that observed in its western counterpart, the Arabian Sea. However, several studies indicated that a suite of physical processes regulate the primary activity in this bay. In the present chapter, we detailed about the physical processes and their impact on nutrient dynamics and biological productivity of the bay on a different spatiotemporal scale. In offshore areas, the cyclonic circulation associated with thermocline is an important process. During northeast monsoon, cyclonic gyre occupies in the southwestern BoB and is associated with high chlorophyll concentration. The cyclonic eddies also have higher chlorophyll concentrations. The Ekman pumping caused by the positive wind stress curl augments the chlorophyll concentration in the southwestern part of the BoB and enhances the chlorophyll concentration within the Sri Lanka dome during the southwest monsoon. The tropical cyclone in the bay narrows down the thermocline and enables the injection of nutrient-enriched water into the photic zone, which leads to phytoplankton blooms. Enhanced productivity due to the coastal upwelling takes place in three different regions, that is, in the south of Sri Lankan coast, the southern part of the east coast, and the northeastern coast of the bay. The first two regions become biologically enriched during the southwest monsoon season while the third one from December to March. The riverine influx also helps to increase the productivity in different parts of the BoB, particularly in coastal and river mouth areas. The flow of water enriched with chlorophyll along the east coast of India depends upon the pattern and magnitudes of coastal currents. The East India coastal current also plays a very significant role in governing the river plumes in the BoB.

References

Amol P, Vinayachandran PN, Shankar D, Thushara V, Vijith V, Chatterjee A, Kankonkar A (2019) Effect of freshwater advection and winds on the vertical structure of chlorophyll in the northern Bay of Bengal. Deep Sea Research Part II: Topical Studies in Oceanography, 104622

Baliarsingh SK, Chandanlal P, Lotliker AA, Suchismita S, Sahu KC, Srinivasa Kumar T (2015) Biological implications of cyclone Hudhud in the coastal waters of northwestern Bay of Bengal. Curr Sci 109(7):1243–1245

Chacko N (2017) Chlorophyll bloom in response to tropical cyclone Hudhud in the Bay of Bengal: bio-Argo subsurface observations. Deep-Sea Res I Oceanogr Res Pap 124:66–72

Choudhury AK, Pal R (2010) Phytoplankton and nutrient dynamics of shallow coastal stations at Bay of Bengal, Eastern Indian coast. Aquat Ecol 44(1):55–71

Emanuel K (2003) Tropical cyclones. Annu Rev Earth Planet Sci 31:75–104

Falkowski PG, Ziemann D, Kolber Z, Bienfang PK (1991) Role of eddy pumping in enhancing primary production in the ocean. Nature 352(6330):55–58

George JV, Vinayachandran PN, Vijith V, Thushara V, Nayak AA, Pargaonkar SM, Matthews AJ (2019) Mechanisms of barrier layer formation and erosion from in situ observations in the Bay of Bengal. J Phys Oceanogr 49(5):1183–1200

Girishkumar MS, Ravichandran M, Pant V (2012) Observed chlorophyll-a bloom in the southern Bay of Bengal during winter 2006–2007. Int J Remote Sens 33(4):1264–1275

Gomes HR, Goes JI, Saino T (2000) Influence of physical processes and freshwater discharge on the seasonality of phytoplankton regime in the Bay of Bengal. Cont Shelf Res 20(3):313–330

Jayaram C, Bhaskar TU, Kumar JP, Swain D (2019) Cyclone enhanced chlorophyll in the Bay of Bengal as evidenced from satellite and BGC-Argo float observations. J Indian Soc Remote Sensing 47(11):1875–1882

Jyothibabu R, Maheswaran PA, Madhu NV, Ashraf TM, Gerson VJ, Haridas PC, Gopalakrishnan TC (2004) Differential response of winter cooling on biological production in the northeastern Arabian Sea and the northwestern Bay of Bengal. Curr Sci 87:783–791

Jyothibabu R, Vinayachandran PN, Madhu NV, Robin RS, Karnan C, Jagadeesan L, Anjusha A (2015) Phytoplankton size structure in the southern Bay of Bengal modified by the summer monsoon current and associated eddies: implications on the vertical biogenic flux. J Mar Syst 143:98–119

Lévy M, Shankar D, André JM, Shenoi SSC, Durand F, de Boyer Montégut C (2007) Basin-wide seasonal evolution of the Indian Ocean's phytoplankton blooms. J Geophys Res Oceans 112(C12):1–14

Lotliker AA, Baliarsingh SK, Sahu KC, Kumar TS (2020) Long-term chlorophyll-a dynamics in tropical coastal waters of the western Bay of Bengal. Environ Sci Pollut Res 27(6):6411–6419

Madhu NV, Jyothibabu R, Maheswaran PA, Gerson VJ, Gopalakrishnan TC, Nair KKC (2006) Lack of seasonality in phytoplankton standing stock (chlorophyll a) and production in the western Bay of Bengal. Cont Shelf Res 26(16):1868–1883

Madhu NV, Jyothibabu R, Maheswaran PA, Gerson VJ, Gopalakrishnan TC, Nair KKC (2003) Lack of seasonality in phytoplankton standing stock (chlorophyll a) and production in the western Bay of Bengal. Cont Shelf Res 26(16):1868–1883

Madhupratap M, Gauns M, Ramaiah N, Kumar SP, Muraleedharan PM, De Sousa SN, Muraleedharan U (2003) Biogeochemistry of the Bay of Bengal: physical, chemical and primary productivity characteristics of the central and western Bay of Bengal during summer monsoon 2001. Deep-Sea Res II Top Stud Oceanogr 50(5):881–896

Martin JM, Eisma D, Burton JD (eds) (1981) River inputs to ocean systems. United Nations Environment Programme

McCreary JP Jr, Kundu PK, Molinari RL (1993) A numerical investigation of dynamics, thermodynamics and mixed-layer processes in the Indian Ocean. Prog Oceanogr 31(3):181–244

McCreary JP, Han W, Shankar D, Shetye SR (1996) Dynamics of the East India coastal current: 2. Numerical solutions. J Geophys Res:Oceans 101(C6):13993–14010

McCreary JP, Murtugudde R, Vialard J, Vinayachandran PN, Wiggert JD, Hood RR, Shetye S (2009) Biophysical processes in the Indian Ocean. Indian Ocean Biogeochemical Processes and Ecological Variability 185:9–32

McGillicuddy DJ Jr, Anderson LA, Doney SC, Maltrud ME (2003) Eddy-driven sources and sinks of nutrients in the upper ocean: results from a 0.1 resolution model of the North Atlantic. Glob Biogeochem Cycles 17(2):1035. https://doi.org/10.1029/2002GB001987

Muraleedharan KR, Jasmine P, Achuthankutty CT, Revichandran C, Kumar PD, Anand P, Rejomon G (2007) Influence of basin-scale and mesoscale physical processes on biological productivity in the Bay of Bengal during the summer monsoon. Prog Oceanogr 72(4):364–383

Narvekar J, Kumar SP (2006) Seasonal variability of the mixed layer in the central Bay of Bengal and associated changes in nutrients and chlorophyll. Deep-Sea Res I Oceanogr Res Pap 53(5):820–835

Neetu S, Lengaigne M, Vincent EM, Vialard J, Madec G, Samson G, Durand F (2012) Influence of upper-ocean stratification on tropical cyclone-induced surface cooling in the Bay of Bengal. J Geophys Res Oceans 117(C12):C12020

Patra PK, Kumar MD, Mahowald N, Sarma VVSS (2007) Atmospheric deposition and surface stratification as controls of contrasting chlorophyll abundance in the North Indian Ocean. J Geophys Res Oceans 112(C5):1–14

Paul JT, Ramaiah N, Sardessai S (2008) Nutrient regimes and their effect on distribution of phytoplankton in the Bay of Bengal. Mar Environ Res 66(3):337–344

Prasad TG (1997) Annual and seasonal mean buoyancy fluxes for the tropical Indian Ocean. Curr Sci 73:667–674

Prasanna Kumar S, Muraleedharan PM, Prasad TG, Gauns M, Ramaiah N, De Souza SN, Madhupratap M (2002) Why is the Bay of Bengal less productive during summer monsoon compared to the Arabian Sea? Geophys Res Lett 29(24):88–81

Prasanna Kumar S, Nuncio M, Narvekar J, Kumar A, Sardesai DS, De Souza SN, Madhupratap M (2004) Are eddies nature's trigger to enhance biological productivity in the Bay of Bengal? Geophys Res Lett 31(7):1–5

Radhakrishna K, Bhattathiri PMA, Devassy VP (1978) Primary productivity of the Bay of Bengal during August-September 1976. Indian J Mar Sci 7(1):94–98

Rao KH, Smitha A, Ali MM (2006) A study on cyclone induced productivity in south-western Bay of Bengal during November-December 2000 using MODIS (SST and chlorophyll-a). IJMS 35(2):153–160

Sanilkumar KV, Kuruvilla TV, Jogendranath D, Rao RR (1997) Observations of the Western boundary current of the Bay of Bengal from a hydrographic survey during march 1993. Deep-Sea Res I Oceanogr Res Pap 44(1):135–145

Sarangi RK, Nayak S, Panigrahy RC (2008) Monthly variability of chlorophyll and associated physical parameters in the southwest Bay of Bengal water using remote sensing data. IJMS 37(3):256–266

Sarangi RK (2011) Impact of cyclones on the Bay of Bengal chlorophyll variability using remote sensing satellites. Indian J Geo-Mar Sci 40(6):794–801

Sarangi RK, Mishra MK, Chauhan P (2014) Remote sensing observations on impact of Phailin cyclone on phytoplankton distribution in northern Bay of Bengal. IEEE J Sel Top Appl Earth Obs Remote Sens 8(2):539–549

Sardessai S, Ramaiah N, Prasanna Kumar S, De Sousa SN (2007) Influence of environmental forcings on the seasonality of dissolved oxygen and nutrients in the Bay of Bengal. J Mar Res 65(2):301–316

Sarma VVSS, Sridevi B, Maneesha K, Sridevi T, Naidu SA, Prasad VR, Kiran BS (2013) Impact of atmospheric and physical forcings on biogeochemical cycling of dissolved oxygen and nutrients in the coastal Bay of Bengal. J Oceanogr 69(2):229–243

Shankar D, McCreary JP, Han W, Shetye SR (1996) Dynamics of the East India coastal current: 1. Analytic solutions forced by interior Ekman pumping and local alongshore winds. J Geophys Res Oceans 101(C6):13975–13991

Shankar D, Vinayachandran PN, Unnikrishnan AS (2002) The monsoon currents in the North Indian Ocean. Prog Oceanogr 52(1):63–120

Shenoi SSC, Shankar D, Shetye SR (2002) Differences in heat budgets of the near-surface Arabian Sea and Bay of Bengal: implications for the summer monsoon. J Geophys Res Oceans 107(C6):5–1

Shetye SR, Shenoi SSC, Gouveia AD, Michael GS, Sundar D, Nampoothiri G (1991) Wind-driven coastal upwelling along the western boundary of the Bay of Bengal during the southwest monsoon. Cont Shelf Res 11(11):1397–1408

Shetye SR, Gouveia AD, Shenoi SSC, Sundar D, Michael GS, Nampoothiri G (1993) The western boundary current of the seasonal subtropical gyre in the Bay of Bengal. J Geophys Res Oceans 98(C1):945–954

Shetye SR, Gouveia AD, Shankar D, Shenoi SSC, Vinayachandran PN, Sundar D, Nampoothiri G (1996) Hydrography and circulation in the western Bay of Bengal during the northeast monsoon. J Geophys Res Oceans 101(C6):14011–14025

Smitha A, Rao KH, Sengupta D (2006) Effect of May 2003 tropical cyclone on physical and biological processes in the Bay of Bengal. Int J Remote Sens 27(23):5301–5314

Thushara V, Vinayachandran PN (2016) Formation of summer phytoplankton bloom in the northwestern Bay of Bengal in a coupled physical-ecosystem model. J Geophys Res Oceans 121(12):8535–8550

Thushara V, Vinayachandran PN, Matthews A, Webber B, Queste B (2019) Vertical distribution of chlorophyll in dynamically distinct regions of the southern Bay of Bengal. Biogeosciences 16(7):1447–1468

UNESCO (1988) River inputs to the ocean systems: status and recommendations for research. UNESCO Technical papers in Marine Science, No. 55, Final Report of the SCOR Working group 46, Paris, 25pp.

Vidya PJ, Das S (2017) Contrasting Chl-a responses to the tropical cyclones Thane and Phailin in the Bay of Bengal. J Mar Syst 165:103–114

Vinayachandran PN (2009) Impact of physical processes on chlorophyll distribution in the Bay of Bengal. Indian Ocean Biogeochemical Processes and Ecological Variability 185:71–86

Vinayachandran PN, Mathew S (2003) Phytoplankton bloom in the Bay of Bengal during the northeast monsoon and its intensification by cyclones. Geophys Res Lett 30(11):1572. https://doi.org/10.1029/2002GL016717

Vinayachandran PN, Yamagata T (1998) Monsoon response of the sea around Sri Lanka: generation of thermal domes and anticyclonic vortices. J Phys Oceanogr 28(10):1946–1960

Vinayachandran PN, Shetye SR, Sengupta D, Gadgil S (1996) Forcing mechanisms of the Bay of Bengal. Curr Sci 70(10):753–763

Vinayachandran PN, Masumoto Y, Mikawa T, Yamagata T (1999) Intrusion of the southwest monsoon current into the Bay of Bengal. J Geophys Res Oceans 104(C5):11077–11085

Vinayachandran PN, Murty VSN, Ramesh Babu V (2002) Observations of barrier layer formation in the Bay of Bengal during summer monsoon. J Geophy Res Oceans 107(C12):SRF-19

Vinayachandran PN, Chauhan P, Mohan M, Nayak S (2004) Biological response of the sea around Sri Lanka to summer monsoon. Geophys Res Lett 31(1):1–4

Vinayachandran PN, McCreary JP Jr, Hood RR, Kohler KE (2005) A numerical investigation of the phytoplankton bloom in the Bay of Bengal during Northeast Monsoon. J Geophys Res Oceans 110(C12):1–14

Chapter 14
A Review of Estuarine CDOM Dynamics of East Coast of India Influenced by Hydrographical Forcing

Sudarsana Rao Pandi, N. V. H. K. Chari, Nittala S. Sarma, Sarat C. Tripathy, G. Chiranjeevulu, and Sourav Das

Abstract Colored Dissolved Organic Matter (CDOM) is a fraction of dissolved organic matter (DOM) that interacts with light and can be detected by satellite ocean color remote sensing. In the estuarine and coastal region, CDOM studies are helpful in the carbon cycle modeling. The Indian summer monsoon causes heavy discharges, enriched in CDOM transporting to the western Bay of Bengal, and the estuaries act as conduits. CDOM-related studies are relatively few in estuaries of the Bay of Bengal. Limited findings suggest a dual behavior, conservative in the 15–31 × 10^{-3} salinity range (October/November to April) during non-lean/lean discharge period and nonconservative behavior outside of this salinity range during monsoon season. Studies integrating the river, estuary, and coastal regions, over temporal cycles, are still lacking. A summary of the present status is expected to be helpful to pin the areas on which to concentrate and as a reference for future studies

S. R. Pandi (✉)
National Centre for Polar and Ocean Research, Ministry of Earth Sciences, Vasco-da-Gama, Goa, India

Marine Chemistry Laboratory, Andhra University, Visakhapatnam, Andhra Pradesh, India

N. V. H. K. Chari
Marine Chemistry Laboratory, Andhra University, Visakhapatnam, Andhra Pradesh, India

Center for Marine Living Resources and Ecology, Ministry of Earth Sciences, Kochi, Kerala, India

N. S. Sarma · G. Chiranjeevulu
Marine Chemistry Laboratory, Andhra University, Visakhapatnam, Andhra Pradesh, India

S. C. Tripathy
National Centre for Polar and Ocean Research, Ministry of Earth Sciences, Vasco-da-Gama, Goa, India

S. Das
School of Oceanographic Studies, Jadavpur University, Kolkata, West Bengal, India

S. Das, T. Ghosh (eds.), *Estuarine Biogeochemical Dynamics of the East Coast of India*, https://doi.org/10.1007/978-3-030-68980-3_14

as we witness changes in the river discharge due to either dam construction or climate change–related perturbations affecting the export of terrigenous DOM, its dynamics, and nutrients that fertilize the sea.

Keywords CDOM · FDOM · Estuaries · Freshwater discharge · Bay of Bengal

14.1 Introduction

Dissolved organic matter (DOM) is operationally defined as aquatic organic matter that passes along with water through filters 0.2 μm in pore size (Morel and Bricaud 1981). The DOM of the global oceans has concentrations between 30 and 300 micromoles (as carbon) litre^{-1} with higher concentrations at surface of the coastal seas where it is introduced by the ex situ terrestrial input and the in situ living processes. The chemically characterized fraction of DOM in natural waters consists of carbohydrates, amino acids, hydrocarbons, lipids, phenolic acids, etc., adding up to 30% of the total DOM (Repeta et al. 2002). The remaining 70% is a chemically unclassified fraction, generally termed as gelbstoff or yellow substance. The gelbstoff is formed in soil by degradation of terrestrial plants followed by polycondensation between different monomers (Kalle 1938) and transported into the sea via river runoff. Humic substances (HS) are also formed in situ from the planktonic exudates. They are the most hydrophobic component of DOM and are composed of organic acids of molecular weight (MW) range of 500–10,000 or more atomic mass units (AMU). HS are generally colored yellow to dark brown and are divided into three categories. Humic acids (HA) are insoluble in water or in acid, but become soluble at alkaline pH conditions, forming metal salts (humates). Fulvic acids (FA) are hydrophilic acids soluble under all pH conditions. Humin is the fraction which is insoluble under all pH conditions (Ishiwatari 1992). The chromophoric DOM, also known as colored DOM (CDOM) is a major fraction of DOM, and amenable to structural study through optical methods due to its absorption and fluorescence properties. Allochthonous CDOM is from terrestrial sources and the autochthonous CDOM is of internal origin.

An exponential law roughly governs the absorption behavior of CDOM in terms of wavelength in the ultraviolet-visible (UV-Vis) region of the electromagnetic radiation: $a(\lambda) = a(\lambda_0)\exp^{-S(\lambda-\lambda_0)}$, where $a(\lambda)$ is the absorption coefficient at wavelength λ and λ_0 is the reference wavelength. The units of $a(\lambda)$ are m^{-1}, an optical unit while that of concentration (of HS or CDOM) is mg CL^{-1} with reference a standard, for example, reference HA. The exponent term in the above relationship is the spectral slope (S, nm^{-1}). The calculation of S is by a chosen curve-fitting method. The globally recommended method is nonlinear exponential curve fitting of $a(\lambda)$ vs λ (Stedmon and Markager 2003).

The absorption of natural waters is due, apart from CDOM, to the particles – detrital and phytoplanktonic. As CDOM absorbs frequencies of the composite white light that are also absorbed by chlorophyll (Chl), in remote sensing applications, the presence of CDOM tends to lead to an overestimation of Chl-a. This necessitates correction for CDOM in the algorithms for Chl-a. As CDOM is heterogeneous in its composition and abundance in different coastal waters, the CDOM of each location has to be studied in detail and relations established specific to the given region.

A fraction of CDOM exhibits fluorescence, and the fluorescent CDOM (FDOM) can be characterized in terms of the fluorescence excitation and emission maxima of the constituting fluorescence components, termed as fluorophores. Initially, fluorescence was measured at single excitation and emission maxima, for example, Excitation/Emission 320–350/430–450 nm. Quinine bisulfate was used as standard, and the concentration expressed as quinine bisulfate units or equivalents (QSU or QSE) (Velapoldi and Mielenz 1981). The QSE of the DOM was used as an index of estuarine water in the coastal and offshore regions, for example, the coastal region away from Godavari estuary (Sarma et al. 1994). When the three-dimensional fluorescence measurement became available (Coble 1996), it soon replaced all fluorescence studies as it became possible to study the differences in the origin of samples such as river, lake, estuary, coastal, and marine water based on peak shifts and intensities in the Excitation Emission Matrix (EEM) spectra (Coble 1996; Romera-Castillo et al. 2011). These spectra are calibrated with the water Raman line using Milli-Q water as the reference and the intensities are converted into the Raman units (RU), which can be compared across different laboratories and sample regimes being globally calibrated (Lawaetz and Stedmon 2009). Further advantage of these spectra is that several indices can be calculated such as humification index (HIX) (Zsolnay 2003), fluorescence index (FI) (McKnight et al. 2001), biological index (BIX) (Huguet et al. 2009), and relative fluorescence efficiency (Downing et al. 2009), which can be useful to distinguish the ex situ versus in situ, bacterial versus photochemical, biotic versus abiotic factors, and one water mass versus another.

Parallel factor (PARAFAC) analysis has become a standard statistical tool to extract information on the common fluorophores (components) embedded in the EEM spectra (Stedmon et al. 2003), which revolutionized the fluorescence studies. The fluorescence components are broadly five, three of humic-like and two of protein-like appearance. Two of the humic-like fluorophores emit blue fluorescence (460–480 nm), one when irradiated by UV light (250–280 nm) and the other when irradiated by UV light bordering the visible region (350–370 nm). The third humic fluorophore has both the maxima of excitation and emission in the UV, namely 300–310 and 380–400 nm, respectively. They are designated as A (UV-humic-like), C (Vis-humic-like), and M (marine/microbial-humic-like), respectively. The proteins like fluorophores are termed as the tryptophan-protein-like (T) and tyrosine-protein-like (B) (Coble 1996) (Fig. 14.1). T and B closely occur, with excitation/emission maxima around 280/330 and 275/310 nm, respectively. Since fluorescence measurement is about three orders of magnitude more sensitive than absorption measurement, and since the FDOM is often related to CDOM by simple linear relationship, FDOM measurement can give a better estimate of DOM particularly when the CDOM is in trace concentration (Sarma et al. 1994).

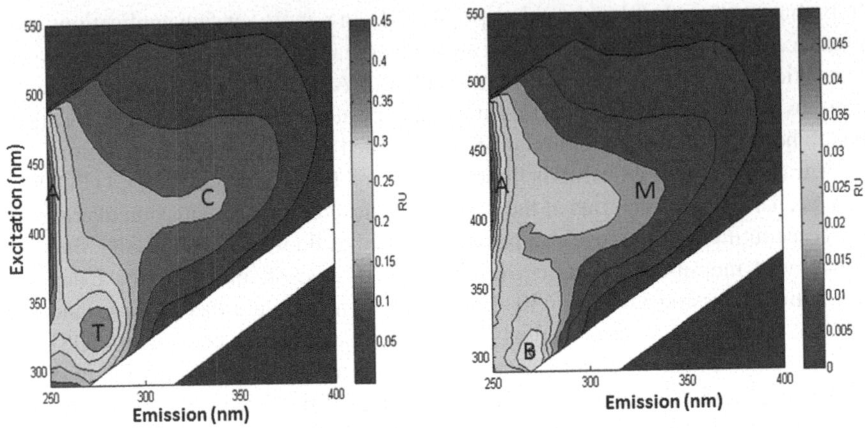

Fig. 14.1 Examples of excitation emission matrix spectra with the fluorophores (where C = visible-humic-like, A = UV-humic-like, M = marine/microbial-humic-like, T = tryptophan-protein-like, and B = tyrosine-protein-like)

14.2 Significance of CDOM and Its Spectral Signature

CDOM, due to its ability to absorb radiation more efficiently the harmful UV-A/B radiation than the visible radiation, can provide protection to the living organisms from DNA damage. Higher concentration of CDOM in the aquatic environments will be an advantage to corals and other light-sensitive organisms because of this protection. CDOM is used as tracer for a wide variety of processes such as mixing of water masses, anthropogenic inputs in the coastal waters, and estuarine mixing processes. CDOM and its absorption and fluorescence spectral analyses were helpful in the study of human wastes, for example, sewage, agricultural waste, and other organic pollutants' distribution in rivers and estuaries (Baker and Spencer 2004). In the estuarine and coastal waters having significant concentration of CDOM, it affects the productivity as it significantly controls the penetration of photosynthetically active radiation (PAR, 400–700 nm).

CDOM plays a significant role in climatology and affects the global temperature. The carbon fixed by phytoplankton by carbon dioxide uptake during photosynthesis or biological pump reduces the global temperature. In the reverse process, the organic matter is degraded by bacteria and carbon dioxide is liberated back to the atmosphere, consequently causing global temperature rise. The DOM determination is an important tool in the coastal and estuarine waters which regulates the global temperature.

The latest addition to the suite of marine pollutants is microplastics, which are changing marine ecosystem. Microplastics are plastic particles with few nanometers to 5 millimeter size. The global release of microplastic into the ocean is estimated as 8 million metric tons per year. The presence of microplastics can lead to an increased production of CDOM along with changes in its molecular weight due either to increased microbial CDOM production or enhanced transformation of lower to higher molecular weight CDOM (Galgani et al. 2018).

The CDOM absorption coefficient has been reported in literature at different wavelengths such as 320, 325, 350, 355, 370, and 375 nm in the UV and 412, 440, and 443 nm, etc., in the visible regions. The more frequently preferred wavelengths are 350/355 nm and 440 nm (Del Castillo et al. 2000; Kowalczuk et al. 2006, 2009; Para et al. 2010). In this review, the CDOM is presented at 350 nm, the preferred wavelength, and also 440 nm. The advantage of 350 nm is that the fluorescence intensities exhibit better relationship with $a_{CDOM}(350)$ than $a_{CDOM}(440)$ as the excitation maxima of fluorophores are closer to it. The advantage of 440 nm is that it can be potentially used directly in remote sensing applications as 440 nm lies close to the major absorption band of chlorophyll a, and a satellite sensor channel.

Peaks in the CDOM absorption spectra indicate the nature of dissolved metabolites released by algae, bacteria, etc. The macroalgae in culture, such as *Fucus spiralis* and *F. serratus* cultures, produced a peak at 270 nm. *Palmaria palmata* produced in addition to 270 nm, a peak at 335 nm. *Enteromorpha intestinalis* produced 280 nm and 395 nm peaks (Hulatt et al. 2009). Microalgae in culture, for example, *Cylindrotheca closterium* and *Tetraselmis suecica* produced peaks at 258 nm and 353 nm, respectively (Chari et al. 2013). Mycosporine-like amino acids (MAA) that are UV-protecting metabolites of algae in natural waters released as extracellular metabolites having absorption maxima in the 320–370 nm range (Singh et al. 2008). Cytochromes are the dominant light absorbers in heterotrophic bacteria (Stanier et al. 1988), particularly cytochrome *c*, which exhibits a typical 415 nm Soret absorption peak (Fruton and Simmonds 1961). A common occurrence of 415 nm absorbing CDOM hydrophilic component has been observed after preconcentration (Röttgers and Koch 2012) and attributed to a degraded porphyrin pigment of heterotrophic (or autotrophic) origin. This degraded pigment may be contributing to a variable fraction of the 412 nm absorption that is measured by various remote sensors. A broad 520–540 nm peak is attributed to a red pigment of the quinone type, which was isolated from culture broth of bacterial isolates from tidal flat sediments (Yi et al. 2003). Quinones are considered ubiquitous in the coastal marine environment (Cory and McKnight 2005).

14.2.1 Absorption Coefficient Ratios

The ratio of absorption at 254–365 nm (called E2:E3) is used to investigate the changes in molecular weight (MW) of the constituent CDOM (Peuravuori et al. 2002). Another combination of wavelengths is 465–665 nm (known as E4:E6) to indicate the CDOM aromaticity (Chin et al. 1994). In many natural waters in which no measurable absorption occurs at 665 nm, Specific Ultraviolet absorption (SUVA) is calculated as a ratio of DOC and absorption coefficient at different wavelengths, for example, 254 nm ($SUVA_{254}$) (Weishaar et al. 2003), 280, 350, or 370 nm ($SUVA_{280}$, $SUVA_{350}$, $SUVA_{370}$, respectively) (Chin et al. 1994). SUVA at these wavelengths is an indicator of aromatic content.

14.2.2 Spectral Slopes

Spectral slopes (*S*) provide insight into the average structural characteristics of CDOM from which the source, reactivity, and fate can be inferred and from a_{CDOM}, the magnitude of those processes (Brown 1977). It was observed that *S* correlates strongly with molecular weight of isolates of fulvic acids. In the case of humic acids, no such correlation was noted (Carder et al. 1989). The *S* has also been known as an indicator of photobleaching, diagenesis of CDOM (Nelson et al. 2007), implying that it can be used as a tracer of source and biogeochemical history of CDOM. In the beginning a wide spectral range, for example, 300–700 nm was preferred for calculating *S* but now the 300–500 nm range is more common. Ever new ranges are proposed, for example, the 330–440 nm range proposed for estuaries (Sarma et al. 2018).

14.2.3 Spectral Slope of Narrower Ranges

A narrow range of 275–295 nm for slope has been proposed to hold better advantage, facilitating comparison among dissimilar water types and as a proxy for DOM MW (Helms et al. 2008). The advantage of $S_{275\text{–}295}$ is that it can be measured with high precision due to stronger absorption of CDOM and steeper fall in this region. The $S_{275\text{–}295}$ was used to retrieve DOC concentration, and as a tracer of terrestrial DOC in river influenced ocean margins (Fichot and Benner 2011, 2012; Fichot et al. 2013). The ratio of the spectral slopes (S_R) obtained from two narrow wavelength ranges, that is, *S* at 275–295 nm to 350–400 nm, has been proposed as an index of MW (of the DOM) and photobleaching (Helms et al. 2008). The S_R has since gained wide application in evaluating environmental dynamics of DOM in terrestrial aquatic (Yamashita et al. 2010; Osburn et al. 2011) as well as coastal environments (Guéguen et al. 2011; Shank and Evans 2011).

14.3 Estuarine CDOM Dynamics

14.3.1 CDOM Estimation

For CDOM absorption measurements, Teflon (inner wall)-coated Niskin bottles with external opening spring system were preferred and for storing of samples, amber-colored corning bottles (to prevent photo decay) with Teflon caps and inner walls free of any organic matter, for example, grease. Storing may be done of filtered samples in an ice box (4 °C) immediately after the sampling. The samples should not be frozen as that would lead to formation of CDOM flocs that will not redissolve upon thawing. Measurements should be done without delay as bacterial metabolism would change the CDOM concentration and profile, preferably within less than 3 h. The samples should be filtered under gentle vacuum through 0.22 μm Millipore membrane filters immediately following sampling. They should be equilibrated for constant duration (15 min) to reach constant temperature (22 ± 1 °C), transferred to 10 cm pathlength quartz cuvettes gently along the walls, and absorbance measured on a precision double beam UV-Vis spectrophotometer, and using (organic-free) Milli-Q water in the reference cell. The absorbance spectra are measured preferably between 200 to 800 nm at 1 nm intervals. The estuarine stations of the east coast studied are marked in Fig. 14.2.

14.3.2 Godavari Estuary

Systematic studies of CDOM in estuarine regions of the east coast started around the year 2010 (Pandi et al. 2021). The excessive freshwater input through rivers into the Bay of Bengal leads to the entire northern and western Bay to be estuary-like at surface and stratification at surface. The Godavari estuary experiences a large input of CDOM of terrestrial origin through the DOC-rich (2–3 mg/L) river water, >90% of which discharges during the 4-month flood season (July–October). In the year 2010, during the monsoon season, CDOM (a_{350} throughout, unless otherwise stated) was 2.75 m^{-1} and non-conservative on salinity. The CDOM behaved conservatively in the 15–31 × 10^{-3} salinity range during the non-monsoon. During this time, CDOM was related to silicate, and a laboratory-leaching experiment at a pH of 8.2 on estuarine mud showed that CDOM and silicate were co-leached. From the linear relationship of CDOM with salinity, it was inferred that freshwater entering the estuary during non-monsoon time had a CDOM of 2.2 m^{-1}. A similar concentration of CDOM was reported for Atchafalaya River water (D'Sa 2008). Again during the summertime, when the estuary was filled with the coastal water and evaporative condition prevailed, the conservative character was lost. The spectral slope ($S_{300-500}$) of CDOM during the discharge season was 0.016 nm^{-1}, which is similar to that of major world rivers water (Bowers and Brett 2008). The slope was 0.015 nm^{-1} for the conservative CDOM during the non-monsoon season, which agreed with a higher MW of DOM due to phytoplankton production. The flood water fluorescence was lower than that of the non-discharge season. Periods of peak abundance of diatoms versus nano and picoplankton were associated with higher or lower M:A fluorophore (from PARAFAC analysis) ratio and this ratio was proposed as an ecological indicator. First flood of the discharge season was distinct with highest CDOM concentration, for example, 3.9 m^{-1}. The slope which was lowest (0.011 nm^{-1}) then suggested that photochemically degraded DOM constituted it (Chiranjeevulu et al. 2014; Pandi et al. 2014).

Further observations on similar lines were reported by Chari et al. (2019) by absorption and fluorescence measurements during a 24-h tidal cycle in April (2015) and January (2016) at a station in the lower estuary. The 2 days of sampling were during the non-discharge season. April is considered as pre-monsoon season

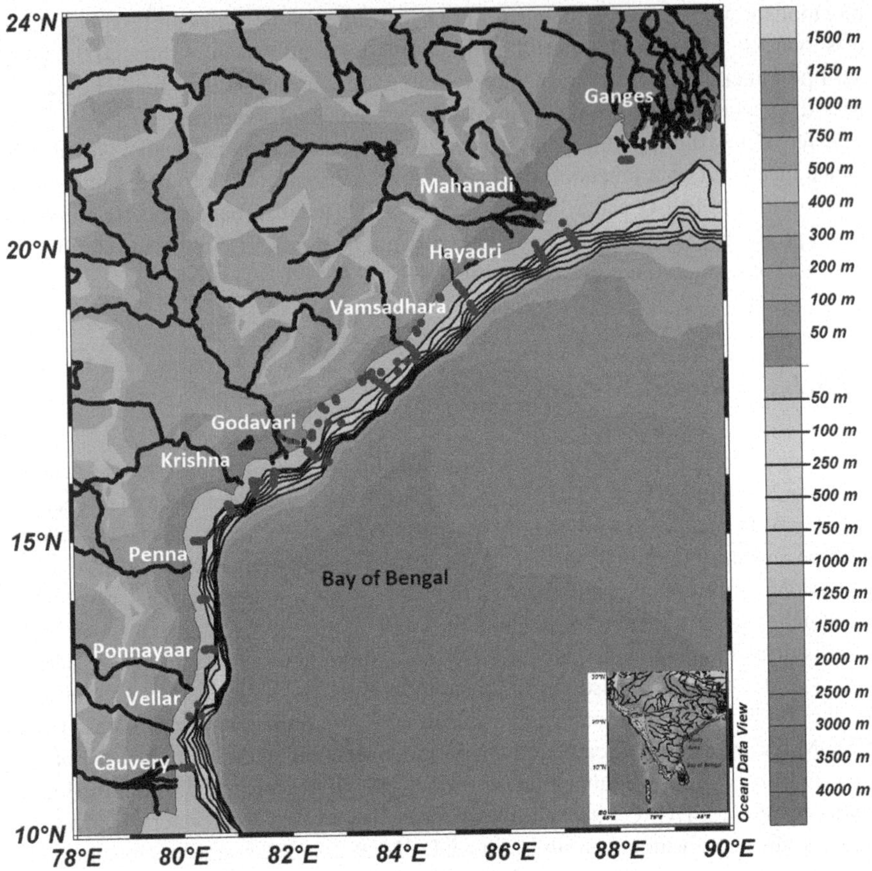

Fig. 14.2 The study region and stations (blue-filled circles) where samples were collected in the coastal Bay of Bengal

(referred as summer in the paper) when spring blooms occur. The spectral slope ratio S_R, which is the ratio of $S_{275:295}$ and $S_{350–400}$, an inverse DOM MW proxy was higher in April (1.28 ± 0.09) than in winter (1.10 ± 0.18); so also the $SUVA_{254}$ (DOC:a_{254} ratio) (3.95 ± 0.21) and (1.91 ± 0.35) $LmgC^{-1}$ L^{-1}, respectively), which indicated that the pre-monsoon CDOM was of lower MW with lower aromatic content due to the in situ production being a major contributor of CDOM, than the winter CDOM for which the MW was higher and the aromatic content too because of a significant fraction of terrestrial CDOM was contributed by the dam-stored water that was released at the upstream (Chari et al. 2019).

Four fluorophores were identified by PARAFAC analysis of the EEM spectra, three humic (A, C and M), and the other protein-like (T). For April samples, the humic fluorophores correlated with Chl-a indicating that river input through a ground water discharge was supporting phytoplankton production in the estuary. The protein (T) fluorophore correlated with DOC, that is, phytoplankton was acting as DOC source within the estuary (Chari et al. 2019).

The humic acids (HA) which are a major fraction of CDOM occur as humate anions. By chromatography over anion-exchange resins they can be fractionated due to their variable charge density. The aromatic richer HA of larger MW owing their origin to terrestrial OM are more ionized than the in situ produced aliphatic HA of lower MW (Chari and Pandi 2017), by the use of multiple spectral indices, for example, E2:E3, and S_R of the resin fractionated HAs showed that compared to XAD-4, the XAD-8 resin retained HA of higher aromaticity (and higher molecular weight of the constituent DOM). Seasonally, they showed that during post-monsoon season the biological process produced fresh DOM of higher MW which underwent bacterial decay and photodegradation during pre-monsoon season, completing the annual cycle (Chari and Pandi 2017).

14.3.3 Chilika Lake

The Chilika Lake on the east coast of India is a shallow brackish water body. It is the world's second largest lagoon and behaves like an estuary due to a narrow mouth connection to the Bay of Bengal, and fed by a large number of monsoonal rivers. Its salinity fluctuates between close to nil in the monsoon season at the northern part where these rivers impound to higher than the coastal water in summer at the mouth due to excessive evaporation, a characteristic of a negative/inverse estuary. CDOM was high in September (3.6 m^{-1}) at the time of peak monsoon freshwater flux and again in March–April (3.9 m^{-1}) when chlorophyll was also at its peak but salinity near to that of the coastal water. The lake constitutes a unique (bacterial) metabolism dominated ecosystem in which CDOM maxima occur twice a year due to opposite mechanisms, viz., ex situ and in situ input maxima. The $S_{300-500}$ was 0.018 and 0.017 nm^{-1} respectively of the two types of CDOM (Chiranjeevulu et al. 2014).

The spectral slope, for example, $S_{300-500}$, is an essential parameter for the estimation of CDOM by remote sensing and must be known precisely for a given ecosystem. A small error in slope can result in large errors in the CDOM retrieval by the algorithms. By individually examining about 800 UV-Vis spectra of CDOM of the Godavari estuary and the Chilika lagoon during different seasons of an year each, Sarma et al. (2018) noted that $S_{300-500}$ was not appropriate for CDOM retrieval because of the absorption peaks at 270–280 and 350–360 nm regions of the spectra which distort the slope. Nitrate's absorption maximum also coincides with the latter region particularly at nitrate exceeding 30 μM. They proposed that $S_{330-440}$ is a better option to calculate CDOM at any λ with minimum error (Sarma et al. 2018).

14.3.4 Rushikulya Estuary

Rushikulya is a minor river joining east coast at Gopalpur, Odisha. Freshwater fluxing through the river during monsoon season makes the estuary a source of CDOM of terrestrial origin and the anthropogenic material. During pre-monsoon season when there is no river flux, and salinity increased, and due to the warm winds that cause evaporation at surface followed by upwelling and consequent nutrient enrichment, diatoms followed by a bloom of the Red *Noctiluca*, the bioluminescent dinoflagellate, was observed. The mixotroph soon experiences death due to exhaustion of their preferred food (diatom) followed by degradation which enhanced the CDOM concentrations in the region (Parida et al. 2019). The blooms were restricted to the nearshore waters in the vicinity of Rushikulya mouth and hence were attributed to a possible source from the estuary (Lotliker et al. 2020).

14.3.5 Hooghly Estuary

Unlike the peninsular rivers in which the discharge differential between the monsoon and non-monsoon seasons was extremely high, the Ganga–Brahmaputra (G–B) system has a more even discharge pattern. From an average discharge of 65,000 m^3 s^{-1} during the 5-month monsoon season (June–October) at the Ganga mouth, it falls to an average of about 9000 m^3 s^{-1} for the remaining 7-month period. The Sundarbans mangrove forest, one of the largest of its kind in the world, is formed at the delta of the Ganga–Brahmaputra (G–B) river system in the Bay of Bengal.

In a study that covered sampling from October (2014) when the SW monsoon driven freshwater flux was still at its near peak through April (2015) when the flux was lowest, the CDOM was shown to be conservative with a highly significant negative correlation with salinity. The CDOM was the highest 0.23 m^{-1} (reported as a_{440}), which corresponds to 0.4641 m^{-1} of a_{350} calculated using 0.0078 nm^{-1} of spectral slope ($S_{400–700}$). The CDOM (a_{440}) decreased in the subsequent months up to the lowest observed of 0.0987 m^{-1}, that is, a_{350}, 0.4166 m^{-1} using the slope observed of 0.018 nm^{-1}. The implication is that CDOM concentration remained similar, but the MW decreased drastically from October (terrestrial CDOM) to April CDOM (in situ CDOM). This underscores the necessity of an accurate measure of spectral slope. Due to the high relief of the drainage basin, the Ganga water is rich in suspended sediment. CDOM also showed a very significant, but less strong correlation to suspended sediment compared to salinity (Das et al. 2016, 2017).

It may be noted that the salinity range in the above study was 18–29 × 10^{-3}, which is similar to the range in which conservative behavior was seen in the Godavari estuary (see above) in the corresponding non-monsoon season. Whether or not the CDOM of Ganga estuary shows conservative relationship in the 0–15 × 10^{-3} salinity range is not known.

Sanyal et al. (2020) have recently observed higher CDOM in the Sundarbans mangrove forest ca 0.4–2.26 m^{-1} than observed by Das et al. (2017) (see above) during the post-monsoon season, due probably to the significant effect of the mangrove litterfall. The CDOM was nonconservative, but correlated significantly with DOC yielding the relationship: DOC (μM) = 36.35 $a_{CDOM}(350)$ + 125.72, $r^2 = 0.44$ ($p < 0.0001$) which can facilitate using CDOM as a DOC proxy.

14.3.6 Coastal Waters

In the western Bay of Bengal, during the monsoon season, estuarine conditions spread into the coastal region. It was shown that CDOM and $SUVA_{254}$ had the power to distinguish Godavari water from Ganga water as the former's CDOM is more aged and humified than the CDOM originated from the much younger Himalayan drainage basin. The northern coastal Bay of Bengal closer to river Ganga mouth is mainly populated by low absorbing (per unit Chl a) microphytoplankton while the southern coastal Bay of Bengal, which is influenced by peninsular rivers, is populated by less abundant but higher absorbing nano- and picophytoplankton. In the northern Bay, there is an inverse power relationship for chl-a with specific absorption coefficient of phytoplankton. This changed in the nano- and picoplankton dominated (>60%) coastal regions off peninsular rivers to an inverse linear model (Pandi et al. 2014).

During a seasonal study of FDOM in the Godavari estuary and the adjacent coastal waters of Bay of Bengal, it was observed that the ratio of UV-humic to visible-humic like fluorophores (A/C) varied seasonally. They were higher during the monsoon season. The change of the ecosystem from one of bacteria-replete during the pre-monsoon to one of bacteria-free during the monsoon season was very well captured by FDOM. The bacterially degraded organic matter was indicated by the tyrosine-protein-like fluorophore (Chari et al. 2012). When the freshwater discharge was high, the humic-like fluorophores showed significant enrichment and a relationship to pH. Increase of pH (salinity) and an accompanying increase in the activity of Ca and Mg caused the externally sourced humates and fulvates to give flocs of the insoluble Ca and Mg humates and fulvates respectively, and thereby get removed to sediment (Sholkovitz et al. 1978; Zepp et al. 2004; Chari et al. 2012). During the nil-discharge period, the CDOM which is newly introduced by the in situ biological processes is spectrally distinguishable (Chari et al. 2019).

The western Bay of Bengal adjoining the estuaries is visited by meso-scale eddies throughout the year, of which a few reach the surface. The formation of eddies may be attributed to the northward moving East India Coastal Current (EICC) that is surrounded by warm Indian Ocean water on the offshore and cool (upwelled) water toward the inshore region (Shetye et al. 1993). In April 2010, when an anticyclonic eddy transited along the mid-western Bay of Bengal, it caused an upwelling of cold, nutrient-rich water to the surface, increased production, and a strong enrichment of protein (T)-like fluorescence. A new petroleum hydrocarbon-like fluorescence, designated as P fluorophore was also noticed in low abundance and attributed to petroleum hydrocarbons possibly from exploration activity along the coast (Chiranjeevulu et al. 2014).

14.4 Future Studies

The coastal ecosystem is exposed to the influence of mangroves, aquaculture, and industries surrounding the estuaries as well as the hydrocarbon exploration and urban and agricultural inputs along the coast. A long-term Coastal Ocean Monitoring and Prediction System (COMAPS) study of the Ministry of Earth Sciences, Government of India, executed by the National Centre of Coastal Research (NCCR) did not detect any significant amounts of anthropogenic inputs in the Indian coastal water beyond 5 km from the shore. However, since CDOM is a relatively easily measurable parameter and FDOM since it is about three orders of magnitude higher in sensitivity over CDOM, if supplemented, may reveal gray areas for further research. The movement of the East India Coastal Current along the coast creates estuary-like condition along its path. The southward spreading Ganga water from the north and northward drifting Indian Ocean water of the south create a front at ~19°N latitude which is an important region from a fishery point of view and because of its distinct hydrographic conditions and seasonal variations. CDOM studies may explain some of the questions, for example, changes in fishery resource, algal production, etc., in these areas. CDOM studies of the Bay of Bengal water at different depths have the potential to distinguish different sources of water/water masses, for example, the Arabia Sea High Saline Water (ASHSW), Persian Gulf Water (PGW), and Indian Ocean Water (IOW) in the Bay of Bengal. Size-fractionated CDOM studies are extremely limited so far and may help delineate these different sources. CDOM structural characteristics may be used for distinguishing subtle changes in river inputs. Collecting large volumes of water followed by isolation and spectral investigation (including NMR, Mass, and XRD spectra) of its constituent CDOM is a gap area for Indian seas. A relatively new area that has come up is the presence of microplastics by the dissolution of the dumped debris. It is not known to what extent the phthalates of the microplastics can be detected/estimated reliably via absorption/fluorescence measurements (and gas chromatography) on water samples and needs to be explored as an easier alternative to laborious experimental practices.

Acknowledgments Authors thank the Directors of Indian National Centre for Ocean Information Services (INCOIS) and National Centre for Polar and Ocean Research (NCPOR) for their encouragement, the National Institute of Ocean Technology (NIOT) (MoES) for the ship facility, and the crew for onboard cooperation. The INCOIS provided project grant (to NSS, F&A:XII:D2:023), DST-SERB Fast Track Young Scientist (to NVHKC, SR/FTP/ES-56/2013), and Human Resource Development Group Council of Scientific and Industrial Research (CSIR-HRDG) Research Associateship (to SRP, 09/002 (510)/2016 EMR-I). NSS thanks Andhra University for awarding Honorary Professor. SD thanks Global Challenges Research Fund (GCRF)-Living Delta Research Hub for providing him the Postdoctoral Research Associateship. This is NCPOR contribution number B-11/2020-21.

References

Baker A, Spencer RGM (2004) Characterization of dissolved organic matter from source to sea using fluorescence and absorbance spectroscopy. Sci Total Environ 333:217–232. https://doi.org/10.1016/j.scitotenv.2004.04.013

Bowers DG, Brett HL (2008) The relationship between CDOM and salinity in estuaries: an analytical and graphical solution. J Mar Syst 73:1–7. https://doi.org/10.1016/j.jmarsys.2007.07.001

Brown M (1977) Transmission spectroscopy examinations of natural waters. C. Ultraviolet spectral characteristics of the transition from terrestrial humus to marine yellow substance. Estuar Coast Mar Sci 5:309–317. https://doi.org/10.1016/0302-3524(77)90058-5

Carder KL, Steward RG, Harvey GR, Ortner PB (1989) Marine humic and fulvic acids: their effects on remote sensing of ocean chlorophyll. Limnol Oceanogr 34:68–81. https://doi.org/10.4319/lo.1989.34.1.0068

Chari NVHK, Pandi SR (2017) Characterization of humic substances and their distribution of XAD fractions by absorption spectroscopy in the Godavari estuary, India. Curr Sci 113:299–303. https://doi.org/10.18520/cs/v113/i02/299-303

Chari NVHK, Sarma NS, Pandi SR, Murthy KN (2012) Seasonal and spatial constraints of fluorophores in the midwestern Bay of Bengal by PARAFAC analysis of excitation emission matrix spectra. Estuar Coast Shelf Sci 100:162–171. https://doi.org/10.1016/j.ecss.2012.01.012

Chari NVHK, Keerthi S, Sarma NS et al (2013) Fluorescence and absorption characteristics of dissolved organic matter excreted by phytoplankton species of western Bay of Bengal under axenic laboratory condition. J Exp Mar Biol Ecol 445:148–155. https://doi.org/10.1016/j.jembe.2013.03.015

Chari NVHK, Pandi SR, Kanuri VV, Basuri CK (2019) Structural variation of coloured dissolved organic matter during summer and winter seasons in a tropical estuary: a case study. Mar Pollut Bull 149:110563. https://doi.org/10.1016/j.marpolbul.2019.110563

Chin YP, Alken G, O'Loughlin E (1994) Molecular weight, polydispersity, and spectroscopic properties of aquatic humic substances. Environ Sci Technol 28:1853–1858. https://doi.org/10.1021/es00060a015

Chiranjeevulu G, Murty KN, Sarma NS et al (2014) Colored dissolved organic matter signature and phytoplankton response in a coastal ecosystem during mesoscale cyclonic (cold core) eddy. Mar Environ Res 98:49–59. https://doi.org/10.1016/j.marenvres.2014.03.002

Coble PG (1996) Characterization of marine and terrestrial DOM in seawater using excitation-emission matrix spectroscopy. Mar Chem 51:325–346. https://doi.org/10.1016/0304-4203(95)00062-3

Cory RM, McKnight DM (2005) Fluorescence spectroscopy reveals ubiquitous presence of oxidized and reduced quinones in dissolved organic matter. Environ Sci Technol 39:8142–8149. https://doi.org/10.1021/es0506962

D'Sa E (2008) Colored dissolved organic matter in coastal waters influenced by the Atchafalaya River, USA: effects of an algal bloom. J Appl Remote Sens 2:023502. https://doi.org/10.1117/1.2838253

Das S, Hazra S, Lotliker AA et al (2016) Delineating the relationship between chromophoric dissolved organic matter (CDOM) variability and biogeochemical parameters in a shallow continental shelf. Egypt J Aquat Res 42:241–248. https://doi.org/10.1016/j.ejar.2016.08.001

Das S, Hazra S, Giri S et al (2017) Light absorption characteristics of chromophoric dissolved organic matter (CDOM) in the coastal waters of northern Bay of Bengal during winter season. Indian J Geomarine Sci 46:884–892

Del Castillo CE, Gilbes F, Coble PG, Müller-Karger FE (2000) On the dispersal of riverine colored dissolved organic matter over the West Florida Shelf. Limnol Oceanogr 45:1425–1432. https://doi.org/10.4319/lo.2000.45.6.1425

Downing BD, Boss E, Bergamaschi BA et al (2009) Quantifying fluxes and characterizing compositional changes of dissolved organic matter in aquatic systems in situ using combined acoustic and optical measurements. Limnol Oceanogr Methods 7:119–131. https://doi.org/10.4319/lom.2009.7.119

Fichot CG, Benner R (2011) A novel method to estimate DOC concentrations from CDOM absorption coefficients in coastal waters. Geophys Res Lett 38:1–5. https://doi.org/10.1029/2010GL046152

Fichot CG, Benner R (2012) The spectral slope coefficient of chromophoric dissolved organic matter (S 275-295) as a tracer of terrigenous dissolved organic carbon in river-influenced ocean margins. Limnol Oceanogr 57:1453–1466. https://doi.org/10.4319/lo.2012.57.5.1453

Fichot CG, Kaiser K, Hooker SB et al (2013) Pan-Arctic distributions of continental runoff in the Arctic Ocean. Sci Rep 3:1–6. https://doi.org/10.1038/srep01053

Fruton JS, Simmonds S (1961) General biochemistry, 2nd edn. Wiley, New York

Galgani L, Engel A, Rossi C et al (2018) Polystyrene microplastics increase microbial release of marine Chromophoric Dissolved Organic Matter in microcosm experiments. Sci Rep 8:1–11. https://doi.org/10.1038/s41598-018-32805-4

Guéguen C, Granskog MA, McCullough G, Barber DG (2011) Characterisation of colored dissolved organic matter in Hudson Bay and Hudson Strait using parallel factor analysis. J Mar Syst 88:423–433. https://doi.org/10.1016/j.jmarsys.2010.12.001

Helms JR, Stubbins A, Ritchie JD et al (2008) Absorption spectral slopes and slope ratios as indicators of molecular weight, source, and photobleaching of chromophoric dissolved organic matter. Limnol Oceanogr 53:955–969. https://doi.org/10.4319/lo.2008.53.3.0955

Huguet A, Vacher L, Relexans S et al (2009) Properties of fluorescent dissolved organic matter in the Gironde Estuary. Org Geochem 40:706–719. https://doi.org/10.1016/j.orggeochem.2009.03.002

Hulatt CJ, Thomas DN, Bowers DG et al (2009) Exudation and decomposition of chromophoric dissolved organic matter (CDOM) from some temperate macroalgae. Estuar Coast Shelf Sci 84:147–153. https://doi.org/10.1016/j.ecss.2009.06.014

Ishiwatari R (1992) Macromolecular material (humic substance) in the water column and sediments. Mar Chem 39:151–166. https://doi.org/10.1016/0304-4203(92)90099-V

Kalle K (1938) Zum Problem der Meerwasserfarbe. Ann Hydrol Mar Mitt 66:1–13

Kowalczuk P, Stedmon CA, Markager S (2006) Modeling absorption by CDOM in the Baltic Sea from season, salinity and chlorophyll. Mar Chem 101:1–11. https://doi.org/10.1016/j.marchem.2005.12.005

Kowalczuk P, Durako MJ, Young H et al (2009) Characterization of dissolved organic matter fluorescence in the South Atlantic Bight with use of PARAFAC model: interannual variability. Mar Chem 113:182–196. https://doi.org/10.1016/j.marchem.2009.01.015

Lawaetz AJ, Stedmon CA (2009) Fluorescence intensity calibration using the Raman scatter peak of water. Appl Spectrosc 63:936–940. https://doi.org/10.1366/000370209788964548

Lotliker AA, Baliarsingh SK, Sahu KC, Kumar TS (2020) Long-term chlorophyll-a dynamics in tropical coastal waters of the western Bay of Bengal. Environ Sci Pollut Res 27:6411–6419. https://doi.org/10.1007/s11356-019-07403-0

McKnight DM, Boyer EW, Westerhoff PK et al (2001) Spectrofluorometric characterization of dissolved organic matter for indication of precursor organic material and aromaticity. Limnol Oceanogr 46:38–48. https://doi.org/10.4319/lo.2001.46.1.0038

Morel A, Bricuad A (1981) Theoretical results concerning light absorption in a discrete medium, and application to specific absorption of phytoplankton. Deep-Sea Res I 28(11): 1375-1393. https://doi.org/10.1016/0198-0149(81)90039-X.

Nelson NB, Siegel DA, Carlson CA et al (2007) Hydrography of chromophoric dissolved organic matter in the North Atlantic. Deep Sea Res I Oceanogr Res Pap 54:710–731. https://doi.org/10.1016/j.dsr.2007.02.006

Osburn CL, Wigdahl CR, Fritz SC, Saros JE (2011) Dissolved organic matter composition and photoreactivity in prairie lakes of the U.S. Great Plains. Limnol Oceanogr 56:2371–2390. https://doi.org/10.4319/lo.2011.56.6.2371

Pandi SR, Kiran R, Sarma NS et al (2014) Contrasting phytoplankton community structure and associated light absorption characteristics of the western Bay of Bengal. Ocean Dyn 64:89–101. https://doi.org/10.1007/s10236-013-0678-1

Para J, Coble PG, Charrière B et al (2010) Fluorescence and absorption properties of chromophoric dissolved organic matter (CDOM) in coastal surface waters of the northwestern Mediterranean Sea, influence of the Rhône River. Biogeosciences 7:4083–4103. https://doi.org/10.5194/bg-7-4083-2010

Parida C, Baliarsingh SK, Lotliker AA et al (2019) Seasonal variation in optically active substances at a coastal site along western Bay of Bengal. SN Appl Sci 1:1–8. https://doi.org/10.1007/s42452-019-1257-y

Peuravuori J, Koivikko R, Pihlaja K (2002) Characterization, differentiation and classification of aquatic humic matter separated with different sorbents: synchronous scanning fluorescence spectroscopy. Water Res 36:4552–4562. https://doi.org/10.1016/S0043-1354(02)00172-0

Repeta DJ, Quan TM, Aluwihare LI, Accardi AM (2002) Chemical characterization of high molecular weight dissolved organic matter in fresh and marine waters. Geochim Cosmochim Acta 66:955–962. https://doi.org/10.1016/S0016-7037(01)00830-4

Romera-Castillo C, Sarmento H, Alvarez-Salgado XAÁ et al (2011) Net production and consumption of fluorescent colored dissolved organic matter by natural bacterial assemblages grow-

ing on marine phytoplankton exudates. Appl Environ Microbiol 77:7490–7498. https://doi.org/10.1128/AEM.00200-11
Röttgers R, Koch BP (2012) Spectroscopic detection of a ubiquitous dissolved pigment degradation product in subsurface waters of the global ocean. Biogeosciences 9:2585–2596. https://doi.org/10.5194/bg-9-2585-2012
Sarma NS, Nageswara Rao I, Annapurna K (1994) Ammonium ion and organic phosphorus as major in-situ contributors to dissolved fluorescence of the near northwestern Bay of Bengal. Mar Chem 47:255–267. https://doi.org/10.1016/0304-4203(94)90024-8
Sarma NS, Pandi SR, Chari NVHK et al (2018) Spectral modelling of estuarine coloured dissolved organic matter. Curr Sci 114:1762–1767. https://doi.org/10.18520/cs/v114/i08/1762-1767
Shank GC, Evans A (2011) Distribution and photoreactivity of chromophoric dissolved organic matter in northern Gulf of Mexico shelf waters. Cont Shelf Res 31:1128–1139. https://doi.org/10.1016/j.csr.2011.04.009
Shetye SR, Gouveia AD, Shenoi SSC et al (1993) The western boundary current of the seasonal subtropical gyre in the Bay of Bengal. J Geophys Res 98:945–954. https://doi.org/10.1029/92JC02070
Sholkovitz ER, Boyle EA, Price NB (1978) The removal of dissolved humic acids and iron during estuarine mixing. Earth Planet Sci Lett 40:130–136. https://doi.org/10.1016/0012-821X(78)90082-1
Stanier RY, Ingraham JL, Wheelis ML, Painter PR (1988) The microbial world, 5th edn. Prentice-Hall, London
Stedmon CA, Markager S (2003) Behaviour of the optical properties of coloured dissolved organic matter under conservative mixing. Estuar Coast Shelf Sci 57:973–979. https://doi.org/10.1016/S0272-7714(03)00003-9
Stedmon CA, Markager S, Bro R (2003) Tracing dissolved organic matter in aquatic environments using a new approach to fluorescence spectroscopy. Mar Chem 82:239–254. https://doi.org/10.1016/S0304-4203(03)00072-0
Velapoldi RA, Mielenz KD (1981) A fluorescence standard reference material: quinine sulfate dihydrate. Appl Opt 20:1718. https://doi.org/10.1364/ao.20.001718
Weishaar JL, Aiken GR, Bergamaschi BA et al (2003) Evaluation of specific ultraviolet absorbance as an indicator of the chemical composition and reactivity of dissolved organic carbon. Environ Sci Technol 37:4702–4708. https://doi.org/10.1021/es030360x
Yamashita Y, Cory RM, Nishioka J et al (2010) Fluorescence characteristics of dissolved organic matter in the deep waters of the Okhotsk Sea and the northwestern North Pacific Ocean. Deep Sea Res II Top Stud Oceanogr 57:1478–1485. https://doi.org/10.1016/j.dsr2.2010.02.016
Yi H, Chang YH, Oh HW et al (2003) Zooshikella ganghwensis gen. nov., sp. nov., isolated from tidal flat sediments. Int J Syst Evol Microbiol 53:1013–1018. https://doi.org/10.1099/ijs.0.02521-0
Zepp RG, Sheldon WM, Moran MA (2004) Dissolved organic fluorophores in southeastern US coastal waters: correction method for eliminating Rayleigh and Raman scattering peaks in excitation-emission matrices. Mar Chem 89:15–36. https://doi.org/10.1016/j.marchem.2004.02.006
Zsolnay Á (2003) Dissolved organic matter: artefacts, definitions, and functions. Geoderma 113:187–209. https://doi.org/10.1016/S0016-7061(02)00361-0
Sanyal P, Ray R, Paul M, Gupta VK, Acharya A, Bakshi S, Jana TK, Mukhopadhyay SK (2020) Assessing the Dynamics of Dissolved Organic Matter (DOM) in the Coastal Environments Dominated by Mangroves, Indian Sundarbans. Front Earth Sci 8: 1–21. https://doi.org/10.3389/feart.2020.00218
Singh SP, Kumari S, Rastogi RP, Singh KL, Sinha, RP (2008) Mycosporine-like amino acids (MAAs): Chemical structure, biosynthesis and significance as UV-absorbing/screening compounds. Indian J Exp Biol 46:7–17. PMID: 18697565
Pandi SR, Chari NVHK, Sarma NS, Chiranjeevulu G, Kiran R, Murthy KN, Venkatesh P, Lotliker AA, Tripathy SC (2021) Characteristics of conservative and non-conservative CDOM of a tropical monsoonal estuary in relation to changing biogeochemistry. Reg Stud Mar Sci (Article in Press)

Chapter 15
The Indian Sundarbans: Biogeochemical Dynamics and Anthropogenic Impacts

Andrew C. G. Henderson, Sourav Das, Tuhin Ghosh, Virginia N. Panizzo, Heather L. Moorhouse, Lucy R. Roberts, Richard E. Walton, Ying Zheng, Adrian M. Bass, and Suzanne McGowan

Abstract The Sundarbans region is one of the richest ecosystems in the world and is located on one of the world's largest deltas – the Ganges–Brahmaputra–Meghna system. The Indian Sundarbans have exceptional biodiversity, including rare and globally threatened species, and is made up of a mangrove forest ecosystem with an interconnected network of rivers. The hydrology of the Sundarbans underpin ecosystem health and the potential impact of humans on the region, as the tidal cycle changes water salinity diurnally and freshwater supply changes seasonally with the monsoon. The Indian Sundarbans face multiple pressures with both a reduction in freshwater supply and rising relative sea-level, leading to increased salinization of the mangrove forest. Human-driven alteration of the Sundarbans river catchments is reducing sediment flow, and when coupled with land-use change, is leading to subsidence, deforestation, nutrient enrichment, and heavy metal pollutants impacting the health of the ecosystem. All of these impacts have important ramifications for carbon fluxes that could exacerbate climate change and ecosystem health. In this chapter, we present an overview of our current understanding of biogeochemical dynamics and anthropogenic impacts on the Indian Sundarbans, with a particular focus on water quality, aquatic ecology, and carbon dynamics.

Keywords Indian Sundarbans · Ganges–Brahmaputra–Meghna delta · Water quality · biogeochemistry · carbon · pollutants · ecology · sediments

A. C. G. Henderson (✉) · R. E. Walton
School of Geography, Politics & Sociology, Newcastle University, Newcastle upon Tyne, UK
e-mail: andrew.henderson@ncl.ac.uk

S. Das · T. Ghosh
School of Oceanographic Studies, Jadavpur University, Kolkata, India

V. N. Panizzo · L. R. Roberts · S. McGowan
School of Geography, University of Nottingham, Nottingham, UK

H. L. Moorhouse
Lancaster Environment Centre, Lancaster University, Lancaster, UK

Y. Zheng · A. M. Bass
School of Geographical & Earth Sciences, University of Glasgow, Glasgow, UK

S. Das, T. Ghosh (eds.), *Estuarine Biogeochemical Dynamics of the East Coast of India*, https://doi.org/10.1007/978-3-030-68980-3_15

15.1 Introduction

The Sundarbans region is one of the richest ecosystems in the world and is located on one of the world's largest deltas – the Ganges–Brahmaputra–Meghna (GBM) system. The Sundarbans is located in the estuarine phases of the Rivers Ganga, Brahmaputra, and Meghna between 21°32′N and 21°40′N and 88°05′E and 89°E, spanning regions in both India and Bangladesh (Spalding et al. 1997) and contains arguably the world's largest remaining area of mangroves (an area of ~2529 km^2, Bhattacharyya 2015). The Indian Sundarbans have exceptional biodiversity, including rare and globally threatened species, for example, the northern river terrapin (*Batagur baska*, Lesson 1831), the Irrawaddy dolphin (*Orcaella brevirostris*, Owen in Gray 1866), the Ganges River dolphin (*Platanista gangetica*, Lebeck 1801), the brown-winged kingfisher (*Pelargopsis amauroptera* Pearson 1841), and the Royal Bengal tiger (*Panthera tigris*, Linneaus 1785) – the only mangrove tiger on Earth (RAMSAR 2019). The mangrove ecosystem, which makes up the Indian Sundarbans, is an interconnected network of rivers, creeks, rivulets, and semi-diurnal tides. The lower delta is dominated by a network of tributary rivers, creeks, and channels, with direct marine influence on the most seaward part of the Indian Sundarbans (Fig. 15.1). As a result, there are a range of hydrological influences (including both freshwater and coastal water) on the mangrove system, and when coupled with its topographic heterogeneity it results in a rich biodiversity (Gopal and Chauhan 2006). This has led to the Sundarbans mangrove forest being designated a World Heritage Site by the International Union for Conservation of Nature (IUCN) in 1987; a Biosphere Reserve by United Nations Educational, Scientific and Cultural Organization (UNESCO) in 1989; and a wetland of international importance according to the RAMSAR convention in the year 2019.

Despite its international designation, the Indian Sundarbans face multiple pressures. As the freshwater discharge originating from the Himalayan uplands has decreased in recent decades (Raha et al. 2012), this has led to increased salinization of soil and groundwater within the Sundarbans, leading to the degradation of mangrove ecosystem health (Chowdhury et al. 2019). In addition, anthropogenic activities continue to alter hydrology and sediment flow, while land-use change is leading to deforestation, nutrient enrichment, and heavy metal pollutants causing many mangrove species to become threatened or extinct (Gopal and Chauhan 2006), triggering an overall degraded ecosystem. This, in turn, has important ramifications for carbon fluxes in the Indian Sundarbans that could further exacerbate climate change and ecosystem health. The following sections aim to explore these different pressures and the impacts they are having on the current and future state of this vital ecosystem.

15.2 Hydrological Regime and Sediment Flow

The Indian Sundarbans landscape has evolved from the subduction of the Asian plate under the Burma plate to neotectonic tilting creating a hydrological gradient leading to river discharge from the highlands (Morgan and McIntire 1959). As a

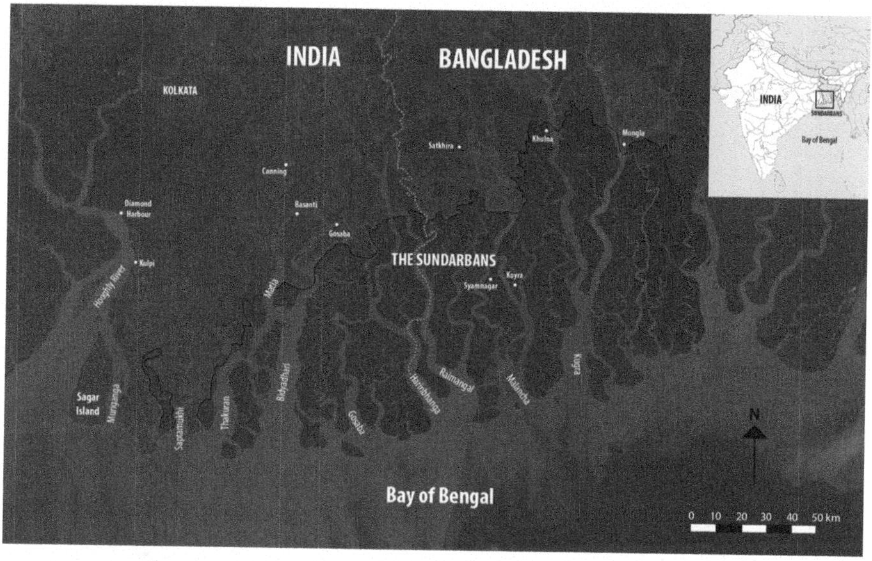

Fig. 15.1 A Sentinel-2 satellite natural color image taken in March 2018 of the Sundarbans region West Bengal, India, generated through the Sentinel Hub. The main rivers that influence the biogeochemistry and anthropogenic impact of the Sundarbans are labelled and major cities and towns are labelled. Inset map shows the location of the Sundarbans within in India

result, there are seven major estuarine rivers flowing through the Indian Sundarbans – the Hooghly, the Muriganga, the Saptamukhi, the Thakuran, the Matla, the Gosaba, and the Harinbhanga (also known as Ichamati and Raimangal) (Fig. 15.1). The combination of freshwater and tidal flow shape the deposition and erosion of sediments across the Sundarbans region, creating the dynamic nature of this deltaic environment. The climate of the Sundarbans is sub-humid and characterized by hot summers and mild winters (Fig 15.2a). The mean monthly temperature varies between 30 °C to 40 °C in the summer (June to September) and 15 °C to 20 °C in winter (October to March). Precipitation from the annual monsoon during June to September is the major freshwater source to the Indian Sundarbans as it represents 80% of all annual rainfall (1750–1800 mm per annum) for the region. As a result, changes in freshwater inputs from monsoon rains, baseline river discharge during the rest of the year and tidal hydrology strongly influence the Sundarban region.

15.2.1 Hydrological Regime

15.2.1.1 Freshwater Hydrology

Despite the Indian Sundarbans being part of the GBM system (Chatterjee et al. 2013), year-round continuous flow is limited to a few river channels. The Hooghly River discharges the most freshwater into the Indian Sundarbans and is the western

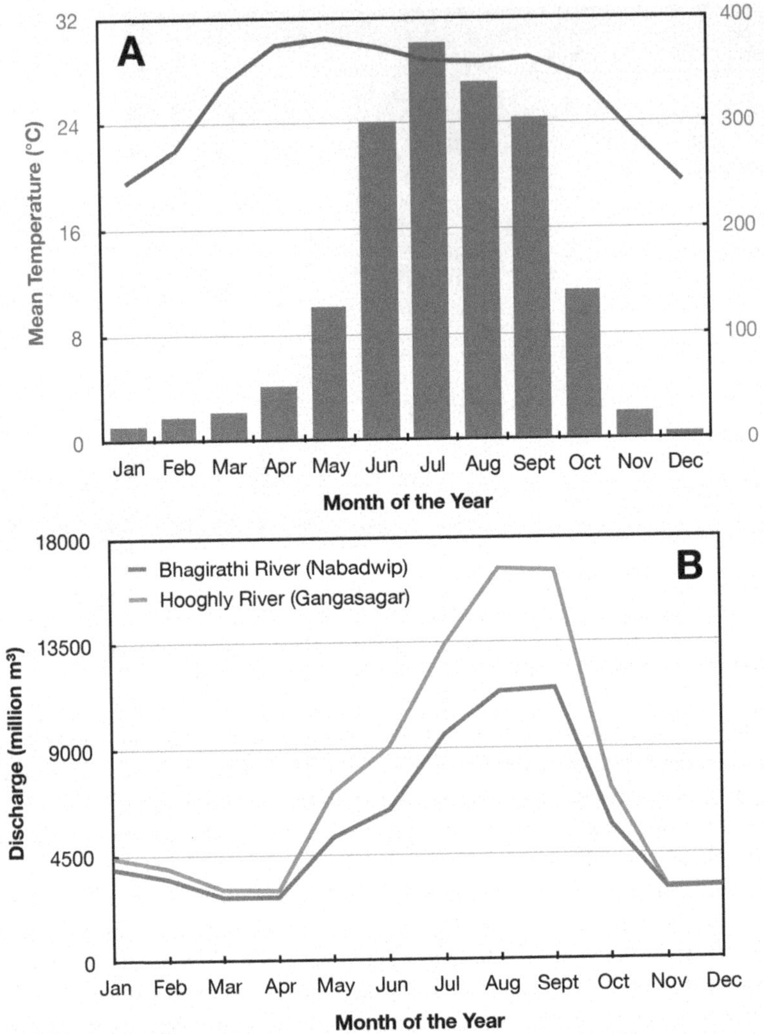

Fig. 15.2 (**a**) Summary of mean monthly temperature and precipitation data from Kolkata, West Bengal, from 1982 to 2012. Data from climate-data.org and is based on an interpolated model of weather station data; (**b**) Mean monthly discharge of the Bhagirathi and Hooghly River systems, West Bengal. (Data from Rudra (2014) and is derived from a rainfall-runoff model)

most branch of the River Ganges reaching the Bay of Bengal (Fig. 15.1) (Rudra 2018). The Raimangal River at the eastern edge of the Indian Sundarbans also brings additional freshwater as a tributary channel of the Ichamati River, and in turn this influences the discharge of the Gosaba, Harinbhanga, and Jhila Rivers (Chatterjee et al. 2013; Sarkar et al. 2013). While the monsoon seasons create variation throughout the year (Fig 15.2b), the Hooghly River has a more consistent input of freshwater than the Raimangal River (Chatterjee et al. 2013; Ghosh et al. 2013) due to the construction of the Farakka Barrage that diverts 7% of the annual flow of the Ganges to provide a regulated stream of freshwater throughout the dry season to support the operation of the Port of Kolkata (Ghosh et al. 2013).

15.2.1.2 Tidal Influence on Hydrology

The Sundarbans are macrotidal (range: 1.8 to 5.2 m between neap and spring high tides) and it experiences a semi-diurnal tide cycle (Gole and Vaidyaraman 1967; Rogers and Goodbred 2014; Sinha et al. 1996). Despite the large volumes of freshwater from the Hooghly and Raimangal Rivers, rising tides still influence the upstream hydrology of the Sundarbans, with tides regularly travelling up to 120 km from the mouth of the Hooghly River during the pre-monsoon season (Gole and Vaidyaraman 1967). In the post-monsoon, the tide can travel 250 km up the Hooghly (Sinha et al. 1996) with the tidal limit at Kalna, West Bengal, during the monsoon (Chatterjee et al. 2013).

As the tides bring saline water with them, they impact both anthropogenic access to freshwater and affect ecological functioning. The incursion of saline waters by flood tides is also controlled by the season in which it happens. For example, the extent of saline waters during the monsoon is low, as the increased freshwater delivered by seasonal rains acts as a barrier to flood tide penetration, with the upper limit typically as far as Nayachar Island in the upper mouth of the Hooghly River (Chatterjee et al. 2013; Ghosh et al. 2013; Sharma et al. 2018). Another effect is the stratification of freshwaters over the saline/brackish waters in the river during the monsoon season (Sadhuram et al. 2005; Chatterjee et al. 2013). In the non-monsoon seasons there is a significant rise in salinity levels within the Hooghly River with 30 ppt (parts per thousand) observed near Diamond Harbour and saline waters reach as far north as Kolkata (Gole and Vaidyaraman 1967), although during ebb tides the limit of saline water moves back down to the mouth of the estuary near Sagar Island (Sinha et al. 1996).

15.2.2 Sediment Flow

All river channels flow into the Indian Sundarbans, including freshwater rivers and tidal inflows, carrying sediments that affect the whole mangrove ecosystem. Sediments carried by the freshwater Hooghly River consist predominantly of sand and silt (Somayajulu et al. 2002; Massolo et al. 2012) and less than 10% of the sediment consists of clay particles. These sediments are predominantly derived from rain-driven terrestrial erosion to the Ganges (Somayajulu et al. 2002; Rudra 2018) and because of the high discharge of the Ganges and Hooghly Rivers these sediments do not experience much water-column weathering before they reach the Sundarbans and the Bay of Bengal (Somayajulu et al. 2002; Flood et al. 2016). While the Hooghly carries a large volume of sediment, there is a notable seasonal variation in sediment loads because of monsoon-driven changes in freshwater discharge (Gole and Vaidyaraman 1967).

Each year flood tides deposit ~12 cm of fresh sediment in to the Indian Sundarbans (Rudra 2018) and tides carry sediments that are more fine-grained than those transported by freshwater rivers (Allison et al. 2003; Flood et al. 2016, 2018). Sediments transported and deposited by flood tides in the Indian Sundarbans also originate from the mouth of the Ganges–Brahmaputra–Meghna River system approximately 275 km to the east (Flood et al. 2016, 2018; Rudra 2018), where >1 billion tons of sediment are discharged each year (Somayajulu et al. 2002; Rogers and Goodbred 2014; Rudra 2018). These sediments are carried by coastal currents westward along the coastline through suspension (Rogers and Goodbred 2014; Flood et al. 2018), where they undergo weathering and degradation in the water column, resulting in fine-grained sediments being transported in suspension by flood tides in to the Sundarbans (Flood et al. 2016, 2018). Sediments are deposited and retained because of lateral accretion along mangrove tree roots (Manna et al. 2012; Flood et al. 2018) and tidal creeks (Rudra 2018).

Resuspension of sediments occurs as a result of bioturbation in intertidal mudflats (Rogers and Goodbred 2014), dredging, winds, and tides. These resuspended sediments are redistributed or carried from the Sundarbans through flooding and wave action. Approximately 430 km^2 of the Indian Sundarbans were eroded between 1917–2016, which is offset by 220 km^2 of sediment accumulation over the same period (Rudra 2018). The dynamics of rivers, tides, and sediment movement means these are key processes that drive Sundarbans water quality, ecology, and overall ecosystem health (Gole and Vaidyaraman 1967; Sinha et al. 1996; Rogers and Goodbred 2014).

15.3 Ecology and Water Quality

The Indian Sundarbans are home to a number of endemic enigmatic and globally vulnerable species. By looking at the biology of these fragile Sundarbans ecosystems and the interface with hydrology and biogeochemistry we can document and understand the threats to the Sundarbans wetland ecosystem and its iconic inhabitants.

15.3.1 Aquatic Ecology

15.3.1.1 Primary Producers

Aquatic primary production in the Sundarbans is a function of nutrient loading and light penetration, with the latter often constrained by river turbidity (Chaudhuri et al. 2012). Large river and estuarine channels are dominated by the

Bacillariophyceae algal group – biosiliceous diatoms, followed by *Pyrrophyceae* – dinoflagellates, and *Chlorophyceae* – chlorophytes (Biswas et al. 2010; Manna et al. 2010; De et al. 2011). There are still large gaps in our knowledge about the role of these primary producers in mangrove ecosystems, especially diatoms (Samanta and Bhadury 2018). However, the biovolume of primary produces is highest in the post-monsoon winter months supporting colonies of long-chain diatoms, whereas there are low biovolumes during the monsoon season because of increased total suspended solids (TSS) (derived from rain-driven catchment erosion), reducing light penetration and photosynthesis (Chaudhuri et al. 2012; Bhattacharjee et al. 2013). Prior to the monsoon season the diatom assemblage is dominated by saline-tolerant species (Manna et al. 2010) and this may become a feature of upstream diatom communities as saline intrusion into the delta region becomes more widespread.

15.3.1.2 Macroinvertebrates

The main consumers of primary producers are the zooplankton, who play an integral role in the transfer of organic matter between trophic levels and export organic carbon to sediments (Bhattacharya et al. 2015a). As macroinvertebrate species occupy distinct trophic levels they respond rapidly to environmental change and are relatively quick and easy to identify, making them effective water-quality indicators (Gannon and Stemberger 1978). Copepods are small cosmopolitan crustaceans, which dominate zooplankton in tidal river systems in the Indian Sundarbans (Bir et al. 2015). Whereas in tidal flats, polychaetes and mollusks are important macrozoobenthic groups, whose spatiotemporal distribution is driven by salinity, the nature of the substrate (e.g., mudflats exhibit greater diversity than sandflats), and anthropogenic activity (Khan 2003; Roy and Nandi 2012).

In the Sundarbans, compositional changes in zooplankton communities are primarily driven by the quantity and quality of primary producer prey, as well as salinity and water transparency, which can vary seasonally and interannually (Bhattacharya et al. 2015a). Much like primary producers, zooplankton biomass is highest during the post-monsoon season when water currents, salinity, and temperature are at their lowest (Bir et al. 2015). However, extreme climate events, such as cyclone "*Aila*" in 2009, can lead to increased suspended particulates and nutrients, reductions in transparency, and primary photosynthesis. As a result there is a decrease in zooplankton diversity, biomass, and abundance (Bhattacharya et al. 2014a). If extreme events across the region worsen, this could modify phytoplankton–zooplankton interactions and threaten the viability of both open-water and aquaculture fisheries, whose stock require good quantity and quality of these prey organisms. Indeed, continued saltwater intrusion may reduce macrozoobenthic diversity due to reductions in decomposition rate of photosynthetic organic matter following higher sediment salinities, which may modify macrozoobenthic feeding behaviors and consequently impact the higher organisms which they support, for example, wading birds (Bandopadhyay and Burman 2006; Roy and Nandi 2012).

15.3.1.3 Microbial Biodiversity

Mangrove environments are hotspots of microbial diversity because of the complexity of habitats they provide and the fluxes in salinity, nutrients, labile organic compounds, and water levels across daily to seasonal timescales (Chakraborty et al. 2015). Seasonal variations in freshwater flow are an important determinant of community diversity, specifically of the bacterioplankton, where diversity is found to be greater in monsoon seasons compared to post-monsoon (Ghosh and Bhadury 2018). These microbes play a profound role in biogeochemical cycling from metabolizing the considerable allochthonous organic matter inputs of mangrove vegetation (Chakraborty et al. 2015), and therefore the sustenance, productivity, and recovery of this ecosystem (Ghosh et al. 2010; Roy et al. 2002; Santos et al. 2011). While there remains a significant gap in the knowledge of microbial diversity and abundance in the Sundarbans (Ghosh et al. 2010), modifications in microbial abundance, diversity, and community composition have been identified (Ghosh and Bhadury 2018). For example, industrial and boating activity has increased polyaromatic hydrocarbons (PAHs), heavy metal, and nutrient pollution detected by bacterial strains with heavy metal resistance and those involved in hydrocarbon degradation processes (Chakraborty et al. 2015). Eutrophication of these waters has meant bacterial productivity exhibits an exponential relationship to temperature as they are no longer nutrient-limited (Manna et al. 2010, 2012).

15.3.2 Water Quality

15.3.2.1 Nutrients

One of the key factors determining the biodiversity of the Indian Sundarbans is water and the role it plays in transporting nutrients and pollutants in the mangrove ecosystem (Sarkar et al. 2004). In general, phosphorus (P) availability is low in tropical regions where soils have been weathered for millions of years (Yang et al. 2013). Nitrogen (N) can be generated and removed from ecosystems by microbes and so mangroves are important sites for N (and C) cycling with mangrove plants being significant stores of N (Kamruzzaman et al. 2019; Purvaja et al. 2008). In coastal zones, P and N availability changes along the freshwater–marine transition, because sediments retain less P in marine environments, releasing P to the waters (Blomqvist et al. 2004). Primary production in freshwaters tends to be limited by P, whereas marine waters are generally N-limited and P-replete. Therefore, the tidal cycle in the Sundarbans is a key influence on nutrient distribution in estuaries, and the nutrient status of waters change seasonally to become P-limited after the monsoon when the influence of freshwaters increases, and N-limited during the monsoon and pre-monsoon periods (Chaudhuri et al. 2012). The main source of nutrients are from either freshwater runoff, for example, dissolved silica, nitrate, and phosphate, and/or from intertidal flats, for example, ammonium, nitrate/nitrite, and phosphate (Singh et al. 2016). During low tides, there is an increase in freshwater input into the northern Bay of Bengal, which dilutes nutrient concentration across the continental shelf and the mangrove ecosystem and vice versa during high tides, and these tidal dynamics play a crucial role in regulating short-term variability in nutrient concentrations (Das et al. 2015, 2017). Atmospheric deposition of P also constitutes a major source in the Sundarbans mangroves, comprising >50% of the annual P inputs (Ray et al. 2018a). P is hypothesized to be transported from arid regions of western India by pre-monsoonal northwesterly (and westerly) winds (Ray et al. 2018a). This seasonal P transport seems likely to either drive or exacerbate the observed seasonal differences in estuaries, but thus far there has been little research into the interplay of monsoonal rainfall, river discharge, and the consequences of desertification in arid regions on nutrient cycling of the Sundarbans.

In addition to natural variability in nutrients, anthropogenic inputs of nutrient-rich effluent have led to the eutrophication of smaller rivers, tidal creeks, and ponds in the Sundarbans, exacerbated by generally reduced flushing rates. However, such phenomena are being more commonly documented within the main estuarine channels such as the Hooghly River where anthropogenic influences has increased at a faster rate (Manna et al. 2010; De et al. 2011). Eutrophication has led to algal blooms, which reduce light penetration for benthic photosynthesis and deplete oxygen for higher trophic species (due to bloom respiration) (Biswas et al. 2014). In addition, harmful algal blooms (HABs) from toxin-producing cyanobacteria (CyanoHABs) such as *Microcystis* species and dinoflagellates have been recorded in Sundarbans aquatic habitats (Manna et al. 2010; Sen et al. 2015). CyanoHABs outcompete other algal groups due to their ability to regulate buoyancy, adaptation to low light, and higher temperatures, and are often able to fix N from the atmosphere (important in systems that are N-limited relative to P typical in these wetlands) (Paerl and Tucker 1995; Walsby and Schanz 2002; Islam et al. 2004; Paerl and Huisman 2008).

15.3.2.2 Heavy Metals

The primary source of heavy metal contamination in coastal areas of West Bengal is the major rivers that run through the Sundarbans (Mitra 1998) and even though these metals can occur naturally in the Sundarbans biogeochemical cycle (Garrett 2000), they predominantly come from industrial and domestic effluents, storm water runoff, dust, and boating activities. The mineralogy and grain size of sediments of the GBM river system has the potential to trap contaminants with silt and clays, predominantly carrying metal contamination from upstream. However, the textual composition and amount of organic matter in the sediment is critical to the sorption of transition metals (Kumar and Ramanathan 2015; Roy et al. 2018). Consequently, river sediments have become a sink of bioavailable heavy metals, with flooding and dredging leading to the resuspension of sediments, releasing their heavy metal load into the water column. Furthermore, salinity influences the partitioning, physiochemical form, and therefore bioavailability of these metals (Mitra 1998).

The Hooghly River catchment encompasses rural, agricultural, urban, and industrial land uses, including the megacity of Kolkata (population ~ 15 million) before draining into the Bay of Bengal. The metal concentrations of the riverine suspended particulate matter (SPM) ranges 7.9–29μg/g (mean: 19 ± 5.5μg/g) for Co, 17–70μg /g (mean: 49 ± 14μg /g) for Ni, and 12–55μg/g (mean: 36 ± 12μg/g) for Cu, which is higher than the average concentrations for global rivers (Samanta and Dalai 2018). The dissolved concentrations of metals in the Hooghly River estuary range 0.8–24 nM/L (mean: 6.2 ± 6 nM/L) for Co, 3.5–172 nM/L (mean: 50 ± 42 nM/ L) for Ni, and 8–178 nM/L (mean: 60 ± 37 nM/L) for Cu. Annually, these contribute up to 1.8% Co, 2.4% Ni, and up to 1.2% Cu of the global riverine metal fluxes (Samanta and Dalai 2018). The heavy metal concentrations of the Hooghly display seasonal variability with the maximum pollution load pre-monsoon and minimum load during the monsoon (Roy et al. 2018). High concentrations pre-monsoon have been attributed to high temperatures and increased evaporation rates of surface water (Bhattacharya et al. 2014a; Ghosh and Choudhury 1989; Mitra 1998).

Mixing of riverine and marine waters also contributes to changes in the speciation of metals, as well as the resuspension of sediments. Mukherjee et al. (2009) argue physicochemical changes limit the enrichment of heavy metals in river sediments and the high concentrations in the Hooghly compared to other regional rivers is because of a large sediment contribution from a bigger catchment area. The elevated concentrations in the Hooghly River are an important mechanism for elevating the amount of dissolved Ba in the river estuary via desorption with mixing of waters (Samanta and Dalai 2018). Similarly, Hg concentrations are positively correlated with pH ($r = 0.58–0.68$, $p < 0.01$) and salinity ($r = 0.52–0.79$, $p < 0.01$) (Bhattacharya et al. 2014b), and some metal concentrations in the waters of the middle and lower Hooghly estuary are significantly higher than other global estuaries in dissolved Ni and Cu (Samanta and Dalai 2018). However, upstream anthropogenic activities are still important in contributing widescale pollution across the Sundarbans.

Anthropogenic pollution within the Sundarbans itself has led to elevated levels of Cd, Cu, Zn, As, Ni, Pb, and Hg, which can cause impacts on biology (Sarkar et al. 2004; Chatterjee et al. 2007, 2009; Chowdhury et al. 2017; Mitra and Ghosh 2014). The source of these contaminants come from a mixture of industrial effluents, boat anti-fouling paint, sewage, fertilizers, and storm water drainage (Chowdhury and Maiti 2016; Mitra and Ghosh 2014; Chatterjee et al. 2007; Mitra et al. 2009; Kumar and Ramanathan 2015). Sediments within the Sundarbans have higher levels of contamination compared with sediments in the Hooghly estuary because of lower tidal energy and finer-grained sediments (Banerjee et al. 2012). Hooghly River inputs of Cu and Zn are a critical source of heavy metal pollution to the Sundarbans (Chakrabarti et al. 1993; Bhattacharya et al. 2015). Moreover, the metal concentration of fine-gained sediment in the Indian Sundarbans is higher than those in the Bangladesh Sundarbans (Kumar and Ramanathan 2015) (Table 15.1). The industrialization of the upper catchment in India compared to Bangladesh has been suggested as the primary reason for this difference (Rahman et al. 2011).

Table 15.1 Comparison of heavy metal concentrations (Fe, Mn, Cu, Zn) across the Indian and Bangladesh Sundarbans

Heavy metal	Indian Sundarban	Bangladesh Sundarban
Fe ($\mu g\ g^{-1}$)	38,760–52,829	29,081–45,025
Mn ($\mu g\ g^{-1}$)	424–770	342–792
Cu ($\mu g\ g^{-1}$)	36–82	12–45
Zn ($\mu g\ g^{-1}$)	55–83	29–75

Data from Kumar and Ramanathan (2015)

Table 15.2 Heavy metal concentrations in water, sediments, and macrobenthos from the Sundarbans. Concentrations in the macrobenthos exceed toxic levels

Heavy metal	Water	Sediment	Macrobenthos
Cd ($\mu g\ g^{-1}$)	0.04–0.10	6.25–7.38	14.63
Zn ($\mu g\ g^{-1}$)	0.01–9.66	24.91–62.0	268.91
Pb ($\mu g\ g^{-1}$)	0.03–0.16	33.7–50.33	174.84
Fe ($\mu g\ g^{-1}$)	14.3–170.0		
Cr ($\mu g\ g^{-1}$)		46.8–78.50	18.76
Cu ($\mu g\ g^{-1}$)		20.38–42.01	90.02

Data from Rahman et al. (2009)

15.3.2.3 Bioaccumulation and Health

Heavy metal pollution of the Sundarbans has important implications for the health of the ecosystem, aquatic organisms, and the local communities (Bhattacharya et al. 2015). River water in the region is largely unpotable due to the dissolved concentrations of Mn, Pb, and Ni (Bhattacharya et al. 2015). River water is also not suitable for irrigation due to the high concentration of Mn (Bhattacharya et al. 2015) and the large-scale metal pollution in riverine water and sediments is a serious concern as fish, prawns, and crabs have been reported to contain significant toxic metals (Dutta et al. 2017a; Mitra et al. 2012). Bioaccumulation of metals in these organisms occurs through the food chain until top level predators accumulate ions at a level that can develop neuronal, abdominal, and cardiovascular diseases. Table 15.2 shows the increase in metal accumulation between water, sediment, and macro benthos. At low concentrations, effects such as diarrhea, vomiting, and skin irritation are common. However, at high concentrations and continued exposure, there are serious health considerations with the International Agency for Research on Cancer (IARC) classifying Cd as a human carcinogen, Pb as possible human carcinogen, and Cr to be the cause of a rare sino-nasal cancer (Dayan and Paine 2001; Järup 2003).

15.4 Carbon Biogeochemistry

The Sundarbans contain nearly 3% of the total area of the world's mangrove ecosystems and have been an important region for understanding carbon cycle dynamics in estuarine delta ecosystems over the past 20 years. In particular, the biogeochemical cycling of different carbon species including dissolved organic carbon (DOC), particulate organic carbon (POC), dissolved inorganic carbon (DIC), and dissolved greenhouse gases (CO_2; CH_4) in different environments including estuarine water (e.g., Biswas et al. 2004; Dutta et al. 2019a, b), sediment (e.g. Dutta et al. 2013; Dutta et al. 2017b), mangrove soil and forests (e.g. Rahman et al. 2015; Chanda et al. 2016; Das et al. 2016).

15.4.1 Carbon Fluxes in the Sundarbans

Mangrove estuaries have been recognized as important organic C sources for the ocean and atmosphere (Rosentreter et al. 2018; Ray et al. 2015), with an estimated flux of 55 Tg C yr.$^{-1}$ (Sippo et al. 2017) derived from plant litter, phytoplankton, and microphytobenthos (Ray et al. 2015). However, the Sundarbans have a conspicuous lack of data related to its carbon budget. In particular, measurements of POC and DOC have only been taken in the last few years (Ray et al. 2018b; Dutta et al. 2019a, b). Ray et al. (2018b) provide the first baseline data of C export (DOC, POC, and DIC) from the Sundarbans mangroves into the Bay of Bengal, which accounts for 3.03 Tg C yr.$^{-1}$, 0.58 Tg C yr.$^{-1}$, and 3.69 Tg C yr.$^{-1}$, respectively. DIC is the major form of C exported in the Sundarbans region, contributing to >50% of the fluvial C budget (Fig. 15.3), with DIC concentration ([DIC]) varying between 1.92 to 2.19 mM

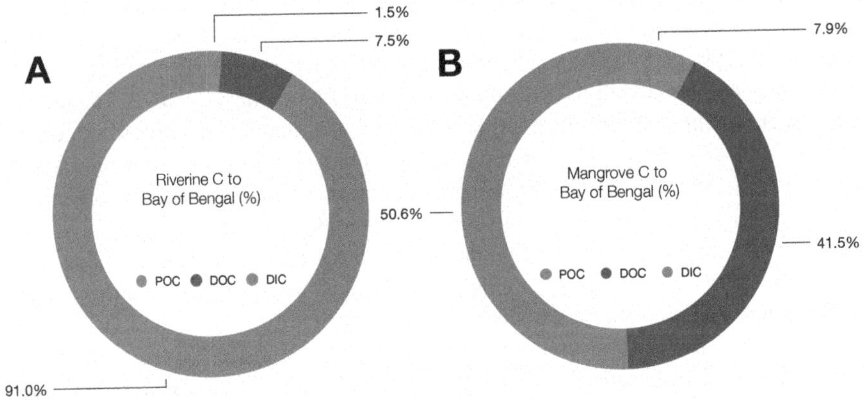

Fig. 15.3 Percentage contribution of the different carbon fractions – dissolved organic carbon (DOC), particulate organic carbon (POC), and dissolved inorganic carbon (DIC) – (**a**) riverine C export from the Hooghly River into the Bay of Bengal during the monsoon season with maximum discharge; (**b**) mangrove-derived C export into the Bay of Bengal. (Data from Ray et al. 2018b)

during a 24-hour period (Dutta et al. 2019a). However, compared to the Hooghly estuary, the major river draining into the Bay of Bengal, the percent contribution and flux of [DIC] from the Sundarbans is much smaller and it has a greater amount of organic-C flux (as DOC and POC). DOC concentration ([DOC]) was monitored in different seasons, with similar values observed during the pre- and post-monsoon (pre-monsoon: 294.3 ± 34 uM; post-monsoon: 262.5 ± 48.2 uM) (Ray et al. 2018b), 235 ± 49 (Dutta et al. 2019b). [POC] is much smaller than [DOC], ranging from 28.0 ± 8.6 uM during pre-monsoon to 45.4 ± 7.5 uM post-monsoon (Ray et al. 2018b). When more locations were monitored, a higher post-monsoon [POC] of 173 ± 111 uM was observed by Dutta et al. (2019b), reflecting the spatial variability of C flux within the Sundarbans region. DIC removal in the Sundarbans is facilitated by phytoplankton uptake, CO_2 outgassing and export to the adjacent continental shelf (Ray et al. 2018b), although significant uncertainty remains.

Much of the work on C biogeochemistry in the Sundarbans has focused on the CO_2 flux from river surface waters (Mukhopadhyay et al. 2002a, b; Biswas et al. 2004; Akhand et al. 2016; Vinh et al. 2019) (Table 15.3). While the mangrove forest is an autotrophic ecosystem and acts as a net C sink (Rodda et al. 2016), more temporal and spatial C flux data is needed to understand its potential to be a large C store. This is important as mangroves can export C to adjacent water bodies, increasing the fraction of CO_2 in water, which can control water-to-air emissions (Akhand et al. 2012) (Table 15.3). A study of the outer part of the Sundarbans found this area to be a CO_2 sink at a rate of 16×10^6 kg C year^{-1} (Mukhopadhyay et al. 2000), while other studies suggest mangrove estuaries are a net CO_2 source at a rate of 13.8 kg C ha^{-1} year^{-1} (Biswas et al. 2004), or it varies between a net source and sink through the seasons as influenced by the monsoon (Mukhopadhyay et al. 2002a, b) (Table 15.3).

The significance of CH_4 production and export from the Sundarbans has been recently documented (e.g., Mukhopadhyay et al. 2002a; Jha et al. 2014; Dutta et al. 2017b). Its importance to the Sundarbans biogeochemical cycle lies in the nature of intertidal mangrove sediments, which are generally anoxic and rich in organic carbon, and therefore favorable environments for methanogenesis (Dutta et al. 2017b). While in the riverine and standing waters, the production of CH_4 is linked to the stratification of the water column and anoxic bottom waters (Koné et al. 2010; Borges and Abril 2011). As a result, dissolved CH_4 concentrations ([CH_4]) are 11.0–129.0 nM throughout the year (Biswas et al. 2007) (Table 15.3), with a distinct increase in CH_4 in the post-monsoon period and maximum recorded values in December across all Sundarbans sites. Higher mean concentration for CH_4 is found in surface waters (69.9 nM) compared to subsurface (56.1 nM) (Dutta et al. 2017b) (Table 15.3).

In mangroves and wetlands, sedimentary-derived CH_4 can escape to the adjacent water and/or atmosphere via plant vascular system–mediated transport, ebullition, and molecular diffusion (Chanton and Dacey 1991), among which ebullition is the dominant pathway (Maher et al. 2019) and is rarely accounted for in the water–air CH_4 budget (Jeffrey et al. 2019). While the [CH_4] in water columns can be partly oxidized to CO_2 via physical and biochemical processes (Hanson and Hanson 1996), this will be limited in well-mixed water bodies, allowing for CH_4 to be emit-

Table 15.3 A summary of CO_2 and CH_4 fluxes and concentration estimates from the Sundarban ecosystem

Place	CO_2 flux (mmol m^{-2} d^{-1})	[CH_4] (nM)	CH_4 flux (umol m^{-2} d^{-1})	Reference
Sundarbans	-3.65×10^9			a.
Hooghly estuary	−2.78–84.4			a.
Sundarbans	0.315			b.
Hooghly-Matla	−0.337			c.
Sundarbans (inner)	0.675			d.
Sundarbans (middle)	0.536			
Sundarbans (outer)	−0.759			
Hooghly estuary	88.8			e.
Matla estuary	6.3			
Saptamukhi estuary (surface water)		69.9		f.
Saptamukhi estuary (sub-surface water)		56.1		
Sundarbans (Muriganga, Saptamukhi, Thakuran)		11.0–129.0	1.97–134.6	g.
Hooghly estuary		10.3–59.3	0.88–148.6	

Data sources: a. Mukhopadhyay et al. 2000, 2002b. Biswas et al. 2004; c. Akhand et al. 2012; d. Akhand et al. 2013; e. Akhand et al. 2016; f. Dutta et al. 2017b; g. Biswas et al. 2007

ted (Abril et al. 2007). The CH_4 emission rate from surface waters was between 1.97 and 134.6μmol m^{-2} d^{-1} in three distributaries in the Sundarbans with clear seasonal variation – minimum during the monsoon and maximum in the post-monsoon (Biswas et al. 2007) (Table 15.3).

15.4.2 *Temporal and Spatial Variations of C Flux*

The biogeochemical processes in the Sundarbans can be significantly different in the monsoon seasons compared to the dry periods of the year. For example, water in the Matla River was found to be marginally oversaturated in CO_2 throughout the year, but transitioned to a CO_2 sink during the post-monsoon season (Akhand et al. 2016). The difference results from the high discharge during the monsoon seasons creating a well-mixed water column, meaning that CO_2 diffusion was limited and there was little organic-rich sediment deposition. Furthermore, the concentrations and fluxes of different forms of C in the Sundarbans are often compared to the Hooghly River estuary, the main artery to the Sundarbans mangroves meaning you are comparing freshwater with coastal saline/brackish waters, which provides a different C dynamic. For [DOC], [POC], and [DIC] there is no distinct or consistent spatial pattern between three Sundarbans estuaries (Dutta et al. 2019b), although

[DIC] and [DOC] were both lower on average than the Hooghly River. Akhand et al. (2013) shows water in the inner and middle Sundarbans regions is oversaturated in CO_2, but undersaturated in the outer region during the summer. As a result, the inner and middle Sundarbans act as a CO_2 source (29.7 and 23.6 mg CO_2 m^{-2} day^{-1}, respectively) while the outer Sundarbans is a net sink (−33.4 mg CO_2 m^{-2} day^{-1}). This change of carbon sink and source results from higher nutrient availability and chlorophyll *a* concentrations, reflecting primary productivity in the outer mangrove system. Variations in the fluxes of CO_2 also demonstrate the heterotrophic nature of the inner mangrove ecosystems at the land–ocean interface and the C-sink character of the outer mangrove on continental shelves (Chen and Borges 2009).

15.4.3 Source of C in the Indian Sundarbans

Very few studies have explored the source of different C species in the Sundarbans water, but a modelling study of the Hooghly–Matla river system by Ray et al. (2015) shows plant litter production and the breakdown of detritus from adjacent Sundarbans mangrove forests are a major source of dissolved inorganic N and C to river waters, and potentially C exports to the continental shelf. In addition, phytoplankton is a leading source of C near Sagar Island, while this is not the case for the Saptamukhi estuary in the Sundarbans, where POC is mainly sourced from riverine suspended sediments and soils, but less from marine plankton, as indicated by their C/N ratios (Dutta et al. 2019a). Higher carbon isotope values in POC ($\delta^{13}C_{POC}$) in estuarine waters compared to mangroves indicate the modification of POC. Ray et al. (2018b) also demonstrate mangrove forests (including plant litter, eroded soil) are the major source of C exported from Sundarbans to the Bay of Bengal, compared to upstream C-inputs and marine phytoplankton. In addition, the negative relationship between [DIC] and its carbon isotope value ($\delta^{13}C_{DIC}$) during low tide, highlights respiration of marine plankton-derived organic carbon may be an important source of DIC rather than exchange of C-rich porewaters derived from terrestrial sources.

15.4.4 Influence of Salinity and Tide on C

In general, CO_2 flux decreases with increasing salinity toward the open sea (Akhand et al. 2012). In the Matla River, the highest fraction of aqueous CO_2 (fCO_2) coincides with the lowest neap tide, overriding CO_2 uptake by photosynthesis. The hydrological change during the ebb and low tide leads to the mixing of sediment porewater and groundwater with brackish/saline estuary waters. The subsequent biogeochemical interaction that leads to increasing fCO_2 and the extent of CO_2 efflux highlights the role of salinity in C-dynamics over the Sundarbans (Akhand et al. 2016). The importance of tidal stage in controlling dissolved greenhouse gas efflux from water is also demonstrated by Padhy et al. (2020), who show the concentrations of dissolved CH_4 and CO_2 are higher in stagnant water during low tide compared to high tide water. This implies the effect of stagnation and lower salinity and therefore less SO_4^{2-} availability, which increases CO_2 emissions. Apart from high pCO_2 during low tide, Dutta et al. (2019b) also suggest there is a strong influence from estuarine mixing on DIC and $\delta^{13}C_{DIC}$ during the low tide, both of which correlate with salinity. This can be explained by the impacts of this biogeochemistry on denitrification, sulfate reduction, and aerobic organic matter mineralization to DIC, along with possible organic contributions from porewater.

15.5 Conclusions

This overview of biogeochemical dynamics and anthropogenic impacts on the Indian Sundarbans highlights the importance of the hydrological regime in driving variability in ecosystem health. Diurnal and seasonal changes in salinity, which are driven by the tides and monsoon-driven freshwater availability, influence biological responses, biogeochemical cycling, and carbon dynamics. Also, high concentrations of heavy metals mean they are bioavailable within the major rivers running into the Sundarbans, but there is little evidence of the short- and long-term implications of this pollution for mangrove health, aquatic organisms, and local communities. Overall, there remains a paucity of research into water-quality impacts on aquatic ecology, including nutrient enrichment and heavy metal pollution, carbon cycling through the mangrove system, and how climate change has and will continue to affect the Indian Sundarbans.

Acknowledgments The writing of this chapter would not have been possible without the funding received from the UK Research and Innovation – Global Challenges Research Fund (UKRI GCRF) Living Deltas Hub (NE/S008926/1) www.livingdeltas.org.

References

Abril G, Commarieu M-V, Guérin F (2007) Enhanced methane oxidation in an estuarine turbidity maximum. Limnol Oceanogr 52:470–475. https://doi.org/10.4319/lo.2007.52.1.0470

Akhand A, Chanda A, Dutta S, Hazra S (2012) Air – water carbon dioxide exchange dynamics along the outer estuarine transition zone of Sundarban, northern Bay of Bengal, India. Indian J Geo-Mar Sci 41:111–116

Akhand A, Chanda A, Dutta S, Manna S, Sanyal P, Hazra S, Rao KH, Dadhwal VK (2013) Dual character of Sundarban estuary as a source and sink of CO2 during summer: an investigation of spatial dynamics. Environ Monit Assess 185:6505–6515. https://doi.org/10.1007/s10661-012-3042-x

Akhand A, Chanda A, Manna S, Das S, Hazra S, Roy R, Choudhury SB, Rao KH, Dadhwal VK, Chakraborty K, Mostofa KMG, Tokoro T, Kuwae T, Wanninkhof R (2016) A comparison of CO2 dynamics and air-water fluxes in a river-dominated estuary and a mangrove-dominated marine estuary. Geophys Res Lett 43(11):726–11,735. https://doi.org/10.1002/2016GL070716

Allison MA, Khan SR, Goodbred SL, Kuehl SA (2003) Stratigraphic evolution of the late Holocene Ganges–Brahmaputra lower delta plain. Sediment Geol 155:317–342. https://doi.org/10.1016/S0037-0738(02)00185-9

Bandopadhyay B, Burman D (2006) Characterization of soil and water of brackish water fisheries of coastal region of Sundarbans, West Bengal. J Indian Soc Coast Agric Res 24(1):78–82

Banerjee K, Senthilkumar B, Purvaja R, Ramesh R (2012) Sedimentation and trace metal distribution in selected locations of Sundarbans mangroves and Hooghly estuary, Northeast coast of India. Environ Geochem Health 34:27–42. https://doi.org/10.1007/s10653-011-9388-0

Bhattacharjee D, Samanta B, Danda A, Bhadbury P (2013) Temporal succession of phytoplankton assemblages in a tidal creek system of the Sundarbans mangroves: an integrated approach. Int J Biodivers 2013:824543

Bhattacharya BD, Hwang JS, Tseng LC, Sarkar SK, Rakshit D, Mitra S (2014a) Bioaccumulation of trace elements in dominant mesozooplankton group inhabiting in the coastal regions of Indian Sundarban mangrove wetland. Mar Pollut Bull 87:345–351. https://doi.org/10.1016/j.marpolbul.2014.07.050

Bhattacharya S, Dubey SK, Dash JR, Patra PH, Das AK, Mandal TK, Bandyopadhyay SK (2014b) Assemblages of Total mercury in the tropical Macrotidal Bidyadhari estuarine stretches of Indian Sundarban mangrove eco-region. J Environ Anal Toxicol 4(6):1

Bhattacharya BD, Nayak DC, Sarkar SK, Biswas SN, Rakshit D, Ahmed MK (2015) Distribution of dissolved trace metals in coastal regions of Indian Sundarban mangrove wetland: a multivariate approach. J Clean Prod 96:233–243. https://doi.org/10.1016/j.jclepro.2014.04.030

Bhattacharya BD, Hwang JS, Sarkar SK, Rakhsit D, Murugan K, Tseng LC (2015a) Community structure of mesozooplankton in coastal waters of Sundarban mangrove wetland, India: a multivariate approach. J Mar Syst 141:112–121

Bhattacharyya S (2015) Study of changing facades of Sundarbans – A Remote Sensing & GIS based management approach, PhD Thesis in Department of Geological Sciences, Jadavpur University

Bir J, Sumon MS, Rahaman S (2015) The effects of different water quality parameters on zooplankton distribution in major river systems of Sundarbans Mangrove. IOSR J Environ Sci Toxicol Food Technol 11:56–63

Biswas H, Mukhopadhyay SK, De TK, Sen S, Jana TK (2004) Biogenic controls on the air-water carbon dioxide exchange in the Sundarban mangrove environment, northeast coast of Bay of Bengal, India. Limnol Oceanogr 49:95–101. https://doi.org/10.4319/lo.2004.49.1.0095

Biswas H, Mukhopadhyay SK, Sen S, Jana TK (2007) Spatial and temporal patterns of methane dynamics in the tropical mangrove dominated estuary, NE coast of Bay of Bengal, India. J Mar Syst 68:55–64. https://doi.org/10.1016/j.jmarsys.2006.11.001

Biswas H, Dey M, Ganguly D, De TK, Ghosh S, Jana TK (2010) Comparative analysis of phytoplankton composition and abundance over a two-decade period at the land–ocean boundary of a tropical mangrove ecosystem. Estuar Coasts 33(2):384–394

Biswas SN, Rakshit D, Sarkar SK, Sarangi RK, Satpathy KK (2014) Impact of multispecies diatom bloom on plankton community structure in Sundarban mangrove wetland, India. Mar Pollut Bull 85(1):306–311

Blomqvist S, Anneli G, Ragnar E (2004) Why the limiting nutrient differs between temperate coastal seas and freshwater lakes: A matter of salt. Limnol Oceanogr 49(6):2236–2241

Borges AV, Abril G (2011) 5.04 – carbon dioxide and methane dynamics in estuaries. In: Wolanski E, McLusky D (eds) Treatise on estuarine and coastal science. Academic Press, Waltham, pp 119–161. https://doi.org/10.1016/B978-0-12-374711-2.00504-0

Chakrabarti C, Kundu S, Ghosh P, Choudhury A (1993) A preliminary study on certain trace metals in some plant and animal organisms from mangroves of Sundarbans, India. Mahasagar 26:17–20

Chakraborty A, Bera A, Mukherjee A, Basak P, Khan I, Mondal A, Roy A, Bhattacharyya A, SenGupta S, Roy D, Nag S (2015) Changing bacterial profile of Sundarbans, the world heritage mangrove: impact of anthropogenic interventions. World J Microbiol Biotechnol 31(4):593–610

Chanda A, Mukhopadhyay A, Ghosh T, Akhand A, Mondal P, Ghosh S, Mukherjee S, Wolf J, Lazar AN, Rahman MM, Salehin M, Chowdhury SM, Hazra S (2016) Blue carbon stock of the Bangladesh Sundarban Mangroves: what could be the scenario after a century? Wetlands 36:1033–1045. https://doi.org/10.1007/s13157-016-0819-7

Chanton JP, Dacey JWH (1991) Effects of vegetation on methane flux, reservoirs, and carbon isotopic composition. In: Sharkey T, Holland E, Mooney H (eds) Trace gas emissions from plants. Academic Press, San Diego, pp 65–92

Chatterjee M, Silva Filho EV, Sarkar SK, Sella SM, Bhattacharya A, Satpathy KK, Prasad MVR, Chakraborty S, Bhattacharya BD (2007) Distribution and possible source of trace elements in the sediment cores of a tropical macrotidal estuary and their ecotoxicological significance. Environ Int 33:346–356. https://doi.org/10.1016/j.envint.2006.11.013

Chatterjee M, Massolo S, Sarkar SK, Bhattacharya AK, Bhattacharya BD, Satpathy KK, Saha S (2009) An assessment of trace element contamination in intertidal sediment cores of Sunderban mangrove wetland, India for evaluating sediment quality guidelines. Environ Monit Assess 150:307–322

Chatterjee M, Shankar D, Sen GK et al (2013) Tidal variations in the Sundarbans estuarine system, India. J Earth Syst Sci 122:899–933. https://doi.org/10.1007/s12040-013-0314-y

Chaudhuri K, Manna S, Sarma KS, Naskar P, Bhattacharyya S, Bhattacharyya M (2012) Physicochemical and biological factors controlling water column metabolism in Sundarbans estuary, India. Aquat Biosyst 8(1):26

Chen CTA, Borges AV (2009) Reconciling opposing views on carbon cycling in the coastal ocean: continental shelves as sinks and near-shore ecosystems as sources of atmospheric CO2. Deep Sea Res Part II Top Stud Oceanogr, Surface Ocean CO2 Variability and Vulnerabilities 56:578–590. https://doi.org/10.1016/j.dsr2.2009.01.001

Chowdhury A, Maiti SK (2016) Assessing the ecological health risk in a conserved mangrove ecosystem due to heavy metal pollution: a case study from Sundarbans Biosphere Reserve, India. Hum Ecol Risk Assess Int J 22:1519–1541. https://doi.org/10.1080/10807039.2016.1190636

Chowdhury R, Favas PJC, Jonathan MP, Venkatachalam P, Raja P, Sarkar SK (2017) Bioremoval of trace metals from rhizosediment by mangrove plants in Indian Sundarban Wetland. Mar Pollut Bull 124:1078–1088. https://doi.org/10.1016/j.marpolbul.2017.01.047

Chowdhury R, Sutradhar T, Begam MM, Mukherjee C, Chatterjee K, Basak SK, Ray K (2019) Effects of nutrient limitation, salinity increase, and associated stressors on mangrove forest cover, structure, and zonation across Indian Sundarbans. Hydrobiologia 842:191–217

Das S, Chanda A, Giri S, Akhand A, Hazra S (2015) Characterizing the influence of tide on the physico-chemical parameters and nutrient variability in the coastal surface water of the northern Bay of Bengal during the winter season. Acta Oceanol Sin 34(12):102–111

Das S, Mukherjee A, De TK, De M, Jana TK (2016) Influence of microbial composition on the carbon storage capacity of the mangrove soil at the land-ocean boundary in the Sundarban Mangrove Ecosystem, India. Geomicrobiol J 33:743–750. https://doi.org/10.1080/01490451.2015.1093567

Das S, Giri S, Das I, Chanda A, Ghosh A, Mukhopadhyay A, Akhand A, Choudhury SB, Dadhwal VK, Maity S, Kumar TS (2017) Nutrient dynamics of northern Bay of Bengal (nBoB)—emphasizing the role of tides. Reg Stud Mar Sci 10:116–134

Dayan AD, Paine AJ (2001) Mechanisms of chromium toxicity, carcinogenicity and allergenicity: review of the literature from 1985 to 2000. Hum Exp Toxicol 20:439–451. https://doi.org/10.1191/096032701682693062

De TK, De M, Das S, Chowdhury C, Ray R, Jana TK (2011) Phytoplankton abundance in relation to cultural eutrophication at the land-ocean boundary of Sunderbans, NE Coast of Bay of Bengal, India. J Environ Stud Sci 1(3):169

Dutta MK, Chowdhury C, Jana TK, Mukhopadhyay SK (2013) Dynamics and exchange fluxes of methane in the estuarine mangrove environment of the Sundarbans, NE coast of India. Atmos Environ 77:631–639. https://doi.org/10.1016/j.atmosenv.2013.05.050

Dutta S, Chakraborty K, Hazra S (2017a) Ecosystem structure and trophic dynamics of an exploited ecosystem of Bay of Bengal, Sundarban Estuary, India. Fish Sci 83:145–159. https://doi.org/10.1007/s12562-016-1060-2

Dutta MK, Bianchi TS, Mukhopadhyay SK (2017b) Mangrove methane biogeochemistry in the Indian Sundarbans: a proposed budget. Front Mar Sci 4. https://doi.org/10.3389/fmars.2017.00187

Dutta MK, Kumar S, Mukherjee R, Sanyal P, Mukhopadhyay SK (2019a) The post-monsoon carbon biogeochemistry of the Hooghly–Sundarbans estuarine system under different levels of anthropogenic impacts. Biogeosciences 16:289–307

Dutta MK, Kumar S, Mukherjee R, Sharma N, Acharya A, Sanyal P, Bhusan R, Mukhopadhyay SK (2019b) Diurnal carbon dynamics in a mangrove-dominated tropical estuary (Sundarbans, India). Estuar Coast Shelf Sci 229:106426. https://doi.org/10.1016/j.ecss.2019.106426

Flood RP, Bloemsma MR, Weltje GJ et al (2016) Compositional data analysis of Holocene sediments from the West Bengal Sundarbans, India: geochemical proxies for grain-size variability in a delta environment. Appl Geochem 75:222–235. https://doi.org/10.1016/j.apgeochem.2016.06.006

Flood RP, Barr ID, Weltje GJ et al (2018) Provenance and depositional variability of the Thin Mud Facies in the lower Ganges-Brahmaputra delta, West Bengal Sundarbans, India. Mar Geol 395:198–218. https://doi.org/10.1016/j.margeo.2017.09.001

Gannon JE, Stemberger RS (1978) Zooplankton (especially crustaceans and rotifers) as indicators of water quality. Trans Am Microsc Soc 97(1):16–35

Garrett RG (2000) Natural sources of metals to the environment. Hum Ecol Risk Assess Int J 6:945–963. https://doi.org/10.1080/10807030091124383

Ghosh A, Bhadury P (2018) Investigating monsoon and post-monsoon variabilities of bacterioplankton communities in a mangrove ecosystem. Environ Sci Pollut Res 25(6):5722–5739

Ghosh P, Choudhury A (1989) Copper, zinc and lead in the sediment of Hooghly estuary. Environ Ecol 7:427–430

Ghosh A, Dey N, Bera A, Tiwari A, Sathyaniranjan K, Chakrabarti K, Chattopadhyay D (2010) Culture independent molecular analysis of bacterial communities in the mangrove sediment of Sundarban, India. Saline Syst 6(1):1

Ghosh P, Chakrabarti R, Bhattacharya SK (2013) Short-and long-term temporal variations in salinity and the oxygen, carbon and hydrogen isotopic compositions of the Hooghly Estuary water, India. Chem Geol 335:118–127. https://doi.org/10.1016/j.chemgeo.2012.10.051

Gole CV, Vaidyaraman PP (1967) Salinity distribution and effect of fresh water flows in the Hooghly River. Coast Eng 1966:1412–1434

Gopal B, Chauhan M (2006) Biodiversity and its conservation in the Sundarban mangrove ecosystem. Aquat Sci 68:338–354

Hanson RS, Hanson TE (1996) Methanotrophic bacteria. Microbiol Mol Biol Rev 60:439–471

Islam MS, Sarkar MJ, Yamamoto T, Wahab MA, Tanaka M (2004) Water and sediment quality, partial mass budget and effluent N loading in coastal brackish water shrimp farms in Bangladesh. Mar Pollut Bull 48(5–6):471–485

Järup L (2003) Hazards of heavy metal contamination. Br Med Bull 68:167–182. https://doi.org/10.1093/bmb/ldg032

Jeffrey LC, Maher DT, Johnston SG, Kelaher BP, Steven A, Tait DR (2019) Wetland methane emissions dominated by plant-mediated fluxes: contrasting emissions pathways and seasons within a shallow freshwater subtropical wetland. Limnol Oceanogr 64:1895–1912. https://doi.org/10.1002/lno.11158

Jha CS, Rodda SR, Thumaty KC, Raha AK, Dadhwal VK (2014) Eddy covariance based methane flux in Sundarbans mangroves, India. J Earth Syst Sci 123:1089–1096. https://doi.org/10.1007/s12040-014-0451-y

Kamruzzaman M, Basak K, Paul SK, Ahmed S, Osawa A (2019) Litterfall production, decomposition and nutrient accumulation in Sundarbans mangrove forests, Bangladesh. For Sci Technol 15(1):24–32

Khan RA (2003) Biodiversity of macrobenthos on the intertidal flats of Sunderban estuarine region, India. Rec Zool Surv India 101(3–4):181–205

Koné YJM, Abril G, Delille B, Borges AV (2010) Seasonal variability of methane in the rivers and lagoons of Ivory Coast (West Africa). Biogeochemistry 100:21–37

Kumar A, Ramanathan A (2015) Speciation of selected trace metals (Fe, Mn, Cu and Zn) with depth in the sediments of Sundarban mangroves: India and Bangladesh. J Soils Sediments 15:2476–2486. https://doi.org/10.1007/s11368-015-1257-5

Maher DT, Drexl M, Tait DR, Johnston SG, Jeffrey LC (2019) iAMES: an inexpensive, automated methane ebullition sensor. Environ Sci Technol 53:6420–6426. https://doi.org/10.1021/acs.est.9b01881

Manna S, Chaudhuri K, Bhattacharyya S, Bhattacharyya M (2010) Dynamics of Sundarban estuarine ecosystem: eutrophication induced threat to mangroves. Saline Syst 6(1):8

Manna S, Chaudhuri K, Sarma KS, Naskar P, Bhattacharyya S, Bhattacharyya M (2012) Interplay of physical, chemical and biological components in estuarine ecosystem with special reference to Sundarbans, India. In: Voudouris K, Voutsa D (eds) Ecological water quality—water treatment and reuse. InTech, Rijeka, Croatia, pp 206–238

Massolo S, Bignasca A, Sarkar SK, Chatterjee M, Bhattacharya BD, Alam A (2012) Geochemical fractionation of trace elements in sediments of Hooghly River (Ganges) and Sundarban wetland (West Bengal, India). Environ Monit Assess 184:7561–7577. https://doi.org/10.1007/s10661-012-2519-y

Mitra A (1998) Status of coastal pollution in West Bengal with special reference to heavy metals. J Indian Ocean Stud 5:135–138

Mitra A, Ghosh R (2014) Bioaccumulation pattern of heavy metals in commercially important fishes on and around Indian Sundarbans 12

Mitra A, Banerjee K, Sengupta K, Gangopadhyay A (2009) Pulse of climate change in Indian Sundarbans: a myth or reality. Natl Acad Sci Lett 32:1–7

Mitra A, Barua P, Zaman S, Banerjee K (2012) Analysis of trace metals in commercially important crustaceans Collected from UNESCO protected world heritage site of Indian Sundarbans. Turk J Fish Aquat Sci 12. https://doi.org/10.4194/1303-2712-v12_1_07

Morgan JP, McIntire WG (1959) Quaternary geology of the Bengal Basin, East Pakistan and India. Geolog Soci Am Bull 70:319–342

Mukherjee D, Mukherjee A, Kumar B (2009) Chemical fractionation of metals in freshly deposited marine estuarine sediments of Sundarban ecosystem, India. Environ Geol 58:1757–1767. https://doi.org/10.1007/s00254-008-1675-4

Mukhopadhyay SK, Jana TK, De TK, Sen S (2000) Measurement of exchange of CO2 in mangrove forest of Sundarbans using micrometeorological method. Trop Ecol 41:57–60

Mukhopadhyay SK, Biswas H, De TK, Sen BK, Sen S, Jana TK (2002a) Impact of Sundarban mangrove biosphere on the carbon dioxide and methane mixing ratios at the NE Coast of Bay of Bengal, India. Atmos Environ 36:629–638. https://doi.org/10.1016/S1352-2310(01)00521-0

Mukhopadhyay SK, Biswas H, De TK, Sen S, Jana TK (2002b) Seasonal effects on the air-water carbon dioxide exchange in the Hooghly estuary, NE coast of Bay of Bengal, India. J Environ Monit 4:549–552. https://doi.org/10.1039/b201614a

Padhy SR, Bhattacharyya P, Dash PK, Reddy CS, Chakraborty A, Pathak H (2020) Seasonal fluctuation in three mode of greenhouse gases emission in relation to soil labile carbon pools in degraded mangrove, Sundarban, India. Sci Total Environ 705:135909. https://doi.org/10.1016/j.scitotenv.2019.135909

Paerl HW, Huisman J (2008) Blooms like it hot. Science 320(5872):57–58

Paerl HW, Tucker CS (1995) Ecology of blue-green algae in aquaculture ponds. J World Aquacult Soc 26(2)

Purvaja R, Ramesh R, Ray AK, Rixen T (2008) Nitrogen cycling: a review of the processes, transformations and fluxes in coastal ecosystems. Curr Sci 94(11):1419–1438

Raha A, Das S, Banerjee K, Mitra A (2012) Climate change impacts on Indian Sunderbans: a time series analysis (1924–2008). Biodivers Conserv 21:1289–1307

Rahman MM, Chongling Y, Islam KS, Haoliang L (2009) A brief review on pollution and ecotoxicologic effects on Sundarbans mangrove ecosystem in Bangladesh. Int J Environ Sci Eng 1(4):369–383

Rahman MT, Rahman S, Quraishi SB, Ahmad JU, Choudhury TR, Mottaleb MA (2011) Distribution of heavy metals in water and sediments in Passur River. Sundarban Mangrove Forest, Bangladesh 6:11

Rahman MM, Khan MNI, Hoque AKF, Ahmed I (2015) Carbon stock in the Sundarbans mangrove forest: spatial variations in vegetation types and salinity zones. Wetl Ecol Manag 23:269–283. https://doi.org/10.1007/s11273-014-9379-x

Ray R, Rixen T, Baum A, Malik A, Gleixner G, Jana TK (2015) Distribution, sources and biogeochemistry of organic matter in a mangrove dominated estuarine system (Indian Sundarbans) during the pre-monsoon. Estuar Coast Shelf Sci 167:404–413

Ray R, Majumder N, Chowdhury C, Das S, Jana TK (2018a) Phosphorus budget of the Sundarban mangrove ecosystem: box model approach. Estuar Coasts 41(4):1036–1049

Ray R, Baum A, Rixen T, Gleixner G, Jana TK (2018b) Exportation of dissolved inorganic and organic and particulate carbon from mangroves and its implication to the carbon budget in the Indian Sundarbans. Sci Total Environ 621:535–547. https://doi.org/10.1016/j.scitotenv.2017.11.225

Rodda SR, Thumaty KC, Jha CS, Dadhwal VK (2016) Seasonal variations of carbon dioxide, water vapor and energy fluxes in tropical Indian mangroves. Forests 7:35. https://doi.org/10.3390/f7020035

Rogers KG, Goodbred SL (2014) The Sundarbans and Bengal Delta: the world's largest tidal mangrove and delta system. In: Kale VS (ed) Landscapes and landforms of India. Springer Netherlands, Dordrecht, pp 181–187

Rosentreter JA, Maher DT, Erler DV, Murray R, Eyre BD (2018) Seasonal and temporal CO2 dynamics in three tropical mangrove creeks–a revision of global mangrove CO2 emissions. Geochim Cosmochim Acta 222:729–745

Roy M, Nandi NC (2012) Distribution pattern of macrozoobenthos in relation to salinity of Hooghly-Matla estuaries in India. Wetlands 32(6):1001–1009

Roy S, Hens D, Biswas D, Biswas D, Kumar R (2002) Survey of petroleum-degrading bacteria in coastal waters of Sunderban Biosphere Reserve. World J Microbiol Biotechnol 18(6):575–581

Roy D, Pramanik A, Banerjee S, Ghosh A, Chattopadhyay D, Bhattacharyya M (2018) Spatio-temporal variability and source identification for metal contamination in the river sediment of Indian Sundarbans, a world heritage site. Environ Sci Pollut Res 25:31326–31345

Rudra K (2014) Changing river courses in the western part of the Ganga-Brahmaputra delta. Geomorphology 227:87–100. https://doi.org/10.1016/j.geomorph.2014.05.013

Rudra K (2018) Rivers of the Ganga – Brahmaputra – Meghna Delta: a fluvial account of Bengal. Springer Internation Publishing AG, part of Springer Nature, Switzerland, pp 1–190

Sadhuram Y, Sarma VV, Murthy TR, Rao BP (2005) Seasonal variability of physico-chemical characteristics of the Haldia channel of Hooghly estuary, India. J Earth Syst Sci 114:37–49. https://doi.org/10.1007/BF02702007

Samanta B, Bhadury P (2018) Study of diatom assemblages in Sundarbans mangrove water based on light microscopy and rbcL gene sequencing. Heliyon 4(6):e00663

Samanta S, Dalai TK (2018) Massive production of heavy metals in the Ganga (Hooghly) river estuary, India: global importance of solute-particle interaction and enhanced metal fluxes to the oceans. Geochim Cosmochim Acta 228:243–258

Santos HF, Carmo FL, Paes JE, Rosado AS, Peixoto RS (2011) Bioremediation of mangroves impacted by petroleum. Water Air Soil Pollut 216(1–4):329–350

Sarkar SK, Frančišković-Bilinski S, Bhattacharya A, Saha M, Bilinski H (2004) Levels of elements in the surficial estuarine sediments of the Hooghly River, Northeast India and their environmental implications. Environ Int 30(8):1089–1098

Sarkar S, Ghosh PB, Das TM et al (2013) Environmental assessment in terms of salinity distribution in the Tropical Mangrove forest of Sundarban, North East Coast of Bay of Bengal, India. Arch Appl Sci Res 5:109–118

Sen A, Saha S, Maita SK, Saha T (2015) A study of planktonic communities along the Bidyadhari River stretch of the Raimangal Estuary of West Bengal, India. Proceedings of the National Conference on Biodiversity – Issues, Concern & Future Strategies, pp 49–53

Sharma M, Chakraborty A, Kuttippurath J, Yadav AK (2018) Potential power production from salinity gradient at the Hooghly Estuary System. Innov Ener Res 7:1–7. https://doi.org/10.4172/2576-1463.1000210

Singh G, Ramanathan AL, Santra SC, Ranjan RK (2016) Tidal control on the nutrient variability in Sundarban mangrove ecosystem. J Appl Geochem 18(4):495–503

Sinha PC, Rao YR, Dubey SK et al (1996) Modeling of circulation and salinity in Hooghly estuary. Mar Geod 19:197–213. https://doi.org/10.1080/01490419609388079

Sippo JZ, Maher DT, Tait DR, Ruiz-Halpern S, Sanders CJ, Santos IR (2017) Mangrove outwelling is a significant source of oceanic exchangeable organic carbon. Limnol Oceanogr Lett 2:1–8. https://doi.org/10.1002/lol2.10031

Somayajulu BLK, Rengarajan R, Jani RA (2002) Geochemical cycling in the Hooghly estuary, India. Mar Chem 79:171–183. https://doi.org/10.1016/S0304-4203(02)00062-2

Spalding M, Blasco F, Field C (1997) World mangrove atlas. International Society for Mangrove Ecosystems, Okinawa

Vinh TV, Allenbach M, Joanne A, Marchand C (2019) Seasonal variability of CO2 fluxes at different interfaces and vertical CO2 concentration profiles within a Rhizophora mangrove forest (Can Gio, Viet Nam). Atmos Environ 201:301–309. https://doi.org/10.1016/j.atmosenv.2018.12.049

Walsby A, Schanz F (2002) Light-dependent growth rate determines changes in the population of Planktothrix rubescens over the annual cycle in Lake Zürich, Switzerland. New Phytol 154(3):671–687

Yang X, Post WM, Thornton PE, Jain A (2013) The distribution of soil phosphorus for global biogeochemical modeling. Biogeosci Discuss (Online) 9(4)

Index

S. Das, T. Ghosh (eds.), *Estuarine Biogeochemical Dynamics of the East Coast of India*, https://doi.org/10.1007/978-3-030-68980-3

D

E

T

U

V

W

Z

The manufacturer's authorised representative in the EU is Springer Nature Customer Service Centre GmbH, Europaplatz 3, 69115 Heidelberg, Germany. If you have any concerns regarding our products, please contact ProductSafety@springernature.com

Printed and bound by CPI Group (UK) Ltd, Croydon, CR0 4YY
23/09/2026
02221542-0008